WSSA/NEWSS 2023 Annual Meeting

Arlington, Virginia, USA
30 January – 2 February 2023

ISBN: 979-8-3313-0917-6

Printed from e-media with permission by:

Curran Associates, Inc.
57 Morehouse Lane
Red Hook, NY 12571

Some format issues inherent in the e-media version may also appear in this print version.

Northeastern Weed Science Society (NEWSS)
P.O. Box 25
Woodstown, NJ 08098
USA

Phone: 315-209-7580

northeasternweedscience@hotmail.com

Additional copies of this publication are available from:

Curran Associates, Inc.
57 Morehouse Lane
Red Hook, NY 12571 USA
Phone: 845-758-0400
Fax: 845-758-2633
Email: curran@proceedings.com
Web: www.proceedings.com

WEED SCIENCE SOCIETY OF AMERICA - NORTHEASTERN WEED SCIENCE SOCIETY JOINT MEETING

63rd Meeting

2023 MEETING PROGRAM

Crystal Gateway Marriott

Arlington, VA

January 30, 2023 – February 2, 2023

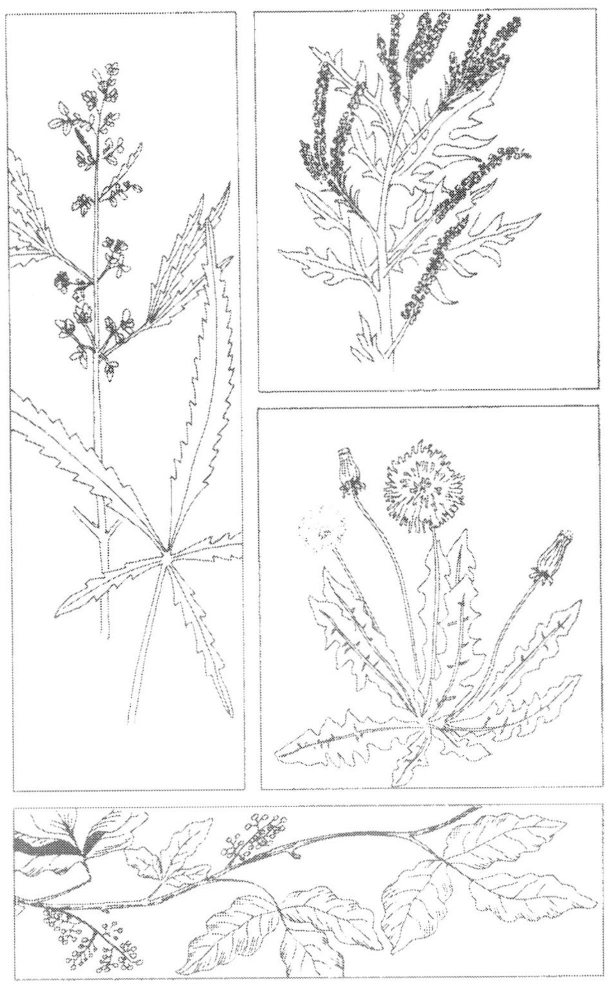

WSSA Sustaining Members

PRESIDENTIAL
BASF Corporation
Bayer Crop Science
Corteva Agrisciences
FMC Corporation
Syngenta Crop Protection
Harpe Bioherbicide

LEADERS
Gowan
UPL NA Inc
ENVU
Valent USA
Winfield United
Helena Agri-Enterprises

PATRONS
Gylling Data Management
Nufarm Americas Inc
Nichino America Inc.
Lehigh Agri & Bio Services, Inc.
Blue River Technology
Precision Laboratories

CONTRIBUTORS
AMVAC Chemical Corp
Greenleaf Technologies
ABG Ag Services
TeeJet Technologies
Nippon Soda Ltd
TKI Novasource
Clariant Corporation
R & D Sprayers
SePRO

63rd Meeting
Weed Science Society of America/
Northeastern Weed Science Society

Local Arrangements Committee 2023 – Arlington

Co-Chairs ....Bill Chism and Sudeep Mathew

The WSSA/NEWSS 2023 Program

Welcome to the 2023 joint Weed Science Society of America (WSSA) and the Northeastern Weed Science Society (NEWSS) meeting at the Crystal Gateway Marriott in Arlington, VA. The hotel is only one mile from Reagan National Airport (DCA), two miles to Washington, DC and directly connected to the Crystal City Metro station providing a home base for your visit to the Washington, DC area. Nearby sites include the new National Landing development and Pentagon City; as well as a quick and convenient Metro ride to downtown Washington D.C. and Old Town Alexandria. Several of my favorite museums include the Smithsonian National Museum of Natural History, National Air and Space Museum, and the National Air and Space Museum Steven F. Udvar-Hazy Center (near Dulles). Don't miss visiting the monuments and memorials – I'm partial to the area encompassed between the Washington Monument and Lincoln Memorial. While you are there, don't miss the National Archives Museum with several incredible historical documents and a new display – All American: The Power of Sports.

Now, back to the meeting! We have an exciting line-up of tours, speakers, symposia, meetings, discussions, a dessert social, and many great papers. I always get "re-charged" by the greatest people involved in Weed Science in the world. Even Washington said, "Agriculture is the most helpful, most useful, and most noble employment." The majority of committee meetings will take place ahead of the annual meeting week. Make sure you are in contact with your committee chair to participate in these meetings. There are some that will take place face to face in Arlington, due to the nature of the meeting. BTW – I am so thankful that this meeting will be taking place in person in 2023!! After going online in 2021 and 2022, we are very fortunate to have this meeting in person. The planning committees for the web meetings did an awesome job switching formats, which basically doubled their workload. Thanks to the Stanley Culpepper and Anita Dille's teams, we were able to pull this off. Meeting in person will bring an even higher excitement to the interactions and progress with weed management that will come from these sessions.

Thanks to Bill Chism and Sudeep Mathew, along with the local arrangements committee, we have two exciting tours scheduled for the meeting. Sunday, there will be a tour of the US Botanic Garden Production Facility and the following day a tour of the Smithsonian Gardens Production Greenhouse.

The General Session and WSSA/WSWS Awards Ceremony will begin Monday at 4:00 PM in Salon III. This will include our Keynote Address by Jake Li, Deputy Assistant Administrator for Pesticide Programs within EPA's Office of Chemical Safety and Pollution Prevention. We will also hear from Stanley Culpepper and Jacob Barney, presidents of WSSA and NEWSS, respectively. We will have a short break after the General Session and head directly into the Society Awards, where we will celebrate the accomplishments of many of our Scientists! Make sure that guests are registered so they can participate in celebrating our WSSA/NEWSS Awardees starting around 7:00 PM at the social.

Tuesday through Thursday will be filled with presentations in the many different sections – including posters, regulatory, integrated weed management, weed biology and ecology, agronomic sections, and many others. Tuesday will include a Graduate Student event, NEWSS Past President's breakfast, Good, Bad, and Ugly symposium about cover crops, Endangered Species Act symposium, SST (Single Slide Talk) contests for students, NEWSS paper contest, and many other exciting presentations. Wednesday will continue with 2 more symposia, Crop-Weed Management in a Rising CO2 and Warming World, and WSSA Research Priorities Survey Results – 13 Federal Agencies Perspectives. This day will also include the President's breakfast, finals of the SST contest, Joint Student and Photo Awards, and a Dessert Social in the evening. Thursday will include a Novel Technologies Symposium, WSSA Business Breakfast, and Board meetings.

We would also like to thank Marty Schraer, and Darrin Dodds for their help in organizing and conducting the student contests. Thanks also to Anita Dille, Stanley Culpepper, and Greg Dahl for their help and guidance preparing the program. A great deal of gratitude is also owed to the Section Chairs, Committee Chairs (over 45 committees), contest organizers, judges, dessert social organizers (thanks Renee Keese, et al.), and other leaders that have pitched in to support the meeting. We had an overwhelming group of committed folks who submitted great symposia ideas this year and some we could not accommodate and hope to have in 2024 in San Antonio. Again, we thank all our Section Chairs for helping to organize the program and especially our WSSA Executive Secretary, Mr. Eric Gustafson, for his commitment to ensure a successful meeting on all levels and the workload he takes on during the month of January. If you run into him at the meeting, please express your gratitude. We are excited about the great participation and the many opportunities to exchange scientific information on weed science research, education, and extension. We look forward to a productive and rewarding meeting.

Carroll Moseley, WSSA Program Chair and President-Elect
Wesley Everman, NEWSS Program Chair and President-Elec

2023 Program Committee

General Program Chair..........	Carroll Moseley
Vice Chair..........................	Gregory Dahl
Agronomic Crops..................	Daljit Singh and Craig Alford
Horticultural Crops................	Kurt Vollmer and Thierry Besancon
Turf and Ornamentals.............	David Hillger and Mayank Milik
Pastures, Rangelands,............. Forests, Rights of Way, Wildlands and Aquatic Invasive Plants	Scott Nolte and Byron Sleugh
Regulatory Aspects................	Amanda Foderaro and Cherilyn Moore
Teaching and Extension..........	Dawn Refsell and Debalin Sarangi
Formulation, Adjuvant, ... & Application Technology	Joe Ikley and Gourav Harma
Weed Biology and Ecology.......	Debalin Sarangi and Shilpa Singh
Biocontrol of Weeds..............	Rin Ragamaihi
Physiology..........................	Chenxi Wu and Chad Brabham
Integrated Weed.................... Management	Sarah Kezar and Ramawater Yadav
Sustaining Member Exhibits......	Kyle Kepner
Poster Sessions......................	Tom Mueller
Student Contest.....................	Marty Schraer
Student Travel Enrichment Experience...........................	Sarah Kezar

Program Booklet and Abstracts

Please also download the meeting app for the most up to date schedules and information.

Posters and Sustaining Member Exhibits

All posters and Sustaining Member Exhibits will be on display from Tuesday at 8am to Thursday at 11am. The times will not be replicated daily throughout the program.

CONDENSED PROGRAM

SUNDAY JANUARY 29

US Botanic Garden Production Facility Tour	09:00 AM - 01:00 PM	
WSSA Board of Directors Meeting	01:00 PM - 05:00 PM	Alexandria

MONDAY JANUARY 30

WSSA Board of Directors Meeting	07:00 AM - 09:30 AM	Alexandria
NEWSS Board of Directors Meeting	08:00 AM - 12:00 PM	McLean
Smithsonian Greenhouse Tour	09:00 AM - 01:00 PM	
Registration	01:00 PM - 04:00 PM	Arlington Reception
Women in Weed Science Meeting	02:00 PM - 03:30 PM	Rosslyn
General Session	04:00 PM - 07:00 PM	Salon III

Awards Reception	07:00 PM - 09:00 PM	Foyer

TUESDAY JANUARY 31

Registration	07:00 AM - 03:00 PM	Arlington Registration
Graduate Student Event - Overcoming Hurdles - Grad Student Only	07:00 AM - 08:00 AM	Salon H
Judges Work Room	07:00 AM - 03:00 PM	Alexandria
NEWSS Past President Breakfast	07:00 AM - 08:00 AM	Salon G
POSTER - Committee Reports	08:00 AM - 10:00 AM	Salon IV-VI
POSTER - 01. Agronomic Crops -	08:00 AM - 10:00 AM	Salon IV-VI
POSTER - 02. Horticultural Crops	08:00 AM - 10:00 AM	Salon IV-VI
POSTER - 03. Turf and Ornamentals	08:00 AM - 10:00 AM	Salon IV-VI
POSTER - 04. Pasture, Range, Forest, & Rights of ways, Wildland, and	08:00 AM - 10:00 AM	Salon IV-VI

Aquatic Invasive plants		
POSTER - 06. Teaching and Extension	08:00 AM - 10:00 AM	Salon IV-VI
POSTER - 07. Formulation, Adjuvant, & Application Technology	08:00 AM - 10:00 AM	Salon IV-VI
POSTER - 08. Weed Biology and Ecology	08:00 AM - 10:00 AM	Salon IV-VI
POSTER - 09. Biocontrol of Weeds	08:00 AM - 10:00 AM	Salon IV-VI
POSTER - 10. Physiology	08:00 AM - 10:00 AM	Salon IV-VI
POSTER - 11. Soil and Environmental Aspects	08:00 AM - 10:00 AM	Salon IV-VI
POSTER - 12. Integrated Weed Management	08:00 AM - 10:00 AM	Salon IV-VI
SYMPOSIUM - The good, the bad, and the ugly the current state of cover crops and weed management	08:00 AM - 12:00 PM	Salon III
ORAL - 08. Weed Biology and Ecology	09:00 AM - 02:00 PM	Salon I

ORAL - 05. Regulatory Aspects	09:00 AM - 01:45 PM	Salon II
ORAL - NEWSS Student Oral	09:00 AM - 04:30 PM	Salon J
SST MS	10:15 AM - 11:30 AM	Salon K
SST PhD	10:15 AM - 12:00 PM	Salon C
SYMPOSIUM - Endangered Species Act: A workshop to determine how the WSSA can provide useful science-based information to regulators	01:00 PM - 05:00 PM	Salon III
ORAL - 13. Student Travel Enrichment Experience	02:00 PM - 04:30 PM	Salon C
NEWSS Business Meeting	05:00 PM - 07:00 PM	Salon II

WEDNESDAY FEBRUARY 1

Presidents Breakfast	07:00 AM - 08:00 AM	Salon J
Judges Work Room	07:00 AM - 03:00 PM	Alexandria
POSTER - NEWSS Student Poster	08:00 AM - 10:00 AM	Salon IV-VI

SYMPOSIUM - Crop-Weed Management in a Rising CO2 and Warming World	08:00 AM - 12:00 PM	Salon II
Oral 01. Agronomic Crops - Day 1	09:00 AM - 05:00 PM	Salon III
ORAL - 07. Formulation, Adjuvant, & Application Technology	09:30 AM - 02:00 PM	Salon H
ORAL - 12. Integrated Weed Management	09:00 AM - 05:00 PM	Salon I
ORAL - 02. Horticultural Crops	02:00 PM - 05:00 PM	Salon K
ORAL - 03. Turf and Ornamentals	09:00 AM - 11:30 AM	Salon F
ORAL - 10. Physiology	09:00 AM - 02:00 PM	Salon B
Joint Graduate Student Luncheon	11:30 AM - 01:00 PM	Salon J
SYMPOSIUM - WSSA Research Priorities Survey Results: 13 Federal Agency-Perspectives and -Funding Opportunities	01:00 PM - 05:00 PM	Salon II

ORAL - 09. Biocontrol of Weeds	02:00 PM - 02:30 PM	Salon B
SST - SST PhD - Finals	02:00 PM - 03:00 PM	Salon C
ORAL - 11. Soil and Environmental Aspects	02:30 PM - 03:00 PM	Salon B
Joint Student/ Photo Awards	05:00 PM - 07:00 PM	Salon III
Joint Dessert Social	09:00 PM - 11:00 PM	Foyer

THURSDAY FEBRUARY 2

WSSA Business Breakfast	06:30 AM - 08:00 AM	Salon H
Registration	08:00 AM - 10:00 AM	Arlington Registration
ORAL - 01. Agronomic Crops - Day 2	08:00 AM - 12:00 PM	Salon III
SYMPOSIUM - Novel Technology for Weed Management: An Update on WSSA Member Activities from Around the US	08:00 AM - 12:00 PM	Salon I
ORAL - 04. Pasture, Range,	09:00 AM - 11:30 AM	Salon II

Forest, & Rights of ways, Wildland, and Aquatic Invasive plants		
ORAL - 06. Teaching and Extension	09:00 AM - 12:00 PM	Salon J
US-HRAC Meeting	12:00 PM - 04:00 PM	Salon K
WSSA Board of Directors Meeting	02:00 PM - 04:00 PM	Alexandria
NEWSS Board of Directors Meeting	02:00 PM - 04:00 PM	Crystal Boardroom

PROGRAM

SUNDAY MORNING JANUARY 29

US Botanic Garden Production Facility Tour

LOCATION:
TIME: 09:00 AM - 01:00 PM Eastern Time

Tours will meet in the Hotel Lobby for transportation

SUNDAY AFTERNOON JANUARY 29

WSSA Board of Directors Meeting

LOCATION:	Alexandria
TIME:	01:00 PM - 05:00 PM Eastern Time
MODERATOR:	A Stanley Culpepper University of Georgia Tifton, GA

MONDAY MORNING JANUARY 30

WSSA Board of Directors Meeting

LOCATION:	Alexandria
TIME:	07:00 AM - 09:30 AM Eastern Time
MODERATOR:	A Stanley Culpepper University of Georgia Tifton, GA

MONDAY MORNING JANUARY 30

NEWSS Board of Directors Meeting

LOCATION:	McLean
TIME:	08:00 AM – 12:00 PM Eastern Time
MODERATOR:	Jacob Barney Virginia Tech Blacksburg, VA

MONDAY MORNING JANUARY 30

Smithsonian Greenhouse Tour

LOCATION:
TIME: 09:00 AM - 01:00 PM Eastern Time

Meet in the hotel Lobby for transportation

MONDAY AFTERNOON JANUARY 30

Registration

LOCATION: Arlington Registration
TIME: 01:00 PM - 04:00 PM Eastern Time

MONDAY AFTERNOON JANUARY 30

Women in Weed Science Meeting

LOCATION: Rosslyn
TIME: 2:00 PM - 03:30 PM Eastern Time
MODERATOR: Cara L. McCauley
Corteva Agriscience
Indianapolis, IN

MONDAY AFTERNOON JANUARY 30

General Session

LOCATION: Salon III
TIME: 04:00 PM - 07:00 PM Eastern Time
MODERATOR: A Stanley Culpepper
University of Georgia

Tifton, GA
CO-MODERATOR: Jacob Barney
Virginia Tech
Blacksburg, VA

***SPEAKER**

04:00 PM **Introduction and Announcements.** Carroll Moseley*[1], Wesley Everman[2]; [1]Syngenta, Brown Summit, NC, [2]North Carolina State University, Raleigh, NC

04:10 PM **IWSC 2024 - Jerusalem and IWSS Update.** Nilda Roma-Burgos*; University of Arkansas, Fayetteville, Fayetteville, AR

04:20 PM **Keynote Presentation - Jake Li - Deputy Assistant Administrator for Pesticide Programs - Protecting Endangered Species from Pesticides: Why it Matters to Your Work.** Jake Li*; US EPA - Deputy Assistant Administrator for Pesticide Programs, Washington, DC

05:05 PM **Presidential Addresses.** A Stanley Culpepper*[1], Jacob Barney[2]; [1]University of Georgia, Tifton, GA, [2]Virginia Tech, Blacksburg, VA

05:30 PM **Awards Presentation.** Mandy Bish*[1], Caren A. Schmidt[2]; [1]University of Missouri, Columbia, MO, [2]BASF, Durham, NC

06:15 PM **Presentation of Honorary Members and Fellows.** Carroll Moseley*[1], Wesley Everman[2]; [1]Syngenta, Brown Summit, Nc, NC, [2]North Carolina State University, Raleigh, NC

MONDAY AFTERNOON JANUARY 30

Awards Reception

LOCATION: Foyer
TIME: 07:00 PM - 09:00 PM Eastern Time

Awardee Reception

TUESDAY MORNING JANUARY 31

Registration

LOCATION: Arlington Registration
TIME: 07:00 AM - 03:00 PM Eastern Time

TUESDAY MORNING JANUARY 31

Graduate Student Event - Overcoming Hurdles - Grad Student Only

LOCATION: Salon H
TIME: 07:00 AM - 08:00 AM Eastern Time
MODERATOR: Sarah E. Kezar
Texas A&M University
College Station, TX

Grad students only

TUESDAY MORNING JANUARY 31

Judges Work Room

LOCATION: Alexandria
TIME: 07:00 AM - 03:00 PM Eastern Time

TUESDAY MORNING JANUARY 31

NEWSS Past President Breakfast

LOCATION: Salon G
TIME: 07:00 AM - 09:00 AM Eastern Time
MODERATOR: Jacob Barney
Virginia Tech
Blacksburg, VA

TUESDAY MORNING JANUARY 31

POSTER - Committee Reports

LOCATION: Salon IV-VI
TIME: 08:00 AM - 10:00 AM Eastern Time

***PRESENTER**

WSSA Public Awareness Committee Update. Theresa Piskackova*[1], Eric Gustafson[2]; [1]Czech University of Life Sciences Prague, Prague 6 - Suchdol, Czechia (Czech Republic), [2]Weed Science Society of America, Westminster, CO

TUESDAY MORNING JANUARY 31

POSTER - 01. Agronomic Crops -

LOCATION: Salon IV-VI
TIME: 08:00 AM - 10:00 AM Eastern Time

WSSA Student Competition Posters will be judged Tuesday only
NEWSS Student Competition Posters will be judged Wednesday only

***PRESENTER † STUDENT CONTEST**

Taxonomic Pattern in Seed Dormancy and its Breaking Methods in Globally Important Weed Species. Aniruddha Maity*, Akashdeep Singh; Auburn University, Auburn, AL **(1)**

The Interaction of Pyroxasulfone and Flumioxazin Applied Preemergence for the Control of Multiple-herbicide-resistant Waterhemp [Amaranthus Tuberculatus (Moq) J.D. Sauer] in Soybean. James C. Ferrier, Nader Soltani*, Peter H. Sikkema; University of Guelph, Ridgetown, ON, Canada **(2)**

†Planting Green in Soybean: Evaluation of Residual Herbicide Interaction, Weed Control Programs, and Soybean Yield. Trey P. Stephens*, Amit J. Jhala; University of Nebraska-Lincoln, Lincoln, NE **(3)**

Control of Larger Herbicide-Resistant Common Ragweed (*Ambrosia artemisiifolia*) in Soybean. Kurt M. Vollmer*; University of Maryland, Queenstown, MD **(4)**

†Urea Ammonium Nitrate as the Carrier for Preplant Burndown Herbicides. Kaone Mookodi*[1], Jared Spackman[2], Albert T. Adjesiwor[1]; [1]University of Idaho, Kimberly, ID, [2]University of Idaho, Aberdeen, ID **(5)**

Brake, Valor, and Their Tank-Mixture Combinations with Residual Herbicides for Broad- Spectrum Weed Management Programs in Mississippi Peanut. Taghi Bararpour*; Mississippi State University, Stoneville, MS **(6)**

†Weed Control Efficacy and Dry Bean Response to Preemergence Herbicide Activation Timing. Prayusha Bhattarai*[1], Albert T. Adjesiwor[2]; [1]University of Idaho, Moscow, ID, [2]University of Idaho, Kimberly, ID **(7)**

Postharvest Control of Russian Thistle (*Salsola kali*) in Spring Wheat Stubble. Fernando H. Oreja*[1], Drew J. Lyon[2], Jennifer A. Gourlie[3], Henry C. Wetzel[2], Judit Barroso[1]; [1]Oregon State University, Adams, OR, [2]Washington State University, Pullman, WA, [3]Oregon State University, Pendleton, OR **(8)**

†Identification of Palmer Amaranth Using Machine Learning. Akhilesh Sharma*[1], Vipan Kumar[2], Ajay Sharda[1], William Hsu[1]; [1]Kansas State University, Manhattan, KS, [2]Kansas State University, Hays, KS **(9)**

Drill-Spacing And Bed Width Influence on Canopy Coverage, Weed Control and Yield in Furrow-Irrigated Rice. Koffi Badou-Jeremie Kouame*[1], Thomas R. Butts[1], Noah H. Reed[2], Brad M. Davis[1], Leah M. Collie[3], Troy W. Dillon[1], Jason K. Norsworthy[2], Tom Barber[1], Jarrod Hardke[4], Jason A. Bond[5], Hunter D. Bowman[6]; [1]University of Arkansas System Division of Agriculture, Lonoke, AR, [2]University of Arkansas,

Fayetteville, AR, [3]University of Arkansas, Lonoke, AR, [4]University of Arkansas System Division of Agriculture, Stuttgart, AR, [5]Mississippi State University, Stoneville, MS, [6]Mississippi State University, Starkville, MS **(10)**

†Evaluating the Metabolic Response of Cloquintocet-Mexyl in Winter Wheat. William Kramer*, Todd A. Gaines, Franck E. Dayan; Colorado State University, Fort Collins, CO **(11)**

Confirmation of Glyphosate Resistance in New York Palmer Amaranth (*Amaranthus palmeri*) Populations and Response to Alternative Chemistries. Aleah L. Butler-Jones*[1], Elizabeth C. Maloney[2], Lynn M. Sosnoskie[2]; [1]Cornell University, Ithaca, NY, [2]Cornell University, Geneva, NY **(12)**

†Does Row Spacing Matter for Weed Suppression in Corn and Soybean Cropping Systems? A Meta-analysis. Mandeep Singh*[1], Resham Thapa[2], Navdeep Singh[3], Steven Brian Mirsky[4], Amit J. Jhala[1]; [1]University of Nebraska-Lincoln, Lincoln, NE, [2]USDA-ARS Beltsville Agricultural Research Center, Beltsville, NC, [3]Punjab Agricultural University, Ludhiana, India, [4]USDA ARS, Beltsville, MD **(13)**

Response of Glyphosate-Resistant Kochia (*Bassia scoparia*) to Dichlorprop-P, 2,4-D, Dicamba Alone or Their Premixture. Sachin Dhanda*[1], Vipan Kumar[1], Augustine Obour[1], Elizabeth Yeager[2], John Holman[3], Anita Dille[2]; [1]Kansas State University, Hays, KS, [2]Kansas State University, Manhattan, KS, [3]Kansas State University, Garden City, KS **(14)**

†Impacts of Florpyrauxifen-benzyl on Rice and Smallflower Umbrella Sedge, *Cyperus difformis*, at Different Growth Stages. Deniz Inci*, Kassim Al-Khatib; University of California, Davis, Davis, CA **(15)**

Late-Season Control of Glyphosate-Resistant Palmer Amaranth with Glufosinate in XtendFlex Soybean. Vipan Kumar*[1], Sachin Dhanda[1], Rui Liu[2], Amit J. Jhala[3]; [1]Kansas State University, Hays, KS, [2]Washington State University, Prosser, WA, [3]University of Nebraska Lincoln, Lincoln, NE **(16)**

†Weed Survey in Colorado Sugarbeet Farms Reveals Resistance to Dicamba and Glyphosate. André Lucas Simões Araujo*, Eric P. Westra, Todd A. Gaines; Colorado State University, Fort Collins, CO **(17)**

†Impact of Soil-Applied Herbicides on Early Planted Soybean. Matthew S. Goddard*, Christy L. Sprague; Michigan State University, East Lansing, MI **(18)**

Estimating Harvest Loss of Corn from Growers' Fields in Nebraska and Implication for Volunteerism. Trey P. Stephens[1], Vipin Kumar*[2], Jenny Rees[3], Amit J. Jhala[1];

[1]University of Nebraska-Lincoln, Lincoln, NE, [2]Virginia Polytechnic Institute and State University, Painter, VA, [3]University of Nebraska-Lincoln, York, NE **(19)**

†Molecular Basis for Resistance to ACCase-inhibiting Herbicides in Downy Brome (*Bromus tectorum*) Populations. Victor Ribeiro*[1], Caio A. Brunharo[2], Carol Mallory-Smith[1], Darrin Walenta[3], Judit Barroso[4]; [1]Oregon State University, Corvallis, OR, [2]Penn State University, University Park, PA, [3]Oregon State University, La Grande, OR, [4]Oregon State University, Adams, OR **(20)**

Expression Levels of Selected GSTs in Palmer Amaranth Resistant to Trifluralin. Fidel Gonzalez Torralva*, Jason K. Norsworthy; UNIVERSITY OF ARKANSAS, Fayetteville, AR **(21)**

†Impact of Spring-Applied Herbicides on Cover Crops Planted After Wheat. Claudia R. Walz*, Christy L. Sprague; Michigan State University, East Lansing, MI **(22)**

PSII Inhibitor Resistance in Italian Ryegrass (Lolium Multiflorum) Populations in Louisiana Sugarcane. Alice A. Wright*[1], Gerald A. McCollam[1], Albert J. Orgeron[2]; [1]USDA-ARS, Houma, LA, [2]LSU AgCenter, Baton Rouge, LA **(23)**

†Optimizing Dicamba and Glufosinate Application Timings for Weed Control in XtendFlex Soybeans in Central Kansas. Monica R. Marrs*[1], Vipan Kumar[2], Anita Dille[1], Rui Liu[3], Augustine Obour[2]; [1]Kansas State University, Manhattan, KS, [2]Kansas State University, Hays, KS, [3]Washington State University, Prosser, WA **(24)**

Optimizing Cover Crop Termination and Herbicide Application Timings to Manage Herbicide-Resistant Waterhemp in Soybean. Ramawatar Yadav*[1], Edward S. Dearden[2], Ryan Hamberg[1], Alexis L. Meadows[1], Austin H. Schleich[1], Prashant Jha[1]; [1]Iowa State University, Ames, IA, [2]Iowa State University, Crop Production & Physiology, Ames, IA **(25)**

†Palmer Amaranth (*Amaranthus palmeri*) Control Options in the Absence of Dicamba. Tyler S. Soignier*, Brian K. Pieralisi, Darrin M. Dodds, Jason Krutz, Dave Spencer, Whitney S. Crow, Bradley J. Norris, William J. Rutland, Eli B. Hobbs; Mississippi State University, Starkville, MS **(26)**

Investigation of Fluroxypyr Resistance in a Kochia Population from Colorado. Eduardo Souza de Amorim*, Olivia E. Todd, Eric P. Westra, André Lucas Simões Araujo, Todd A. Gaines; Colorado State University, Fort Collins, CO **(27)**

†Fitness of Different Ploidy Plants of F1 Progeny Derived from an Interspecific Cross Between Cultivated Sorghum and Johnsongrass. Usha R. Pedireddi*, Rana Afshang, Maximillian Chung, Kayla Buenaventura, Aleena Jiby, Nithya K. Subramanian, Muthukumar V. Bagavathiannan; Texas A&M University, College Station, TX **(28)**

Differential Control of ALS- and PPO-inhibiting Herbicide-resistant Amaranthus Retroflexus Populations with Other Postemergence Herbicides. Eric A. Jones*, Diego J. Contreras, Charlie W. Cahoon, Ramon G. Leon, Wesley Everman; North Carolina State University, Raleigh, NC **(29)**

†Evaluation of Fall-Applied Residual Herbicides for Weed Management in Chickpeas(Cicer Arietinum). Akamjot S. Brar*[1], Qasim Khan[1], Fabian Menalled[1], Zach Miller[1], Clint Beiermann[1], Kent McVay[1], Lovreet S. Shergill[2]; [1]Montana State University, Bozeman, MT, [2]Montana State University, Huntley, MT **(30)**

What Are We Missing? Undetected Changes to Fiber Quality from Low Rates of 2,4-D Using AFIS. Kyle R. Russell*[1], Peter A. Dotray[2], Brendan R. Kelly[1]; [1]Texas Tech University, Lubbock, TX, [2]Texas Tech University, Texas A&M AgriLife Research, Texas A&M AgriLife Extension, Lubbock, TX **(31)**

~~**†Effect of Als and 4-hppd Inhibitors Herbicides on Maize Genotypes.** José I. Saavedra-ávila*; COLPOS, Texcoco, EM, Mexico **(32)**~~

Developer of Novel Adjuvant Systems for Herbicides. Jim T. Daniel[1], Chase T. Boman*[2]; [1]Daniel Ag Consulting, Keenesburg, CO, [2]AgraSyst, Spokane, WA **(33)**

†Seed-Shattering Phenology of Grass and Broadleaf Weeds at Soybean (*Glycine max*) and Corn (*Zea mays*) Harvest in SE Montana. Het Samir Desai*[1], Fabian Menalled[2], Lovreet S. Shergill[3]; [1]Southern Agricultural Research Center, Montana State University, Huntley, MT, [2]Montana State University, Bozeman, MT, [3]Montana State University, Huntley, MT **(34)**

Confirmation of a Three-way (Thifensulfuron, Atrazine and Glyphosate) Herbicide Resistant Population of Palmer Amaranth (*Amaranthus palmeri*) in North Carolina. Ronel J. Argueta*, Eric A. Jones, Diego J. Contreras, Charlie W. Cahoon, Wesley Everman; North Carolina State University, Raleigh, NC **(35)**

†Axant™ Flex Cotton Response and Residual Weed Management Using HPPD Inhibiting Herbicides. Megan Mills*[1], Peter A. Dotray[2], Gregory B. Baldwin[3], Scott Asher[4], Adam C. Hixson[4], Maxwell E. Smith[1], Blaine A. Patton[1], Kyle

R. Russell[1], Bobby Rodriguez[5]; [1]Texas Tech University, Lubbock, TX, [2]Texas Tech University and Texas A&M AgriLife Research & Extension Service, Lubbock, TX, [3]BASF, Research Triangle Park, NC, [4]BASF, Lubbock, TX, [5]Texas A&M AgriLife Extension, Lubbock, TX **(36)**

Hybrid Rice Tolerance with Various Fenclorim Seed Treatments and Microencapsulated Acetochlor. Tristen H. Avent*, Jason K. Norsworthy, Leonard B. Piveta, Casey H. Arnold, Pamela Carvalho-Moore; University of Arkansas, Fayetteville, AR **(37)**

†Transgenes Flow from Corn Confer Resistance to Glyphosate and Glufosinate-Ammonium in Teosinte (__Zea Mays__ Subsp. __Mexicana__). Luan Cutti*[1], Paula Sinigaglia Angonese[1], Guilherme Menegol Turra[1], Estéfani Sulzbach[1], Catarine Markus[1], Todd A. Gaines[2], Aldo Merotto[1]; [1]Federal University of Rio Grande do Sul, Porto Alegre, Brazil, [2]Colorado State University, Fort Collins, CO **(38)**

Comparing Remotely-Sensed Data for Estimating Palmer Amaranth and Soybean Canopy Cover. Byron Evers[1], Hannah Buessing[2], Christopher Weber[2], Sarah Lancaster*[2]; [1]Bayer CropScience, Manhattan, KS, [2]Kansas State University, Manhattan, KS **(39)**

†A Metabolization System Affected by Temperature is Associated with Florpyrauxifen-benzyl Resistance in Barnyardgrass (__Echinochloa__ __Crus-galli__). Luan Cutti*[1], Crystal D. Sparks[2], André Lucas Simões Araujo[2], Angela Da Cas Bundt[3], Catarine Markus[1], Todd A. Gaines[2], Aldo Merotto[1]; [1]Federal University of Rio Grande do Sul, Porto Alegre, Brazil, [2]Colorado State University, Fort Collins, CO, [3]Corteva Agriscience, Santa Maria, Brazil **(40)**

Evaluation of the Axant™ Flex Cotton System. Zachary R. Treadway*[1], Jennifer Dudak[1], Todd A. Baughman[1], Adam C. Hixson[2], Gregory B. Baldwin[3]; [1]Oklahoma State University, Ardmore, OK, [2]BASF, Lubbock, TX, [3]BASF, Research Triangle Park, NC **(41)**

†Use of New Sorghum Herbicide Technologies to Control Annual Grass Weeds. Blaine A. Patton*[1], Peter A. Dotray[2]; [1]Texas Tech University, Lubbock, TX, [2]Texas Tech University, Texas A&M AgriLife Research, Texas A&M AgriLife Extension, Lubbock, TX **(42)**

Distribution of Target Site Mutations in PPO2 Gene in Waterhemp (Amaranthus Tuberculatus) and Palmer Amaranth (A. Palmeri) Populations in the USA Conferring Resistance to PPO Inhibiting Herbicides. Daljit Singh[1], Jacob S. Montgomery[2], Jaishree Chitoor[3], Chandrashekar Aradhya[3], Alejandro Perez-Jones*[3]; [1]Bayer, Chesterfield, MO, [2]Colorado

State University, Fort Collins, CO, [3]Bayer Crop Science, Chesterfield, MO **(43)**

Palmer Amaranth (*Amaranthus palmeri*) Management in Kansas Cotton (*Gossypium hirsutum*). Tyler P. Meyeres*[1], Lucas Haag[2], Gaylon Morgan[3], Sarah Lancaster[1]; [1]Kansas State University, Manhattan, KS, [2]Kansas State University, Colby, KS, [3]Cotton Incorporated, Cary, NC **(44)**

†Changes in Nutrient Partitioning Following Low Rates of 2,4-D. Kyle R. Russell*[1], Peter A. Dotray[2], Irish L. B. Pabuayon[3], Glen L. Ritchie[1]; [1]Texas Tech University, Lubbock, TX, [2]Texas Tech University, Texas A&M AgriLife Research, Texas A&M AgriLife Extension, Lubbock, TX, [3]LSU AgCenter, Rayne, LA **(45)**

First Report of Multiple-Herbicide Resistance in Walter's Barnyardgrass in Texas Rice. Nithya K. Subramanian*, Tamara T. Mundt, Isidor Ceperkovic, Gabriella Elizarraras, Muthukumar V. Bagavathiannan; Texas A&M University, College Station, TX **(46)**

†Preemergence Control of *Amaranthus palmeri* and Axant™ Flex Cotton Weed Management Systems Using Isoxaflutole. Maxwell E. Smith*[1], Peter A. Dotray[2], Adam C. Hixson[3]; [1]Texas Tech University, Lubbock, TX, [2]Texas Tech University, Texas A&M AgriLife Research, Texas A&M AgriLife Extension, Lubbock, TX, [3]BASF, Lubbock, TX **(47)**

Evaluation of Axant Flex Systems in Mid-South Cotton. Bradley J. Norris*, Brian K. Pieralisi, Darrin M. Dodds, Tyler S. Soignier, William J. Rutland; Mississippi State University, Starkville, MS **(48)**

†Crabgrass and Texas Panicum Control with Group 15 Herbicides Applied Preemergence. Livia Pereira*, Steve Li, Ryan D. Langemeier, Justin T. McCaghren; Auburn University, Auburn, AL **(49)**

Effect of Grassy Weed Herbicides on Cereal Rye. Ethan B. Whitmoyer*, Dwight Lingenfelter, John M. Wallace; Penn State University, University Park, PA **(50)**

†Evaluating Weed Control and Crop Response of Pyridate in Field Pea. Adam Leise*[1], Amit J. Jhala[1], Dennis Long[2]; [1]University of Nebraska Lincoln, Lincoln, NE, [2]Belchim, Manchester, TN **(51)**

Evaluation of Biological Nitrification Inhibition (BNI) Potential in Cover Crops. Nithya K. Subramanian*, Nithya Rajan, Sanjay Antony-Babu, Tamara T. Mundt, Megan L. Schill, Dinesh Phuyal, Muthukumar V. Bagavathiannan; Texas A&M University, College Station, TX **(52)**

†**Long-term Palmer Amaranth (*Amaranthus palmeri*) Management in Dicamba-Resistant Cotton Using Herbicide Programs, Crop Rotations, and Tillage.** Rohith Vulchi*[1], Joshua A. McGinty[2], Scott A. Nolte[3]; [1]Texas A&M University, College Station, TX, [2]Texas A&M AgriLife Extension, Corpus Christi, TX, [3]Texas A&M AgriLife Extension, College Station, TX **(53)**

Spot Treatment of Clethodim on Weedy Rice. Rasim Unan*, Liberty B. Galvin, Aaron Becerra-Alvarez; University of California, Davis, Davis, CA **(54)**

†**Utilizing Novel Application Methods to Mitigate Auxin + Graminicide Herbicide Antagonism.** Jake A. Patterson*[1], Darrin M. Dodds[1], Antonio Augusto Correa Tavares[2], Gresham A. Stephens[1], Caleb K. Meyer[1], Sydney C. Baker[1]; [1]Mississippi State University, Starkville, MS, [2]University of Nebraska-Lincoln, North Platte, NE **(55)**

Early Season Weed Control in Rice on Organic Soils with Preemergence Herbicides. D Calvin Odero*; University of Florida, Belle Glade, FL **(56)**

†**Impact of Cereal Rye Cover Crop Seeding Rate and Termination Timing on Weed Management in Soybean.** Eric Y. Yu*[1], Axel Garcia y Garcia[2], Lizabeth Stahl[3], Ryan P. Miller[4], Gregg A. Johnson[1], Ce Yang[1], William Lazarus[1], Debalin Sarangi[1]; [1]University of Minnesota, St. Paul, MN, [2]University of Minnesota, Lamberton, MN, [3]University of Minnesota Extension, Worthington, MN, [4]University of Minnesota, Rochester, MN **(57)**

Intercropping Soybean and Corn into the New Winter Oilseed Crop Golden Pennycress (*Thlaspi arvense*). Brent S. Heaton*, Tad L. Wesley, Winthrop B. Phippen, Mark L. Bernards; Western Illinois University, Macomb, IL **(58)**

Quantifying the Impact of Herbicide Drift on Peanuts Using Unmanned Aerial System. Pratap Devkota*, Navjot Singh, Joseph E. Iboyi; University of Florida, Jay, FL **(59)**

Weed Seed Fate at Corn Harvest. Haylee E. Schreier*[1], Jacob E. Vaughn[1], Kevin W. Bradley[1], Bryan G. Young[2], Jason K. Norsworthy[3], Prashant Jha[4], Travis Legleiter[5], Bill Johnson[6]; [1]University of Missouri, Columbia, MO, [2]Purdue University, Brookston, IN, [3]University of Arkansas, Fayetteville, AR, [4]Iowa State University, Ames, IA, [5]University of Kentucky, Princeton, KY, [6]Purdue University, Brooksone, IN **(60)**

Seed Germination Responses of Branched Broomrape (*Phelipanche ramosa*) to Different Sanitation Chemicals and

Exposure Durations. Pershang Hosseini*; Brad Hanson, University of California, Davis, Davis, CA **(61)**

TUESDAY MORNING JANUARY 31

POSTER - 02. Horticultural Crops

LOCATION:	Salon IV-VI
TIME:	08:00 AM - 10:00 AM Eastern Time

WSSA Student Competition Posters will be judged Tuesday only
NEWSS Student Competition Posters will be judged Wednesday only

***PRESENTER † STUDENT CONTEST**

Sensitivity of Selected mid-Atlantic Vegetables to Simulated Dicamba Drift Rates. Maggie H. Wasacz[1], Baylee L. Carr[2], Mark VanGessel[3], Thierry E. Besancon*[4]; [1]Rutgers University, Wall Township, NJ, [2]Rutgers University - Weed Science, Chatsworth, NJ, [3]University of Delaware, Georgetown, DE, [4]Rutgers University, Chatsworth, NJ **(62)**

†Integrating Pre-emergence Herbicides with Potential Safeners for Purple Nutsedge (*Cyperus rotundus* L.) Management in Florida's Tomato Plasticulture Production. Ruby Tiwari*[1], Nathan Boyd[2], Pamela Roberts[1], Samira Daroub[3], Ramdas Kanissery[1]; [1]University of Florida, Immokalee, FL, [2]University of Florida, Balm, FL, [3]University of Florida, Everglades, FL **(63)**

Carolina Redroot: Long Term Management with Sequential Residual Herbicide Applications in Cranberry. Thierry E. Besancon*[1], Baylee L. Carr[2], Wesley M. Bouchelle[1]; [1]Rutgers University, Chatsworth, NJ, [2]Rutgers University - Weed Science, Chatsworth, NJ **(64)**

†Evaluation of Weed Removal Timing for Three Sweetpotato (*Ipomoea batatas* L.) Cultivars. Emmanuel G. Cooper*[1], Stephen L. Meyers[1], Jeanine Arana[2]; [1]Purdue University, West Lafayette, IN, [2]Purdue University/Horticulture, West Lafayette, IN **(65)**

Assessing the Risk of Injury Induced by Chloroacetamide and Diphenylether Mixing in Cole Crops Production. Wesley M. Bouchelle*[1], Elizabeth C. Maloney[2], Lynn M. Sosnoskie[2],

Thierry E. Besancon[1]; [1]Rutgers University, Chatsworth, NJ, [2]Cornell University, Geneva, NY **(66)**

†Evaluation of Weed Control Efficacy and Crop Safety of the PPO-inhibiting Herbicide, Tiafenacil in California Orchard Cropping Systems. Guelta Laguerre*; Brad Hanson, UC Davis Department of Plant Sciences, Davis, CA **(67)**

Evaluating Vision-Guided Spraying for Use in Vineyards. Bethany M. Schulteis*[1], Elizabeth C. Maloney[2], Emily J. Sausser[3], Terry Bates[1], Lynn M. Sosnoskie[2]; [1]Cornell University, Ithaca, NY, [2]Cornell University, Geneva, NY, [3]Lebanon Valley College, Annville, PA **(68)**

†Interference of Palmer Amaranth (*Amaranthus palmeri*) in Stevia (*Stevia rebaudiana*). Stephen Ippolito*[1], Katherine M. Jennings[1], David W. Monks[1], Sushila Chaudhari[2], David L. Jordan[1], Levi D. Moore[1], Colton D. Blankenship[1], Kira C. Sims[1], Fnu Chitra[1]; [1]North Carolina State University, Raleigh, NC, [2]Michigan State University, East Lansing, MI **(69)**

Influence of Growth Stage and Dicamba Rate on Eggplant, Cucumber, and Snap Bean Tolerance and Yield Response. Maggie H. Wasacz[1], Lynn M. Sosnoskie[2], Mark VanGessel[3], Thierry E. Besancon*[4]; [1]Rutgers University, Wall Township, NJ, [2]Cornell University, Geneva, NY, [3]University of Delaware, Georgetown, DE, [4]Rutgers University, Chatsworth, NJ **(70)**

†Effect of S-metolachlor and Flumioxazin Herbicides on Sweetpotato Treated with and without Activated Charcoal Applied Through Transplant Water. Colton D. Blankenship*[1], Katherine M. Jennings[1], David W. Monks[1], David L. Jordan[1], Stephen L. Meyers[2], Jonathan R. Schultheis[1], David H. Suchoff[1]; [1]North Carolina State University, Raleigh, NC, [2]Purdue University, West Lafayette, IN **(71)**

Tiafenacil Alone and in Combination with Preemergence Herbicides for Burndown and Residual Weed Control in Grapes. Elizabeth C. Maloney*, Lynn M. Sosnoskie; Cornell University, Geneva, NY **(72)**

In Search of New Weed Management Tools in Spinach, Cilantro, and Parsley for Georgia Farmers. A Stanley Culpepper*, Jenna C. Vance, Taylor M. Randell-Singleton; University of Georgia, Tifton, GA **(73)**

Current Status of Weed Management in Washington State Vegetable Production. Rui Liu*[1], Tim Waters[2]; [1]Washington State University, Prosser, WA, [2]Washington State University, Pasco, WA **(74)**

Defoliation Timing and Growth Responses of Seedling Field Bindweed and Honeyvine Milkweed. Melissa McClements*[1],

Lynn M. Sosnoskie[2]; [1]Cornell, Geneva, NY, [2]Cornell University, Geneva, NY **(75)**

Hemp and Weed Competition in New York. Lynn M. Sosnoskie*, Elizabeth C. Maloney; Cornell University, Geneva, NY **(76)**

Dicamba Residue Persistence in Processing Tomato. Stephen L. Meyers*[1], Jeanine Arana[2], Brandi C. Woolam[3], Nathaly Belen Vargas Arroyo[1], Laura D. Rodriguez Baquero[4], Luz Aide Cardona Giraldo[1]; [1]Purdue University, West Lafayette, IN, [2]Purdue University/Horticulture, West Lafayette, IN, [3]LSU AgCenter, Alexandria, LA, [4]University of Wisconsin-Madison, Madison, WI **(77)**

Weed Surveys of Lima Bean (*Phaseolus lunatus*) Fields in United States Production Systems. Yudai K. Takenaka*[1], Jed Colquhoun[2], Scott Lukas[3], Pavle Pavlovic[4], Barbara Scott[5], Mark VanGessel[5], Martin Williams[6]; [1]University of Illinois at Urbana-Champaign, Urbana, IL, [2]University of Wisconsin-Madison, Madison, WI, [3]Oregon State University, Hermiston, OR, [4]University of Illinois, Champaign, IL, [5]University of Delaware, Georgetown, DE, [6]USDA-ARS, Urbana, IL **(78)**

Michigan's Experience with Sugarbeet Tolerance to Acifluorfen. Brian J. Stiles II*, Gary Edward Powell, Christy L. Sprague; Michigan State University, East Lansing, MI **(79)**

TUESDAY MORNING JANUARY 31

POSTER - 03. Turf and Ornamentals

LOCATION:	Salon IV-VI
TIME:	08:00 AM - 10:00 AM Eastern Time

WSSA Student Competition Posters will be judged Tuesday only
NEWSS Student Competition Posters will be judged Wednesday only

***PRESENTER † STUDENT CONTEST**

Efficacy of Organic Turf and Landscape Herbicides in Montgomery County, Maryland. Kelly M. Nichols*; University of Maryland Extension, Derwood, MD **(80)**

†Ethofumesate Resistant Annual Bluegrass in Oregon Grass Seed Production Systems. Vera Vukovic*, Aaron J. Patton; Purdue University, West Lafayette, IN **(81)**

Glufosinate and Clethodim Efficacy on Bermudagrass. Joe C. Neal*, Christopher D. Harlow; North Carolina State University, Raleigh, NC **(82)**

†Site-Specific Treatment of Kyllinga in Golf Fairways Using a Remotely Piloted Aerial Application System. Ubaldo Torres*[1], Daniel E. Martin[2], Bholuram Gurjar[1], Bishwa B. Sapkota[1], Andrew W. Osburn[1], Chase M. Straw[1], Muthukumar V. Bagavathiannan[1]; [1]Texas A&M University, College Station, TX, [2]United States Department of Agriculture, College Station, TX **(83)**

Non-Target Site Resistance Mechanisms in Annual Bluegrass (*Poa annua*). Ronald R. Rogers*, Travis Gannon, Khalied Ahmed, Mathieu LeCompte; North Carolina State University, Raleigh, NC **(84)**

†Characterization of Potential Quinclorac-Resistant Smooth Crabgrass (*Digitaria ischaemum*) Population. Claudia A. Rutland*[1], Bridgette C. Johnson[2], James D. McCurdy[3], Joseph S. McElroy[2]; [1]Auburn University Department of Crop, Soil, and Environmental Sciences, Auburn, AL, [2]Auburn University, Auburn, AL, [3]Mississippi State University, Starkville, MS **(85)**

Halosulfuron and Sulfentrazone Combinations for False-green Kyllinga (Kyllinga Gracillima) Control. Daniel P. Tuck*[1], Matthew T. Elmore[1], Charles A. Silcox[2]; [1]Rutgers University, New Brunswick, NJ, [2]Amvac Chemical, Lincoln University, PA **(86)**

†Phenotypic and Reproductive Trait Diversity of *Poa annua* L. (Annual Bluegrass) in Common Garden Experiments Across Multiple Ecoregions in the United States. Andrew W. Osburn*, Muthukumar V. Bagavathiannan; Texas A&M University, College Station, TX **(87)**

Mowing Height Influences False-Green Kyllinga (*Kyllinga gracillima*) Establishment in Perennial Ryegrass Turf. Matthew T. Elmore*, Daniel P. Tuck; Rutgers University, New Brunswick, NJ **(88)**

Injury to Nursery Crops from Herbicide Residues in a Contaminated Irrigation Water. Ada C. Montoya*, Joe C. Neal, Wesley Everman, Christopher D. Harlow, Travis Gannon; North Carolina State University, Raleigh, NC **(89)**

POSTER - 04. Pasture, Range, Forest, & Rights of ways, Wildland, and Aquatic Invasive plants

LOCATION: Salon IV-VI
TIME: 08:00 AM - 10:00 AM Eastern Time

WSSA Student Competition Posters will be judged Tuesday only
NEWSS Student Competition Posters will be judged Wednesday only

***PRESENTER † STUDENT CONTEST**

Guardrail Bareground Applications in KY: Spring or Fall Timing? Control Next Season? Joseph Omielan*; University of Kentucky, Lexington, KY **(90)**

†Evaluation of Hemp Dogbane (*Apocynum cannabinum*) Response to Auxinic Herbicides. Kayla L. Broster*, John D. Byrd, Jr.; Mississippi State University, Starkville, MS **(91)**

Environmental Variables Influence Native Plant Composition of Virginia Roadside. Shawn Askew*, Clebson G. Goncalves; Virginia Tech, Blacksburg, VA **(92)**

Does PRE Herbicide Timing Affect the Control of Foxtail Species (*Setaria* Spp.) in Southern Forages. Nicholas T. Basinger*[1], Chris Bocz[1], Sambit Shome[2], Hannah C. Lindell[3]; [1]The University of Georgia, Athens, GA, [2]University of Georgia, Athens, GA, [3]The University of Georgia Crop and Soil Science, Athens, GA **(93)**

Do Forage Grasses Tolerate Graminicides? Dwight Lingenfelter*, John M. Wallace; Penn State University, University Park, PA **(94)**

Invasive Aquatic and Native Marine Vegetation on the Mississippi Barrier Islands. Gray Turnage*, Andrew Sampe; Mississippi State University, Starkville, MS **(95)**

Vegetation Damage by Ecosystem Disturbance Plants '*Sicyos angulatus* and *Humulus japonicus*' in Korea. Soo-Hyun Lim*; Korea National Open University, Seoul, South Korea **(96)**

TUESDAY MORNING JANUARY 31

POSTER - 06. Teaching and Extension

LOCATION: Salon IV-VI
TIME: 08:00 AM - 10:00 AM Eastern Time

WSSA Student Competition Posters will be judged Tuesday only
NEWSS Student Competition Posters will be judged Wednesday only

***PRESENTER † STUDENT CONTEST**

P23: Join the WSSA Plant Names Committee. Theresa Piskackova*; Czech University of Life Sciences Prague, Prague 6 - Suchdol, Czechia (Czech Republic) **(97)**

†WeedR: An R Package and Shiny Application for Weed Scientists. Lucas Kopecky Bobadilla*[1], Patrick Tranel[2]; [1]University of Illinois, Champaign, IL, [2]University of Illinois, Urbana, IL **(98)**

Weeds of the Northeast - 2nd Edition. Joe C. Neal*[1], Richard H. Uva[2], Joe M. DiTomaso[3], Antonio DiTommaso[4]; [1]North Carolina State University, Raleigh, NC, [2]Seaberry Farms, LLC, Federalsburg, MD, [3]UC Davis, retired, Davis, CA, [4]Cornell University, Ithaca, NY **(99)**

†Comparing the Accuracy of Available Weather Resources forPesticide Applicators in Texas. Rohith Vulchi*[1], Zachary S. Howard[1], Don Conlee[1], Scott A. Nolte[2]; [1]Texas A&M University, College Station, TX, [2]Texas A&M AgriLife Extension, College Station, TX **(100)**

~~**Things That Work: Integrated Weed Management Teaching Modules That Stood the Test of Time.** David A. Mortensen*; University of New Hampshire, Durham, NH **(101)**~~

The Importance of Communicating Agricultural Science and Technology □ the Role of CAST and You! Anthony L. Witcher*[1], Jill Schroeder[2]; [1]Tennessee State University, Mcminnville, TN, [2]New Mexico State University, Las Cruces, NM **(102)**

Undergraduate Student Research Experience -- from Proposal to Poster Presentation in *AGRN 479 Weed Science.*

Mark L. Bernards*; Western Illinois University, Macomb, IL **(103)**

TUESDAY MORNING JANUARY 31

POSTER - 07. Formulation, Adjuvant, & Application Technology

LOCATION:	Salon IV-VI
TIME:	08:00 AM - 10:00 AM Eastern Time

WSSA Student Competition Posters will be judged Tuesday only
NEWSS Student Competition Posters will be judged Wednesday only

***PRESENTER † STUDENT CONTEST**

†Weed Management Using Unmanned Aerial Systems. Dhiraj Srivastava*[1], Vijay Singh[2], Vipin Kumar[1]; [1]Virginia Polytechnic Institute and State University, Painter, VA, [2]Virginia Tech, Painter, VA **(104)**

†Designing and Testing an Ultra-Low Volume Spray System for Site-Specific Herbicide Application Using Ground Robots. Bholuram Gurjar*, Sunil Kumar, Joe Johnson, Robert Hardin, Muthukumar V. Bagavathiannan; Texas A&M university, College Station, TX **(105)**

†Clethodim Antagonism from Dicamba on Volunteer Corn Control in Soybean Was Resolved by Using a Dual Tank/Boom Sprayer. Marcelo Zimmer*[1], Bryan G. Young[2], William G. Johnson[1], Tristen H. Avent[3], Jason K. Norsworthy[3], Diego J. Contreras[4], Wesley Everman[4], Michael M. Houston[5], Lauren M. Lazaro[6], William L. Patzoldt[6]; [1]Purdue University, West Lafayette, IN, [2]Purdue University, Brookston, IN, [3]University of Arkansas, Fayetteville, AR, [4]North Carolina State University, Raleigh, NC, [5]Blue River Technology, Greenville, MS, [6]Blue River Technology, Sunnyvale, CA **(106)**

TUESDAY MORNING JANUARY 31

POSTER - 08. Weed Biology and Ecology

LOCATION: Salon IV-VI
TIME: 08:00 AM - 10:00 AM Eastern Time

<u>WSSA Student Competition Posters will be judged Tuesday only</u>
<u>NEWSS Student Competition Posters will be judged Wednesday only</u>

***PRESENTER † STUDENT CONTEST**

The Critical Period for Weed 'Seed' Control: Concept and its Application. Charles M. Geddes*[1], Adam S. Davis[2]; [1]Agriculture and Agri-Food Canada, Lethbridge, AB, Canada, [2]University of Illinois, Urbana, IL **(107)**

†Light Quality Manipulation as a Seedbank Management Tool for Palmer Amaranth (*Amaranthus palmeri*). Sarah E. Kezar*, Shuyang Zhen, Muthukumar V. Bagavathiannan; Texas A&M University, College Station, TX **(108)**

Quantifying Long-term Competitiveness of Weed Species in a Cereal-legume Rotation. Jose L. Gonzalez-Andujar*[1], Maria J. Aguilera[2], Rene Van Acker[3]; [1]Instituto de Agricultura Sostenible (CSIC), Cordoba, Spain, [2]Departmento de Física Aplicada, Universidad de Córdoba, Cordoba, Spain, [3]Ontario Agricultural College, University of Guelph, Guelph, ON, Canada **(109)**

†Characteristics of Self-fertilization in *Lilium formosanum* Var. *pricei*. Sawa Seoka*, Hidehiro Inagaki; Shizuoka university, Shizuoka, Japan **(110)**

The Role of Cytochrome P450 in Herbicide Metabolism. Fatemeh Abdollahi*; Postdoctoral researcher, Fort Collins, CO **(111)**

†In the Ruins: The Neglected Link Between Archaeology and Weed Science. Alice C. Wolff*[1], Sophie Westbrook[2], Antonio DiTommaso[2]; [1]Cornell University/Medieval Studies, Ithaca, NY, [2]Cornell University, Ithaca, NY **(112)**

Rational Design of an Artificial Weed for High-Throughput Soil Bin Physical Weed Control Tool Testing. Sonora Ortiz*[1], Eric R. Gallandt[1], Bashir Khoda[1], Margaret R. McCollough[1], Daniel C. Brainard[2]; [1]University of Maine, Orono, ME, [2]Michigan State University, East Lansing, MI **(113)**

†Cereal Rye Residue Effects on the Germination of Troublesome Southeastern Weeds. Annu Kumari*[1], Steve Li[1], Andrew J. Price[2]; [1]Auburn University, Auburn, AL, [2]USDA-ARS, Auburn, AL **(114)**

Elevated CO_2 and Temperature Effects on Growth and Photosynthetic Responses in Purple and Yellow Nutsedge (*Cyperus rotundus* L. and *C. esculentus* L.). Srinivasa Rao Pinnamaneni*, Krishna N. Reddy; Crop Production Systems Research Unit USDA-ARS, Stoneville, MS **(115)**

†Analysis of the Recently Completed Genome of *Salsola tragus* (Russian Thistle), a Tenacious Tetraploid Tumbleweed. John M. Lemas*[1], Eric L. Patterson[2], Sarah Morran[1], Nickolas A. Johnson[2], Jacob S. Montgomery[1], Victor Llaca[3], Kevin Fengler[3], Philip Westra[1], Todd A. Gaines[1]; [1]Colorado State University, Fort Collins, CO, [2]Michigan State University, East Lansing, MI, [3]Corteva Agriscience, Johnston, IA **(116)**

Using Weed Extracts to Protect Sweet Potato from Deer Browsing. Ziming Yue[1], Karina Beneton[1], Presley Lee[2], Mark Hall[3], Callie J. Morris[1], Mark W. Shankle[3], Te-Ming (Paul) Tseng*[1]; [1]Mississippi State University, Starkville, MS, [2]North Pontotoc High School, Pontotoc, MS, [3]Mississippi State University, Pontotoc, MS **(117)**

†Competitive Interactions Between Sorghum and Johnsongrass (*Sorghum halepense*) is Influenced by Multiple Abiotic Stress Factors. Vikas C. Tyagi*, Aniruddha Maity, Nithya K. Subramanian, Muthukumar V. Bagavathiannan; Texas A&M University, College Station, TX **(118)**

Integrating Ecological Tactics to Manage Multiple Herbicide-Resistant Pigweeds in U.S. Soybean Production. Prashant Jha*[1], Jason K. Norsworthy[2], Vipan Kumar[3], Karla L. Gage[4]; [1]Iowa State University, Ames, IA, [2]University of Arkansas, Fayetteville, AR, [3]Kansas State University, Hays, KS, [4]Southern Illinois University Carbondale, Carbondale, IL **(119)**

†CO_2 And Temperature Effects on Herbicide-resistant and Susceptible Palmer Amaranth Biotypes. Juliana de Souza Rodrigues*[1], Timothy L. Grey[2], Donn Shilling[3], Todd A. Gaines[4], Nicholas T. Basinger[5], John Snider[2], Viktor Tishchenko[6]; [1]University of Georgia/Department of Crop and Soil Science, Tifton, GA, [2]University of Georgia, Tifton, GA, [3]University of Georgia, Athens, GA, [4]Colorado State University, Fort Collins, CO, [5]The University of Georgia, Athens, GA, [6]University of Georgia, Griffin, GA **(120)**

Influence of Multiple Abiotic Stress Factors on Pollen Viability and Stigma Receptivity of Johnsongrass (*Sorghum halepense*). Vikas C. Tyagi*, Aniruddha Maity, Nithya K. Subramanian, Muthukumar V. Bagavathiannan; Texas A&M University, College Station, TX **(121)**

†Differentiation Between Yellow Foxtail (*Setaria pumilla*) and Knotroot Foxtail (*Setaria parviflora*) Via Morphological Characteristics and Genetics Markers. Mikerly M. Joseph*[1], Joseph S. McElroy[1], Claudia A. Rutland[2]; [1]Auburn University, Auburn, AL, [2]Auburn University Department of Crop, Soil, and Environmental Sciences, Auburn, AL **(122)**

†Insights into Weed Aboveground and Seedbank Dynamics from a Long-Term Cropping Systems Experiment. Natasha Djuric*[1], Michel A. Cavigelli[2], Steven Brian Mirsky[2], Eugene P. Law[3], Antonio DiTommaso[1], Matthew R. Ryan[1]; [1]Cornell University, Ithaca, NY, [2]USDAARS, Beltsville, MD, [3]University of Delaware, Beltsville, MD **(123)**

†Evapotranspiration of Volunteer Corn in Corn, Soybean, and Sorghum Cropping Systems. Mandeep Singh*[1], John Lindquist[2], Stevan Knezevic[2], Suat Irmak[3], Vipan Kumar[4], Amit J. Jhala[1]; [1]University of Nebraska-Lincoln, Lincoln, NE, [2]University of Nebraska, Lincoln, NE, [3]Penn State University, University Park, PA, [4]Kansas State University, Hays, KS **(124)**

†Predicting Yield Loss Using the Kropff-Spitters Model and Early Season Leaf Area Indices. Avi S. Goldsmith*, Sandra R. Ethridge, Ramon G. Leon, Robert Austin; North Carolina State University, Raleigh, NC **(125)**

†Fate of *Amaranthus palmeri* (Palmer Amaranth) and *A. tuberculatus* (Waterhemp) Seeds Under Cereal Rye Cover Crop Across the North Central Region. Lily A. Woitaszewski*[1], Sarah Lancaster[1], Anita Dille[1], Kevin W. Bradley[2], Mandy Bish[2], William G. Johnson[3], Bryan G. Young[4], Claudia R. Bland[3], Joseph T. Ikley[5], Rodrigo Werle[6], Nicholas J. Arneson[6]; [1]Kansas State University, Manhattan, KS, [2]University of Missouri, Columbia, MO, [3]Purdue University, West Lafayette, IN, [4]Purdue University, Brookston, IN, [5]North Dakota State University, Fargo, ND, [6]University of Wisconsin-Madison, Madison, WI **(126)**

†Building Reference Genome of Canada Thistle (Cirsium Arevense) and Identifying Genomic Regions Behind Sex Determination. Sofia Marques-hill*[1], Todd A. Gaines[2]; [1]Colorado State University - Agricultural Biology, Fort Collins, CO, [2]Colorado State University, Fort Collins, CO **(127)**

†Identifying Genes Associated with Conferring Clopyralid Resistance in Common Ragweed (*Ambrosia artemisiifolia*) Using RNA Sequence Transcriptome Analysis. Nash D. Hart*[1], Erin E. Burns[2], Eric L. Patterson[2]; [1]Michigan State University, Durand, MI, [2]Michigan State University, East Lansing, MI **(128)**

†Chlorophyll Content and Fluorescence Fail to Detect Crop Injury from Physical Weed Control Tools. Anthony J.

Codega*[1], Ruthie Clements[1], Jordan W. Parks[1], Rebecca Champagne[2], Eric R. Gallandt[1]; [1]University of Maine, Orono, ME, [2]The University of Maine, Orono, ME **(130)**

†Effective Anthraquinone Derivative Dose; A Weed-based Deer Repellent for Soybean. Ncomiwe A. Maphalala*[1], Ziming Yue[2], Te-Ming (Paul) Tseng[2]; [1]Mississippi State University-Department of Plant and Soil Sciences, Starkville, MS, [2]Mississippi State University, Starkville, MS **(131)**

†A Novel Allele of *IAA16* is Associated with Dicamba Resistance in *Bassia scoparia*. Jacob S. Montgomery*[1], Neeta Soni[2], Sarah Morran[1], Franck E. Dayan[1], Philip Westra[1], Eric L. Patterson[3], Todd A. Gaines[1]; [1]Colorado State University, Fort Collins, CO, [2]Corteva Agriscience, Indianapolis, IN, [3]Michigan State University, East Lansing, MI **(132)**

TUESDAY MORNING JANUARY 31

POSTER - 09. Biocontrol of Weeds

LOCATION: Salon IV-VI
TIME: 08:00 AM - 10:00 AM Eastern Time

WSSA Student Competition Posters will be judged Tuesday only
NEWSS Student Competition Posters will be judged Wednesday only

***PRESENTER † STUDENT CONTEST**

A Fungal Pathogen (*Colletotrichum shisoi*) Reduces Biomass, Flowering, and Competitive Ability of Invasive Beefsteak Plant (*Perilla frutescens*). Michael R. Fulcher*; United States Department of Agriculture, Agricultural Research Service, Fort Detrick, MD **(133)**

†The Efficacy and Synergistic Effect of Coumarin and Chlorogenic Acid on Glyphosate- Resistant Palmer Amaranth. Alyssa Lea Miller*; Mississippi State University, Starkville, MS **(134)**

Invasive Weeds Serving as Microbial Reservoirs and Their Role in Pathogen Evolution and Emergence. Matthew A. Tancos*, Michael R. Fulcher; United States Department of

Agriculture, Agricultural Research Service, Fort Detrick, MD **(135)**

†Exploiting the Allelopathic Weed-suppressive Potential of Sweetpotato Varieties Under Field Conditions. Varsha Singh*[1], Worlanyo Segbefia[1], Ziming Yue[1], Callie J. Morris[1], Parth Sharma[1], Mark W. Shankle[2], Te-Ming (Paul) Tseng[1]; [1]Mississippi State University, Starkville, MS, [2]Mississippi State University, Pontotoc, MS **(136)**

Mycorrhizae Mediate the Control of Broomrapes (*Orobanche* Spp.) in Lentils and Sunflowers. Cumali Ozaslan*[1], Antonio DiTommaso[2]; [1]Dicle University, Diyarbakir, Turkey, [2]Cornell University, Ithaca, NY **(137)**

†Root Architecture Study for Allelopathic Traits in Cotton for Weed Suppression. Worlanyo Segbefia*[1], Varsha Singh[1], Parth Sharma[1], Ncomiwe A. Maphalala[2], Te-Ming (Paul) Tseng[1]; [1]Mississippi State University, Starkville, MS, [2]Mississippi State University-Department of Plant and Soil Sciences, Starkville, MS **(138)**

†Root System Architectural Traits and Genes Linked to Allelopathic Effects in Weedy Rice (Oryza Sativa f. Spontanea). Ncomiwe A. Maphalala*[1], Auriana Tucker[2], Brooklyn C. Schumaker[2], Swati Shrestha[2], Shandrea D. Stallworth[2], Nilda Roma-Burgos[3], Te-Ming (Paul) Tseng[2]; [1]Mississippi State University-Department of Plant and Soil Sciences, Starkville, MS, [2]Mississippi State University, Starkville, MS, [3]University of Arkansas, Fayetteville, Fayetteville, AR **(139)**

TUESDAY MORNING JANUARY 31

POSTER - 10. Physiology

LOCATION:	Salon IV-VI
TIME:	08:00 AM - 10:00 AM Eastern Time

WSSA Student Competition Posters will be judged Tuesday only
NEWSS Student Competition Posters will be judged Wednesday only

***PRESENTER † STUDENT CONTEST**

†A Tale of 20 Years: What Can be Learned About Glufosinate Tolerance When Comparing Palmer Amaranth Populations from 2001 Vs. Now? Pamela Carvalho-Moore*[1], Jason K. Norsworthy[1], Fidel Gonzalez Torralva[1], Leonard B. Piveta[1], Maria Carolina C R Souza[2], Tom Barber[3], Thomas R. Butts[3]; [1]University of Arkansas, Fayetteville, AR, [2]University of Arkanas, Fayetteville, AR, [3]University of Arkansas System Division of Agriculture, Lonoke, AR **(140)**

†A Mutation in a Transcription Factor Imparts Non-target Site Herbicide Resistance in Rice. Srishti Gupta*; Colorado State University, Fort Collins, CO **(141)**

†CRISPR/Cas9 Editing of a *CYP81A* Homoeolog in *Triticum aestivum*. Olivia A. Landau*[1], Natalya Nersesian[2], Thomas Clemente[2], Dean E. Riechers[3]; [1]University of Illinois, Urbana, IL, [2]University of Nebraska, Lincoln, NE, [3]University of Illinois-Urbana, Urbana, IL **(142)**

†Reactive Oxygen Species in Herbicidal Activity: What We Know. Catherine J. Traxler*[1], Todd A. Gaines[1], Peter Luemmen[2], Franck E. Dayan[1]; [1]Colorado State University, Fort Collins, CO, [2]Bayer Crop Sciences, Frankfurt, Germany **(143)**

Activation of a Singlet Oxygen Signaling Pathway by Competition Cues in Arabidopsis Thaliana. Nicole Berardi[1], Sasan Amirsadeghi[1], Clarence Swanton*[2]; [1]University of Guelph, Guelph, ON, Canada, [2]University of Guelph, Department of Plant Agriculture, Guelph, ON, Canada **(144)**

†Herbicidal Activity of Chalcones. Raphael M. Garrido*[1], Franck E. Dayan[1], Rosana M. Kolb[2], Luis O. Regasini[3]; [1]Colorado State University, Fort Collins, CO, [2]São Paulo State University, Assis, Brazil, [3]São Paulo State University, São José Do Rio Preto, Brazil **(145)**

Phytotoxicity of Sorgoleone Revealed by Expression of the Biosynthetic Pathway in *Nicotiana benthamiana*. Zhiqiang Pan*, Joanna Bajsa-Hirschel; USDA-ARS, University, MS **(146)**

†An Omics Approach to Investigate the Resistance Mechanism to ACCase Inhibitors in Feather Fingergrass (Chloris Virgata*)*. Luiz Augusto Inojosa Ferreira*[1], Rubem Silvério de Oliveira Jr[2], Jamil Contantin[2], Caio A. Brunharo[3]; [1]Penn State University, State College, PA, [2]State University of Maringa, Maringa, Brazil, [3]Penn State University, University Park, PA **(147)**

Resistance of Sagittaria Montevidensis Biotypes to Acetolactate Synthase Inhibitor and Bentazon Herbicides. Lariza Benedetti[1], Marcus Vinícius Fipke[1], Vívian Ebeling Viana*[1], Diogo da Silva Moura[2], Edinalvo Rabaioli Camargo[1], José Alberto Noldin[3], Luis Antonio de Avila[1]; [1]Federal

University of Pelotas, Pelotas, Brazil, [2]Centro de Ensino Superior Riograndense, Marau, Brazil, [3]Empresa de Pesquisa Agropecuária e Extensão Rural de Santa Catarina, Itajaí, Brazil **(148)**

†Protoplast Generation from Cell Suspension Culture in Waterhemp [*Amaranthus tuberculatus* (Moq.) J. D. Sauer]. Peter E. Beerbower*, Robert Sabba, Michael J. Christoffers; North Dakota State University, Fargo, ND **(149)**

†Agrematch AI Platform Leads to the Discovery of AS9057 as a Novel Natural Herbicide. Alyssa Twitty*[1], Yosef Geva[2], Eyal Simonovsky[2], Ifat Shub[2], Yael Phillip[2], Franck E. Dayan[1]; [1]Colorado State University, Fort Collins, CO, [2]Agrematch, Rehovot, Israel **(150)**

†Transcriptomic Response to Glyphosate in Rapid Response Giant Ragweed (*Ambrosia trifida*). Crystal D. Sparks*[1], Christopher Van Horn[2], Roland S. Beffa[3], Todd A. Gaines[1], Franck E. Dayan[1], Philip Westra[1]; [1]Colorado State University, Fort Collins, CO, [2]Bayer Crop Science, St. Louse, MO, [3]Senior Scientist Consultant, Frankfurt, Germany **(151)**

TUESDAY MORNING JANUARY 31

POSTER - 11. Soil and Environmental Aspects

LOCATION: Salon IV-VI
TIME: 08:00 AM - 10:00 AM Eastern Time

WSSA Student Competition Posters will be judged Tuesday only
NEWSS Student Competition Posters will be judged Wednesday only

***PRESENTER † STUDENT CONTEST**

Relative Soybean Response from 4 Auxin Herbicides Applied POST. Thomas C. Mueller*[1], Larry Steckel[2]; [1]University of Tennessee, Knoxville, TN, [2]University of Tennessee, Jackson, TN **(152)**

†Cereal Rye Cover Crop Impacted the Fate of PRE-emergence Herbicides Flumioxazin and Pyroxasulfone in Soybean. Jose J. Nunes*[1], Eric J. Miller[2], Lily A. Woitaszewski[3], Shelby E. Lanz[4], Nicholas J. Arneson[1], John M. Wallace[5], Karla L. Gage[6], Sarah Lancaster[3], Thomas C.

Mueller[4], Rodrigo Werle[1]; [1]University of Wisconsin-Madison, Madison, WI, [2]Southern Illinois University, Carbondale, IL, [3]Kansas State University, Manhattan, KS, [4]University of Tennessee, Knoxville, TN, [5]Penn State University, University Park, PA, [6]Southern Illinois University Carbondale, Carbondale, IL **(153)**

How Does Environment Influence VLCFA-Inhibitor Efficacy? Tyler P. Meyeres*[1], Lucas Haag[2], Peter Tomlinson[1], Sarah Lancaster[1]; [1]Kansas State University, Manhattan, KS, [2]Kansas State University, Colby, KS **(154)**

TUESDAY MORNING JANUARY 31

POSTER - 12. Integrated Weed Management

LOCATION:	Salon IV-VI
TIME:	08:00 AM - 10:00 AM Eastern Time

WSSA Student Competition Posters will be judged Tuesday only
NEWSS Student Competition Posters will be judged Wednesday only

***PRESENTER † STUDENT CONTEST**

†Impact of Blue Light on the Germination Characteristics of Grass Weed Seeds. Sarah A. Chu*, Sarah E. Kezar, Shuyang Zhen, Muthukumar V. Bagavathiannan; Texas A&M University, College Station, TX **(155)**

Integrated Weed Management Using Biosolarization, Cover Cropping and Strip Tillage. Dwayne D. Joseph*[1], Alan W. Leslie[2], Kurt M. Vollmer[3], Cerruti Rr Hooks[2]; [1]University of Maryland, Chestertown, MD, [2]University of Maryland, College Park, MD, [3]University of Maryland, Queenstown, MD **(156)**

†Ecology and Landscape Scale Management of a Globally Invasive Weed; *Nassella trichotoma* (Serrated Tussock). Talia J. Humphries*; Federation University Australia, Ballarat, Australia **(157)**

Integrated Weed Management in Olive Orchards. Veronica Pedraza[1], Juan A. Lezaun[2], Irache Garnica[2], Jose L. Gonzalez-Andujar*[3]; [1]Instituto de Agricultura Sostenible (CSIC), Córdoba, Spain, [2]Instituto Navarro de Tecnologías e Infraestructuras

Agrarias (INTIA), Villava, Spain, [3]Instituto de Agricultura Sostenible (CSIC), Cordoba, Spain **(158)**

†Atrazine Wash Off from Cereal Rye (*Secale cereale* L.) Residue onto the Soil After Simulated Rainfall. Lucas Oliveira Ribeiro Maia*[1], Bryan G. Young[2], William G. Johnson[1], Eileen J. Kladivko[1], Shalamar Armstrong[1]; [1]Purdue University, West Lafayette, IN, [2]Purdue University, Brookston, IN **(159)**

Changing Labor Landscape in Northeast Weed Management for Specialty Crops. Laura Pineda-Bermudez*, Lynn M. Sosnoskie; Cornell University, Geneva, NY **(160)**

†Summer Cover Crop Suppression of Broadleaf, Grass, and Nutsedge Weeds. Moriah Williams*[1], Nathan Boyd[2]; [1]University of Florida, Saint Petersburg, FL, [2]University of Florida, Balm, FL **(161)**

Field Scale Evaluation of the Harrington Seed Destructor in Alberta, Canada. Breanne D. Tidemann*[1], K. Neil Harker[1], Hugh Beckie[2]; [1]Agriculture and Agri-Food Canada, Lacombe, AB, Canada, [2]University of Western Australia, Perth, Australia **(162)**

Weed Suppression Provided by a Second Season Perennial Living Mulch and Self-reseeded Cover Crop. Veronica Yurchak*, Alan W. Leslie, Cerruti Rr Hooks, Bill Phillips; University of Maryland, College Park, MD **(163)**

Can Exogenous Application of Phytohormones Involved in the Stimulus of Seed Germination Become Part of Integrated Waterhemp (*Amaranthus tuberculatus*) Management Programs? Jose J. Nunes, Nicholas J. Arneson, Rodrigo Werle*; University of Wisconsin-Madison, Madison, WI **(164)**

†Integrative Effect of Row Spacing and Herbicide Programs on Photosynthetically Active Radiation (PAR) in Corn Infested with Multiple Herbicide-resistant Palmer Amaranth. Ramandeep Kaur*[1], Stevan Knezevic[2], Yeyin Shi[2], Nevin Lawrence[2], Parminder Chahal[2], Rachana Jhala[2], Amit J. Jhala[1]; [1]University of Nebraska-Lincoln, Lincoln, NE, [2]University of Nebraska, Lincoln, NE **(165)**

Cover Crops as a Weed Seed Bank Management Tool. Cynthia Sias[1], Bethany R. Wolters[2], Mark S. Reiter[3], Michael L. Flessner*[1]; [1]Virginia Tech, Blacksburg, VA, [2]University of Tennessee, Martin, Martin, TN, [3]Virginia Tech, Painter, VA **(166)**

†Impacts of Soil Organic Amendments and Weed Control Practices on Nutritional Status and Performance of Valencia Sweet Orange (*Citrus Sinensis 'Valencia'*) Trees. Ankit

Pokhrel*, Ute Albrecht, Ramdas Kanissery, Sarah Strauss; University of Florida, Immokalee, FL **(167)**

Evaluating Different Deep Learning Models for the Detection of Poison Ivy (*Toxicodendron radicans) in Images.* Renato Herrig Furlanetto*[1], Nathan Boyd[2], Ana Claudia Buzanini[1], Alex G. Rodriguez[3]; [1]University of Florida, Wimauma, FL, [2]University of Florida, Balm, FL, [3]University of Florida, Belle Glade, FL **(168)**

†Multi-year Herbicide Comparison for Italian Ryegrass (Lolium Perenne Ssp. Multiflorum) Control in Wheat. Wyatt J. Stutzman*, Kevin Bamber, Michael L. Flessner; Virginia Tech, Blacksburg, VA **(169)**

Weed Suppression from Frost-seeded *Brassicaceae* Cover Crops. Huong Nguyen*, Olivia Fisher, Amy Fox, Kristen Loria, Kathryn Marini, Chris J. Pelzer, Adam Sharifi, Domenic Varma, Sandra Wayman, Matthew R. Ryan; Cornell University, Ithaca, NY **(170)**

Evaluation of Two-pass Herbicide Programs for Broad-spectrum Weed Control in Conventional Corn in Wisconsin Atrazine Prohibition Areas. Ahmadreza Mobli*[1], Ryan P. DeWerff[2], Nicholas J. Arneson[1], Rodrigo Werle[1]; [1]University of Wisconsin-Madison, Madison, WI, [2]University of Wisconsin, Madison, WI **(171)**

†A Multi-state Evaluation of Weed Seed Fate During Soybean Harvest. Jacob E. Vaughn*[1], Joseph T. Ikley[2], Vipan Kumar[3], Sarah Lancaster[4], Christy L. Sprague[5], Rodrigo Werle[6], Nicholas J. Arneson[6], Kevin W. Bradley[1]; [1]University of Missouri, Columbia, MO, [2]North Dakota State University, Fargo, ND, [3]Kansas State University, Hays, KS, [4]Kansas State University, Manhattan, KS, [5]Michigan State University, East Lansing, MI, [6]University of Wisconsin-Madison, Madison, WI **(172)**

Cover Crop Residue Management Strategies Impact Deposition and Washoff of Soil-applied Residual Herbicides. Cody Smith[1], John M. Wallace*[2], Mark VanGessel[3], Michael L. Flessner[4]; [1]Penn State University, University Park Pa, PA, [2]Penn State University, University Park, PA, [3]University of Delaware, Georgetown, DE, [4]Virginia Tech, Blacksburg, VA **(173)**

†Goosegrass (*Eleusine indica)* Control Options in Bentgrass Putting Greens. Bridgette C. Johnson*, Joseph S. McElroy; Auburn University, Auburn, AL **(174)**

US-Herbicide Resistance Action Committee (HRAC): Who We Are and What We Do? Zahoor A. Ganie*[1], Lowell Sandell[2], Greg Elmore[3], Chandrashekar Aradhya[4]; [1]FMC,

Newark, DE, [2]Valent, Ashland, NE, [3]Bayer Crop Science, Saint Louis, MO, [4]Bayer Crop Science, Chesterfield, MO **(175)**

†Light Quality Alteration is a Mechanism of Weed Suppression by Cereal Rye Cover Crop. Gustavo Camargo Silva*, Muthukumar V. Bagavathiannan; Texas A&M University, College Station, TX **(176)**

Building an Agriculture Weed ID Network for New York: Development of Web Based Resources to Interface with Growers and Beyond. Sharon Bachman*[1], Caroline Marschner[2], Antonio DiTommaso[2], Isabella Colucci[2]; [1]Cornell Cooperative Extension, East Aurora, NY, [2]Cornell University, Ithaca, NY **(177)**

†Weed Control and Rice Response to Florpyrauxifen-benzyl. Rachelle Berger*; University of Florida, Belle Glade, FL **(178)**

To Bury or Uproot? Modelling Selectivity of Physical Weed Control in Sensitive Crops. Daniel C. Brainard*, Noelle Connors; Michigan State University, East Lansing, MI **(179)**

†CANCELLED Impact of Seeding Rate, Nitrogen Availability, and Disease Pressure in *Bromus tectorum* and Winter Wheat Interactions. Laura Berrios-Ortiz*[1], Fabian Menalled[2]; [1]Department of Land Resources and Environmental Sciences- Montana State University, Bozeman, MT, [2]Montana State University, Bozeman, MT **(180)**

†Effect of Tillage, Cover Crop, and Herbicide on Giant Ragweed (*Ambrosia trifida* L.) Management in Soybean. Venkatanaga Shiva Datta Kumar Sharma Chiruvelli*[1], Thomas J. Peters[2], Gregg A. Johnson[1], Debalin Sarangi[1]; [1]University of Minnesota, St Paul, MN, [2]North Dakota State University, Fargo, ND **(181)**

Weed Detection Using Convolutional Neural Networks Under Scenarios of Occlusion. Daniel J. Ginn*[1], Ramon G. Leon[2], Chris Reberg-Horton[2], Steven Brian Mirsky[3], Muthukumar V. Bagavathiannan[1]; [1]Texas A&M University, College Station, TX, [2]North Carolina State University, Raleigh, NC, [3]USDA ARS, Beltsville, MD **(182)**

†A Multi-Tactic Waterhemp Management Approach in Soybean-Sugar Beet Rotation. Navjot Singh*[1], Thomas J. Peters[2], Seth L. Naeve[1], Ryan P. Miller[3], Debalin Sarangi[4]; [1]University of Minnesota, Saint Paul, MN, [2]North Dakota State University, Fargo, ND, [3]University of Minnesota, Rochester, MN, [4]University of Minnesota, St. Paul, MN **(183)**

SYMPOSIUM - 04. The good, the bad, and the ugly the current state of cover crops and weed management

LOCATION:	Salon III
TIME:	08:00 AM - 12:00 PM Eastern Time
CHAIR:	Erin Haramoto University of Kentucky Lexington, KY
MODERATOR:	Erin Haramoto University of Kentucky Lexington, KY
CO-MODERATOR:	Steve Young USDA ARS Beltsville, MD

***SPEAKER**

08:00 AM **Symposium Welcome.** Steve Young*; USDA-ARS, Beltsville, MD **(184)**

08:05 AM **Current State of Cover Crops and Weed Management □ What Do We Know and How Do We Know It?** Erin Haramoto*; University of Kentucky, Lexington, KY **(185)**

08:20 AM **Cover Crop Options and Opportunities in Australian Cropping Systems.** Michael Walsh*[1], Asad Shabbir[2], Saliya Gurusinghe[3], Leslie A. Weston[4]; [1]University of Sydney, Narrabri, Australia, [2]The University of Sydney, Camden, Australia, [3]Charles Sturt University, Bathurst, Australia, [4]Charles Sturt University, Wagga Wagga, Australia **(186)**

08:40 AM **What is the Potential for Actually Managing Weeds with Cover Crops? A Reality Check.** Richard G. Smith*; University of New Hampshire, Durham, NH **(187)**

09:00 AM **Opportunities and Challenges of Growing Cover Crops in the Semi-arid Great Plains.** John Holman*[1], Augustine Obour[2]; [1]Kansas State University, Garden City, KS, [2]Kansas State University, Hays, KS **(188)**

09:20 AM **Weighing Cover Crop Mediated Weed Management Benefits with Other Risks.** Erin Haramoto*; University of Kentucky, Lexington, KY **(189)**

09:40 AM **The Highs and Lows of Working with a Perennial Cover Crop for Weed Suppression.** Cerruti Rr Hooks*, Alan W. Leslie, Veronica Yurchak; University of Maryland, College Park, MD **(190)**

10:00 AM **Coffee Break**

10:15 AM **Cover Crop-Based Integrated Weed Management: Communicating an Adaptive Management Framework.** John M. Wallace*; Penn State University, University Park, PA **(191)**

10:35 AM **Emerging Opportunities to Understand Cover Crop x Weed Management Interactions from Data-Model Mashups.** Armen R. Kemanian*, Yuning Shi, Curt A. McConnell; Penn State University, State College, PA **(192)**

10:55 AM **Assessing Regional Variations of Weed Suppression with Cover Crops.** Steven Brian Mirsky*[1], Victoria Ackroyd[2], Muthukumar V. Bagavathiannan[3], Kevin W. Bradley[4], William S. Curran[5], Adam S. Davis[6], Jeffrey Evans[7], Wesley Everman[8], Michael L. Flessner[9], Nicholas R. Jordan[10], Eugene P. Law[11], Lauren M. Lazaro[12], Ramon G. Leon[8], John Lindquist[13], Jason K. Norsworthy[14], Lovreet S. Shergill[15], Mark VanGessel[16]; [1]USDA ARS, Beltsville, MD, [2]University of Maryland, College Park, MD, [3]Texas A&M University, College Station, TX, [4]University of Missouri, Columbia, MO, [5]Penn State University, Bozeman, MT, [6]University of Illinois, Urbana, IL, [7]Five Rivers Conservation Trust, Concord, NH, [8]North Carolina State University, Raleigh, NC, [9]Virginia Tech, Blacksburg, VA, [10]University of Minnesota, St. Paul, MN, [11]University of Delaware, Beltsville, MD, [12]Blue River Technology, Sunnyvale, CA, [13]University of Nebraska, Lincoln, NE, [14]University of Arkansas, Fayetteville, AR, [15]Montana State University, Huntley, MT, [16]University of Delaware, Georgetown, DE **(193)**

11:15 AM **Panel Discussion.** Steve Young*; USDA-ARS, Beltsville, MD **(194)**

ORAL - 08. Weed Biology and Ecology

LOCATION:	Salon I
TIME:	09:00 AM - 02:00 PM Eastern Time
CHAIR:	Debalin Sarangi University of Minnesota St. Paul, MN
MODERATOR:	Debalin Sarangi University of Minnesota St. Paul, MN
CO-MODERATOR:	Shilpa Singh Texas A&M University College Station, TX

***SPEAKER**

09:00 AM **Directional Selection of Competitive Ability in *Setaria faberi* Confirmed with GWAS.** Sandra R. Ethridge[1], Saket Chandra[1], Wesley Everman[1], David L. Jordan[1], Anna M. Locke[2], Micheal D. Owen[3], Ramon G. Leon*[1]; [1]North Carolina State University, Raleigh, NC, [2]USDA-ARS, Raleigh, NC, [3]Iowa State University, Ames, IA **(195)**

09:15 AM **Impact of Combined Temperature and Moisture Stress on Palmer Amaranth (*Amaranthus palmeri*) Growth and Fecundity: Adaptations and Constraints.** Sarah E. Kezar*, Aniruddha Maity, Muthukumar V. Bagavathiannan; Texas A&M University, College Station, TX **(196)**

09:30 AM **The International Weed Genomics Consortium: a Review of Current Reference Genomes and Future Directions.** Sarah Morran*[1], Todd A. Gaines[1], Dana R. MacGregor[2], Eric L. Patterson[3], Mithila Jugulam[4], Roland S. Beffa[5]; [1]Colorado State University, Fort Collins, CO, [2]Rothamsted Research, Harpenden, United Kingdom, [3]Michigan State University, East Lansing, MI, [4]Kansas State University, Manhattan, KS, [5]Senior Scientist Consultant, Frankfurt, Germany **(197)**

09:45 AM **New Insights in Amaranth PPO Target Site Resistance and PPO1 and PPO2 Biochemistry.** Jens Lerchl*[1], Aimone Porri[2], Inessa De[1], Lucie Meyer[1], Nilda Roma-Burgos[3], Matheus Machado Noguera[4], Bernhard Grimm[5], Boris Hedtke[5], Daniel Wittmann[6], Patrick Tranel[7], Kathryn Seebruck[8]; [1]BASF SE, Limburgerhof, Germany, [2]BASF Global Research & Development, Limburgerhof, Germany, [3]University of Arkansas, Fayetteville, Fayetteville, AR, [4]Univ Arkansas, Fayetteville, AR, [5]Humboldt University, Berlin, Germany, [6]BASF SE, Berlin, Germany, [7]University of Illinois, Urbana, IL, [8]University Illinois, Urbana, IL **(198)**

10:00 AM **Coffee Break**

10:15 AM **Modeling the Effect of Density, Emergence Time, Inter- and Intra-Specific Competition of Barnyard Grass (*Echinochloa Crus-Galli)* on Growth and Yield of Dry-Seeded Rice: Simulations with INTERCOM.** Tahir Hussain Awan*[1], Bhagirath Singh Chauhan[2], John Lindquist[3]; [1]Rice Research Institute, Kala Shah Kaku, Punjab, Pakistan, Lahore, Pakistan, [2]bThe Centre for Plant Science, Queensland Alliance for Agriculture and Food Innovation (QAAFI), The University of Queensland, Gatton 4343, Queensland, Australia, Queensland, Australia, [3]University of Nebraska, Lincoln, NE **(199)**

10:30 AM **A Comparative Genomics Approach to Investigate the Parallel Evolution of Weediness in *Alopecurus myosuroides* (Black Grass) and *Alopecurus aequalis* (Orange Foxtail).** Dana R. MacGregor*[1], Jon Wright[2], Hannah Rees[2], Christian Harrison[1], Jun Li[3], Paul Neve[4], Seanna McTaggart[2], Anthony Hall[2]; [1]Rothamsted Research, Harpenden, United Kingdom, [2]Earlham Institute, Norwich, United Kingdom, [3]Nanjing Agricultural University, Nanjing, China, [4]University of Copenhagen, Copenhagen, Denmark **(200)**

10:45 AM **Does the Emergence Pattern of *Echinochloa crus-gali* (Barnyardgrass) Differ with EPOST Timings in Corn?** Theresa Piskackova*; Czech University of Life Sciences Prague, Prague 6 - Suchdol, Czechia (Czech Republic) **(201)**

11:00 AM **A Single GST Identified Via an Association Mapping and Transcriptomics Approach Endows High Levels of Resistance to Atrazine in**

Amaranthus palmeri. Shiv S. Kaundun*; Syngenta, Bracknell, United Kingdom **(202)**

11:15 AM **Utilizing Thermal Requirements of California Weedy Rice (*Oryza sativa spontanea*) to Predict Seedling Emergence Timing Under Field Conditions.** Liberty B. Galvin*[1], Whitney Brim-DeForest[2], Kassim Al-Khatib[1]; [1]University of California, Davis, Davis, CA, [2]University of California, Davis, Yuba City, CA **(203)**

11:30 AM ***In-vitro* Screening to Phenotype Cereal Rye (*Secale cereale*) Lines for Allelopathic Activity.** Democrito B. Rebong*[1], Ramon G. Leon[1], Chris Reberg-Horton[1], J. Paul Murphy[1], Virginia Moore[2], Steven Brian Mirsky[3], Lisa Kucek[4], Nancy Ehlke[5]; [1]North Carolina State University, Raleigh, NC, [2]Cornell University, Ithaca, NY, [3]USDA ARS, Beltsville, MD, [4]USDA Agricultural Research Service, Madison, WI, [5]University of Minnesota, St. Paul, MN **(204)**

11:45 AM **Demographic Implications of Self-fertilization and Migration for the Maintenance of Genetic Variation and Successful Adaptation to New Environments in Annual Weeds.** Samuel R. Revolinski*, Ian Burke; Washington State University, Pullman, WA **(205)**

12:00 PM **Lunch**

01:00 PM **Mapping of Metabolic Resistance to Tembotrione in Palmer Amaranth.** Carlos Alberto Gonsiorkiewicz Rigon*[1], Jacob S. Montgomery[1], Anita Küpper[2], Roland S. Beffa[3], Franck E. Dayan[1], Todd A. Gaines[1]; [1]Colorado State University, Fort Collins, CO, [2]Bayer AG, Frankfurt, Germany, [3]Senior Scientist Consultant, Frankfurt, Germany **(206)**

01:15 PM **Coupling Seedbank Management and Cultivation Tool Stacking for Improved Weed Control in Organic Vegetables.** Margaret R. McCollough*[1], Rebecca Champagne[2], Eric R. Gallandt[1], Daniel C. Brainard[3]; [1]University of Maine, Orono, ME, [2]The University of Maine, Orono, ME, [3]Michigan State University, East Lansing, MI **(207)**

01:30 PM **Phenotypic Variation Among Populations of the Invasive Weed *Parthenium hysterophorus* L. (Asteraceae) as a Tool to Assess Multiple Introductions.** Maor Matzrafi*[1], Sahar Malka[2],

Hanan Eizenberg[3]; [1]039904412, Ramat-yishay, Israel, [2]The Robert H. Smith Faculty of Agriculture, Food and Environment, The Hebrew University of Jerusalem, Jerusalem, Israel, [3]Department of Plant Pathology and Weed Research, Agricultural Research Organization Volcani Institute, Newe-Ya ar Research Center, Ramat-yishay, Israel **(208)**

TUESDAY MORNING JANUARY 31

ORAL - 05. Regulatory Aspects

LOCATION:	Salon II
TIME:	09:00 AM - 01:45 PM Eastern Time
CHAIR:	Amanda Foderaro North Carolina State University Raleigh, NC
MODERATOR:	Amanda Foderaro North Carolina State University Raleigh, NC
CO-MODERATOR:	Cherilyn Moore Syngenta Crop Protection, LLC Greensboro, NC

***SPEAKER**

09:00 AM **EPA's Plant Assessment Tool: A Non-Target Plant Exposure Model Used in Endangered Species Assessments.** Melissa Bridges*[1], Elizabeth Donovan[1], Gretchen E. Dykes[1], William P. Eckel[1], Frank T. Farruggia[1], Jerrett Fowler[1], Kristina Garber[1], Brian Kiernan[1], Stephan Miller[2], Chuck Peck[1], Dana Spatz[1], Holly Summers[1], Peter Tellez[1], Dirk Young[1]; [1]US Environmental Protection Agency, Washington, DC, [2]US Environmental Protection Agency, Washington, VA **(209)**

09:15 AM **EPA's Consideration of Pesticide Use, Mitigating Environmental Exposure, and Updating Labels in Endangered Species Evaluations.** Frank T. Farruggia*[1], Amy Adams[2], Jennifer Connolly[3], Kristina Garber[1], Stephen Muela[4], Tracy Perry[3]; [1]US Environmental Protection Agency, Washington, DC, [2]United States Environmental Protection Agency,

Washigton, DC, [3]United States Environmental Protection Agency, Washington, DC, [4]United States Environmental Protection Agency, Washtington, DC **(210)**

09:30 AM **Public Comment Periods: How Can WSSA Help EPA? Part 1.** Kelly V. Tindall*, Bradley Kells; US Environmental Protection Agency - Office of Pesticide Programs, Washington, DC **(211)**

09:45 AM **Public Comment Periods: How Can WSSA Help EPA? Part 2.** Bradley Kells*, Kelly V. Tindall; US Environmental Protection Agency - Office of Pesticide Programs, Washington, DC **(212)**

10:00 AM **The Importance of Herbicide Resistance Information for Regulatory Agencies.** Cameron H. Douglass[1], John Orlowski*[2]; [1]US Department of Agriculture, Office of the Chief Economist, Office of Pest Management Policy, Washington, DC, [2]US Environmental Protection Agency, Office of Pesticide Programs, Biological and Economic Analysis Division, Washington, DC **(213)**

10:15 AM **Coffee Break**

10:30 AM **Climate Impacts of Herbicide Regulations: Case Studies with Cover Cropping and No-till.** Laura Dodson[1], Fengxia Dong[2], Cameron H. Douglass*[3], Rebecca Nemec[4], Ryan Olver[2], Michelle Ranville[3]; [1]US Department of Agriculture, Economic Research Service, Washington, DC, [2]US Department of Agriculture, Economic Research Service, Kansas City, MO, [3]US Department of Agriculture, Office of the Chief Economist, Office of Pest Management Policy, Washington, DC, [4]US Department of Agriculture, Office of the Chief Economist, Washington, DC **(214)**

10:45 AM **Pesticide Task Forces: Role, Function, and Benefits.** Montague U. Dixon*; Syngenta Crop Protection, Greensboro, NC **(215)**

11:00 AM **ESA And FIFRA: A Regulatory Perspective on Successes and Challenges.** Reuben D. Baris*; Corteva Agriscience, Indianapolis, IN **(216)**

11:15 AM **Types of Registrations Under FIFRA - Section 3, Special Local Needs, Section 18 Emergency Use: What They Are and How They Are Utilized.** Rachel Hardie*; Syngenta Crop Protection, Greensboro, NC **(217)**

11:30 AM **Paraquat Dichloride Interim Registration Review Decision Update.** Jordan Moseley*; Syngenta Crop Protection, LLC, Greensboro, NC **(218)**

11:45 AM **Regulatory Update on Atrazine.** Mark White*; Syngenta, Macon, MO **(219)**

12:00 PM **Lunch**

01:00 PM **Persistent Herbicides in Compost: A Regulatory Pathway.** Travis Bui*; Corteva Agriscience, Indianapolis, IN **(220)**

01:15 PM **Introduction to the Center for Excellence in Regulatory Science in Agriculture at North Carolina State University.** Keith L. Edmisten*, Danesha Seth Carley; North Carolina State University, Raleigh, NC **(221)**

01:30 PM **The Economic, Nutritional, and Environmental Benefits of International Cooperation for the Establishment of MRLs on Specialty Crops.** Dirk C. Drost*; Minor Use Foundation, Inc, Raleigh, NC **(222)**

TUESDAY MORNING JANUARY 31

ORAL - NEWSS Student Oral

LOCATION: Salon J
TIME: 09:00 AM - 04:30 PM Eastern Time
CHAIR: Arthur E. Gover
Penn State
University Park, PA
MODERATOR: Arthur E. Gover
Penn State
University Park, PA

***SPEAKER † STUDENT CONTEST**

09:00 AM **†Influence of *Sorghum halepense* (Johnsongrass) Density on Early Growth of *Zea mays* (Corn).** Ryan P. O'Briant*[1], Antonio DiTommaso[2], Lynn M. Sosnoskie[3]; [1]Cornell University - Soil and Crop Sciences, Brooktondale,

NY, [2]Cornell University, Ithaca, NY, [3]Cornell University, Geneva, NY **(223)**

09:15 AM **†Competitive and Hormetic Responses of Velvetleaf (*Abutilon theophrasti*) Historical Seed Collection.** Sandra R. Ethridge*[1], Saket Chandra[1], Wesley Everman[1], Anna M. Locke[2], David L. Jordan[1], Micheal D. Owen[3], Ramon G. Leon[1]; [1]North Carolina State University, Raleigh, NC, [2]USDA-ARS, Raleigh, NC, [3]Iowa State University, Ames, IA **(224)**

09:30 AM **†Soybean and Palmer Amaranth Identification Using Machine Learning.** Dhiraj Srivastava*[1], Vijay Singh[2]; [1]Virginia Polytechnic Institute and State University, Painter, VA, [2]Virginia Tech, Painter, VA **(225)**

09:45 AM **†Do Cover Crops Affect Weed Community Structure?** Uriel D. Menalled*[1], Stephane Cordeau[2], Antonio DiTommaso[1], Richard G. Smith[3], Matthew R. Ryan[1]; [1]Cornell University, Ithaca, NY, [2]INRAE, Lyon, France, [3]University of New Hampshire, Durham, NH **(226)**

10:00 AM **Coffee Break**

10:15 AM **†Varietal Tolerance of Cucurbits Crops with Postemergence Applications of *S*-Metolachlor.** Carrie J. Mansue*[1], Thierry E. Besancon[2], Lynn M. Sosnoskie[3], Kurt M. Vollmer[4]; [1]Rutgers University, Mays Landing, NJ, [2]Rutgers University, Chatsworth, NJ, [3]Cornell University, Geneva, NY, [4]University of Maryland, Queenstown, MD **(227)**

10:30 AM **†Neutral Weed Communities: the Junction Point Between Crop Productivity, Biodiversity, and Ecosystem Services.** Marco Esposito*[1], Valerio Cirillo[2], Albino Maggio[2], Antonio DiTommaso[3]; [1]Cornell University & University of Napoli Federico II, Ithaca, NY, [2]University of Naples Federico II, Naples, Italy, [3]Cornell University, Ithaca, NY **(228)**

10:45 AM **†Optimizing Impregnated Pyroxasulfone for Cotton.** Brock A. Dean*, Charlie W. Cahoon, Zackary R. Taylor, Jose H. de Sanctis, Jacob Forehand, Hunter Lee; North Carolina State University, Raleigh, NC **(229)**

11:00 AM **†Anaerobic Soil Disinfestation: A Catalyst for Organic Watermelon Production in South**

Carolina? Sohaib Chattha*, Matthew A. Cutulle; Clemson University, Charleston, SC **(230)**

11:15 AM **†Weed Potential of Roll-crimped Cereal Rye Differs Across Cultural Practices.** Laurel Wellman*[1], John M. Wallace[2]; [1]Penn State University, State College, PA, [2]Penn State University, University Park, PA **(231)**

11:30 AM **†False-Green Kyllinga (*Kyllinga gracillima*) Establishment from Seed is Influenced by Turfgrass Mangement Regime.** Trevor S. Watson*, Matthew T. Elmore, Daniel P. Tuck; Rutgers University, New Brunswick, NJ **(232)**

11:45 AM **†Soybean Response to Simulated Drift of Herbicides Commonly Used to Manage Roadside Weeds.** Estefania Gomiero Polli*, Travis Gannon, Ronald R. Rogers, Mathieu LeCompte, Khalied Ahmed; North Carolina State University, Raleigh, NC **(233)**

12:00 PM **Lunch**

01:00 PM **†Multi-Seed *Zea* Pellets (MSZP) for Increasing Agroecosystem Biodiversity.** Sophie Westbrook*[1], Masoume Amirkhani[2], Alan G. Taylor[2], Michael T. Loos[2], John E. Losey[1], Antonio DiTommaso[1]; [1]Cornell University, Ithaca, NY, [2]Cornell University, Geneva, NY **(234)**

01:15 PM **†Using 3-D Image Reconstruction to Characterize Canopy Architecture and Predict Biomass.** April M. Dobbs*[1], Daniel J. Ginn[2], Søren K. Skovsen[3], Muthukumar V. Bagavathiannan[2], Steven Brian Mirsky[4], Chris Reberg-Horton[1], Prashant Jha[5], Ramawatar Yadav[5]; [1]North Carolina State University, Raleigh, NC, [2]Texas A&M University, College Station, TX, [3]Aarhus University, Aarhus, Denmark, [4]USDA ARS, Beltsville, MD, [5]Iowa State University, Ames, IA **(235)**

01:30 PM **†Cover Crops and Falls Residual Herbicides for Managing Multiple Herbicide-Resistant Italian Ryegrass.** Jose H. de Sanctis*, Charlie W. Cahoon, Wesley Everman, Travis Gannon, Zachary R. Taylor; North Carolina State University, Raleigh, NC **(236)**

01:45 PM **†Herbicides Affect White Clover Bloom Morphology and Pollinator Foraging in Lawn**

Turf. Navdeep Godara*, Shawn Askew; Virginia Tech, Blacksburg, VA **(237)**

02:00 PM **†Effect of Metcamifen on Trifloxysulfuron Absorption, Translocation and Metabolism in Turfgrass Systems.** Ronald R. Rogers*, Travis Gannon, Khalied Ahmed; North Carolina State University, Raleigh, NC **(238)**

02:15 PM **†Confirmation of Glufosinate Resistance and Inheritance in a North Carolina Palmer Amaranth (Amaranthus Palmeri) Population.** Eric A. Jones*, Jeffrey Dunne, Diego J. Contreras, Charlie W. Cahoon, Katherine M. Jennings, Ramon G. Leon, Wesley Everman; North Carolina State University, Raleigh, NC **(239)**

02:30 PM **†Factors That Influence Uniform Droplet Deposition from an Agricultural Spray Drone.** Daewon Koo*[1], Navdeep Godara[1], John M. Peppers[1], Clebson G. Goncalves[2], Shawn Askew[1]; [1]Virginia Tech, Blacksburg, VA, [2]University of California, Agriculture and Natural Resources, Ukiah, CA **(240)**

02:45 PM **†The Confirmation and Characterization of Auxinic Herbicide Resistance in a Population of Green Pigweed (*Amaranthus powellii*) from Ontario, Canada.** Isabelle Aicklen*[1], Peter Smith[1], Brendan Metzger[2], Todd A. Gaines[3], Mithila Jugulam[4], Darren E. Robinson[5], Peter H. Sikkema[5], Francois Tardif[1]; [1]University of Guelph, Guelph, ON, Canada, [2]BASF Canada, Morden, MB, Canada, [3]Colorado State University, Fort Collins, CO, [4]Kansas State University, Manhattan, KS, [5]University of Guelph, Ridgetown, ON, Canada **(241)**

03:00 PM **Coffee Break**

03:15 PM **†Speed, Angle and Spacing Effects on Finger Weeder Performance.** Jordan W. Parks*, Margaret R. McCollough, Eric R. Gallandt; University of Maine, Orono, ME **(242)**

03:30 PM **†Interaction of See and Spray™ Model Sensitivity and Application Timing in Soybean.** Diego J. Contreras*[1], Wesley Everman[1], Jason K. Norsworthy[2], Tristen H. Avent[2], Bryan G. Young[3], Marcelo Zimmer[4], Jasmine M. Mausbach[5], Lauren M. Lazaro[6], William L. Patzoldt[6]; [1]North Carolina State University, Raleigh, NC, [2]University of Arkansas, Fayetteville, AR, [3]Purdue University,

Brookston, IN, [4]Purdue University, West Lafayette, IN, [5]University of Nebraska-Lincoln, Lincoln, NE, [6]Blue River Technology, Sunnyvale, CA **(243)**

03:45 PM **†Residual Control of Palmer Amaranth Varies Across VLCFA-Inhibiting Herbicides.** Jackson W. Alsdorf*[1], Amanda M. Foderaro[2], Eric A. Jones[1], Diego J. Contreras[1], Wesley Everman[1]; [1]North Carolina State University, Raleigh, NC, [2]Syngenta, Raleigh, NC **(244)**

TUESDAY MORNING JANUARY 31

SST - SST MS

LOCATION:	Salon K
TIME:	10:15 AM - 11:30 AM Eastern Time
CHAIR:	Marty Schraer Syngenta Meridian, ID
MODERATOR:	Marty Schraer Syngenta Meridian, ID

***SPEAKER † STUDENT CONTEST**

10:15 AM **†Confirmation of Paraquat Resistance in Horseweed (*Erigeron canadensis*) in New York.** Bethany M. Schulteis*[1], Elizabeth C. Maloney[2], Caio A. Brunharo[3], Lynn M. Sosnoskie[2]; [1]Cornell University, Ithaca, NY, [2]Cornell University, Geneva, NY, [3]Penn State University, University Park, PA **(245)**

10:20 AM **†Preliminary Responses of Weeds, the Weed Soil Seedbank, and Microarthropod Communities to Zasso Electrical Weeding.** Aleah L. Butler-Jones*[1], Elizabeth C. Maloney[2], Charlee R. Smith[3], Emily J. Sausser[4], Lynn M. Sosnoskie[2]; [1]Cornell University, Ithaca, NY, [2]Cornell University, Geneva, NY, [3]The Ohio State University, Columbus, OH, [4]Lebanon Valley College, Annville, PA **(246)**

10:25 AM **†Improved Herbicide Selectivity in Tomato by Safening Action of Benoxacor, Fenclorim, Melatonin, and 2,4,6- Trichlorophenoxyacetic**

Acid. Tabata Raissa De Oliveira*, Antonio Augusto Tavares, Josiane C. Argenta, Varsha Varsha, Te-Ming (Paul) Tseng; Mississippi State University, Starkville, MS **(247)**

10:30 AM **†Sulfentrazone Efficacy and Safety in Cabbage and Broccoli.** Laura Pineda-Bermudez*[1], Elizabeth C. Maloney[1], Baylee L. Carr[2], Wesley M. Bouchelle[3], Thierry E. Besancon[3], Lynn M. Sosnoskie[1]; [1]Cornell University, Geneva, NY, [2]Rutgers University - Weed Science, Chatsworth, NJ, [3]Rutgers University, Chatsworth, NJ **(248)**

10:35 AM **†Weed Impacts on Alfalfa Forage Accumulation and Nutritive Value.** Chandra L. Montgomery*[1], Albert T. Adjesiwor[2]; [1]University of Idaho, Moscow, ID, [2]University of Idaho, Kimberly, ID **(249)**

10:40 AM **†GST Inhibition - Influence on Quizalofop P-ethyl Metabolism in Winter Wheat.** Amber Pelon*[1], Franck E. Dayan[2], Todd A. Gaines[2]; [1]Colorado State University, Berthoud, CO, [2]Colorado State University, Fort Collins, CO **(250)**

10:45 AM **†Weed Suppression Via Cotton Chromosome Substitution Lines at Different Cover Crop Systems.** Alyssa Lea Miller*; Mississippi State University, Starkville, MS **(251)**

10:50 AM **†Surface Ponds as a Source of Weed Contamination in Container Nurseries.** Alisha O. Ray*[1], Anthony LeBude[1], James Altland[2], Christopher D. Harlow[1], Joe C. Neal[1]; [1]North Carolina State University, Raleigh, NC, [2]USDA-ARS, Wooster, OH **(252)**

TUESDAY MORNING JANUARY 31

SST - SST PhD

LOCATION: Salon C
TIME: 10:15 AM - 12:00 PM Eastern Time
CHAIR: Marty Schraer
Syngenta
Meridian, ID
MODERATOR: Marty Schraer

Syngenta
Meridian, ID

***SPEAKER † STUDENT CONTEST**

10:15 AM **†Investigating New Peanut Cultivar Responses to Postemergence Applications of Chlorimuron.** Nicholas J. Shay*[1], Eric P. Prostko[2]; [1]The University of Georgia, Tifton, GA, [2]University of Georgia, Tifton, GA **(253)**

10:20 AM **†Weed Community Shift in Nitrogen Immobilized Soils.** Maria A. Gannett*, Antonio DiTommaso, Jenny Kao-Kniffin; Cornell University, Ithaca, NY **(254)**

10:25 AM **†Effect of Fall- or Spring-Planted Cover Crops on Weed Suppression in High Plains.** Sachin Dhanda*[1], Vipan Kumar[1], Augustine Obour[1], Elizabeth Yeager[2], John Holman[3], Anita Dille[2]; [1]Kansas State University, Hays, KS, [2]Kansas State University, Manhattan, KS, [3]Kansas State University, Garden City, KS **(255)**

10:30 AM **†Utility of John Deere's See & Spray™ Ultimate in Southern U.S. Row Crops.** Tristen H. Avent*[1], Jason K. Norsworthy[1], William L. Patzoldt[2], Lauren M. Lazaro[2]; [1]University of Arkansas, Fayetteville, AR, [2]Blue River Technology, Sunnyvale, CA **(256)**

10:35 AM **†Transgenerational Effect of Weed Competition on Wheat.** Albert O. Kwarteng*[1], Albert T. Adjesiwor[2]; [1]University of Idaho, Moscow, ID, [2]University of Idaho, Kimberly, ID **(257)**

10:40 AM **†Assessing the Efficacy of a Biculture Living and Dead Cover Crop Mixture for Weed Suppression in Sweet Corn.** Veronica Yurchak*, Alan W. Leslie, Bill Phillips, Cerruti Rr Hooks; University of Maryland, College Park, MD **(258)**

10:45 AM **†A Deep Learning Approach for Nutsedge (*Cyperus* Spp.) Detection in Strawberry Plasticulture Production.** Alex G. Rodriguez*[1], Arnold W. Schumann[2], Nathan Boyd[3]; [1]University of Florida, Wimauma, FL, [2]University of Florida, Lake Alfred, FL, [3]University of Florida, Balm, FL **(259)**

10:50 AM **†Weed Seed Fate in the Combine During Wheat and Soybean Harvest.** Eli C. Russell*, Michael L.

Flessner, Matthew P. Spoth, Kevin Bamber; Virginia Tech, Blacksburg, VA **(260)**

10:55 AM **†Evaluating the Effect of Cover Crop Termination Management on Palmer Amaranth (Amaranthus Palmeri) Suppression in Soybean.** Cynthia Sias*, Michael L. Flessner, Kevin Bamber; Virginia Tech, Blacksburg, VA **(261)**

11:00 AM **†Using Black Soldier Fly Larvae Composting to Reduce Weed Seed Germination.** Jeanine Arana*[1], Stephen L. Meyers[2], Luz Aide Cardona Giraldo[2], Nathaly Belen Vargas Arroyo[2], Laura D. Rodriguez Baquero[3]; [1]Purdue University/Horticulture, West Lafayette, IN, [2]Purdue University, West Lafayette, IN, [3]University of Wisconsin-Madison, Madison, WI **(262)**

11:05 AM **†Weed Genetic Control: Potential Application for Controlling Dioecious Amaranths.** Lucas Kopecky Bobadilla*[1], Patrick Tranel[2]; [1]University of Illinois, Champaign, IL, [2]University of Illinois, Urbana, IL **(263)**

11:10 AM **†Determining the Critical Period of Weed Control in Organic Carrots.** Emma Kubinski*[1], Mac Burgess[2], Fabian Menalled[3]; [1]Montana State University/Land Resources and Environmental Science, Bozeman, MT, [2]Montana State University/ Plant Sciences and Plant Pathology, Bozeman, MT, [3]Montana State University, Bozeman, MT **(264)**

11:15 AM **†Allelopathic Weed-suppressive Effect of Cotton Chromosome Substitution Lines Under Field Conditions.** Worlanyo Segbefia*[1], Varsha Singh[1], Ncomiwe A. Maphalala[2], Tabata Raissa De Oliveira[1], Ziming Yue[1], Te-Ming (Paul) Tseng[1]; [1]Mississippi State University, Starkville, MS, [2]Mississippi State University-Department of Plant and Soil Sciences, Starkville, MS **(265)**

11:20 AM **†Biochemical Profiling of Sweetpotato Roots for Weed-Suppressing Allelochemicals.** Varsha Singh*[1], Worlanyo Segbefia[1], Ziming Yue[1], Mark W. Shankle[2], Te-Ming (Paul) Tseng[1]; [1]Mississippi State University, Starkville, MS, [2]Mississippi State University, Pontotoc, MS **(266)**

11:25 AM **†Prediction of Weed Emergence Patterns in a Rice-Wheat Cropping System Using Population-Based Threshold Models.** Faryal Ali*[1], Irfan

Afzal[2]; [1]University of agriculture Faisalabad Pakistan, Mardan, Pakistan, [2]University of agriculture Faisalabad Pakistan, Faisalabad, Pakistan **(267)**

11:30 AM **†Soybean Weed Management with Residual Herbicide Programs.** Zachary R. Treadway*, Jennifer Dudak, Todd A. Baughman; Oklahoma State University, Ardmore, OK **(268)**

11:35 AM **†Evaluation of Pre-emergence Herbicides for Weed Management in Florida's Strawberry Plasticulture Production.** Ruby Tiwari*[1], Nathan Boyd[2], Pamela Roberts[1], Samira Daroub[3], Ramdas Kanissery[1]; [1]University of Florida, Immokalee, FL, [2]University of Florida, Balm, FL, [3]University of Florida, Everglades, FL **(269)**

11:40 AM **†Johnsongrass (*Sorghum halepense*) Promotes Biological Nitrification Inhibition in its Rhizosphere.** Megan L. Schill*, Dinesh Phuyal, Nithya Rajan, Nithya K. Subramanian, Muthukumar V. Bagavathiannan; Texas A&M University, College Station, TX **(270)**

11:45 AM **†Subtelomeric Rearrangements Cause Glyphosate Resistance in *Eleusine indica*.** Nicholas A. Johnson*[1], Nathan D. Hall[2], Qin Yu[3], Chun Zhang[4], Eric L. Patterson[1]; [1]Michigan State University, East Lansing, MI, [2]Michigan State, East Lansing, MI, [3]University of Western Australia, Perth, Australia, [4]Guangdong Academy of Agricultural Sciences, Guangzhou, China **(129)**

TUESDAY AFTERNOON JANUARY 31

SYMPOSIUM - 01. Endangered Species Act: A workshop to determine how the WSSA can provide useful science-based information to regulators

LOCATION: Salon III
TIME: 01:00 PM - 05:00 PM Eastern Time
CHAIR: William J. Chism
US EPA
Point of Rocks, MD

MODERATOR: William J. Chism
US EPA
Point of Rocks, MD

***SPEAKER**

01:00 PM **Introduction to Workshop, Objectives and Format.** William J. Chism*; US EPA, Point Of Rocks, MD **(271)**

01:10 PM **Keynote.** Jake Li*; US EPA - Deputy Assistant Adminsitrator for Pesticide Programs, Washington, DC **(272)**

01:30 PM **Keynote.** Kyle Kunkler*; American Soybean Association, St Louis, MO **(273)**

01:50 PM **Discuss WSSA Comments to EPA Endangered Species Workplan.** William J. Chism*; US EPA, Point of Rocks, MD **(274)**

02:05 PM **Description of Workshop Topics.** William J. Chism*; US EPA, Point Of Rocks, MD **(275)**

02:55 PM **Coffee Break**

03:10 PM **Workgroup Discussion Sessions.** William J. Chism*; US EPA, Point Of Rocks, MD **(276)**

04:40 PM **Post Workshop Discussion.** William J. Chism*; US EPA, Point Of Rocks, MD **(277)**

TUESDAY AFTERNOON JANUARY 31

ORAL - 13. Student Travel Enrichment Experience

LOCATION: Salon C
TIME: 02:00 PM - 04:30 PM Eastern Time

***SPEAKER**

02:00 PM **A Tour Through Corteva Headquarters: A Global Perspective of an Integrated Field Science Organization.** Estefania Gomiero Polli*; North Carolina State University, Raleigh, NC **(278)**

02:15 PM **Travel Enrichment Experience at Beltsville Agricultural Research Center, Maryland.** Sachin Dhanda*; Kansas State University, Hays, KS **(279)**

02:30 PM **Travel Enrichment Experience with the Weed Research Lab at Colorado State University.** Joshua W. Miranda*; Oregon State University, Corvallis, OR **(280)**

02:45 PM **From East to West: Biocontrol Musings from Montana.** Emily Duenk*; University of Guelph, Ridgetown, ON, Canada **(281)**

03:00 PM **Coffee Break**

03:15 PM **Weed Science Research, Policy, and Partnerships with USDA-ARS: 2022 Travel Enrichment Experience.** Sarah E. Kezar*[1], Steve Young[2]; [1]Texas A&M University, College Station, TX, [2]USDA-ARS, Beltsville, MD **(282)**

03:30 PM **Science Policy Fellowship: A Unique Learning Opportunity for Aspiring Weed Scientists.** Taylor M. Randell-Singleton*[1], Navdeep Godara[2], Lee Van Wychen[3]; [1]University of Georgia, Tifton, GA, [2]Virginia Tech, Blacksburg, VA, [3]Weed Science Society of America, Alexandria, VA **(283)**

TUESDAY AFTERNOON JANUARY 31

NEWSS Business Meeting

LOCATION: Salon II
TIME: 05:00 PM - 07:00 PM Eastern Time
MODERATOR: Jacob Barney
Virginia Tech
Blacksburg, VA

WEDNESDAY MORNING FEBRUARY 1

Presidents Breakfast

LOCATION: Salon J

TIME: 07:00 AM - 08:00 AM Eastern Time

MODERATOR: A Stanley Culpepper
University of Georgia
Tifton, GA

WEDNESDAY MORNING FEBRUARY 1

Judges Work Room

LOCATION: Alexandria

TIME: 07:00 AM - 03:00 PM Eastern Time

WEDNESDAY MORNING FEBRUARY 1

POSTER - NEWSS Student Poster

LOCATION: Salon IV-VI

TIME: 08:00 AM - 10:00 AM Eastern Time

WSSA Student Competition Posters will be judged Tuesday only
NEWSS Student Competition Posters will be judged Wednesday only

***PRESENTER † STUDENT CONTEST**

†Increased Soil Temperature Effects on Weed Seed Mortality Vary Across Species. Ashley N. Isaacson*[1], Richard G. Smith[2], Chris Stamplis[3], Alix Contosta[2], Carolyn J. Lowry[4]; [1]Penn State University, State College, PA, [2]University of New Hampshire, Durham, NH, [3]Pennsylvania State University, University Park, PA, [4]The Pennsylvania State University, University Park, PA **(284)**

†Steam Disinfestation of Soil with and without Accelerants. Donald E. Fox*[1], Emma Volk[1], Mark Hoffmann[1], Steve Fennimore[2], Joe C. Neal[1]; [1]North Carolina State University, Raleigh, NC, [2]University of California - Davis, Salinas, CA **(285)**

†Acreage Allocation for Variable Density Planting to Optimize Crop Yield and Weed Suppression. Sandra R. Ethridge*[1], Wesley Everman[1], Anna M. Locke[2], David L. Jordan[1], Ramon G. Leon[1]; [1]North Carolina State University, Raleigh, NC, [2]USDA-ARS, Raleigh, NC **(286)**

†Degradation of Spring Applied Methiozolin and its Effects on Smooth Crabgrass and Goosegrass Control. John M. Peppers*, Shawn Askew; Virginia Tech, Blacksburg, VA **(287)**

†Weed Risk Assessment of *Brassica carinata* in the Southeastern U.S. April M. Dobbs*, Chris Reberg-Horton, Lori Unruh Snyder, Ramon G. Leon; North Carolina State University, Raleigh, NC **(288)**

†Response of Italian Ryegrass Accessions from South-Central North Carolina to Spring Burndown Herbicides. Jose H. de Sanctis*, Charlie W. Cahoon, Wesley Everman, Travis Gannon; North Carolina State University, Raleigh, NC **(289)**

†Optimizing Application Time of Residual Herbicides in a Planted Green Soybean System. Megan R. Czekaj*[1], John M. Wallace[2], Mark VanGessel[3], Michael L. Flessner[4]; [1]Penn State University- Agricultural and Environmental Plant Science, State College, PA, [2]Penn State University, University Park, PA, [3]University of Delaware, Georgetown, DE, [4]Virginia Tech, Blacksburg, VA **(290)**

†Do Plant Growth Regulator Admixtures Influence Annual Bluegrass Control by Methiozolin? Navdeep Godara*, Shawn Askew, John M. Peppers; Virginia Tech, Blacksburg, VA **(291)**

†Evaluation of Various Preemergent Herbicides at Different Rates to Control False-Green Kyllinga (*Kyllinga gracillima*) from Seed. Matthew T. Elmore[1], Brandon C. McNally[1], Aaron J. Patton[2], Daniel P. Tuck[1], Trevor S. Watson*[1]; [1]Rutgers University, New Brunswick, NJ, [2]Purdue University, West Lafayette, IN **(292)**

†Discrimination Between PPO-inhibiting Herbicide-resistant Redroot Pigweed (Amaranthus Retroflexus) Utilizing Spectral Reflectance. Eric A. Jones*, Robert Austin, Diego J. Contreras, Jeffrey Dunne, Charlie W. Cahoon, Ramon G. Leon, Wesley Everman; North Carolina State University, Raleigh, NC **(293)**

†Influence of Tall Fescue Canopy Architecture on Weed Emergence. Daewon Koo*[1], Clebson G. Goncalves[2], Shawn Askew[1]; [1]Virginia Tech, Blacksburg, VA, [2]University of California, Agriculture and Natural Resources, Ukiah, CA **(294)**

†**Control of Common Row Crop Weeds with Dicamba and Glufosinate Applied Alone, Mixed, and Sequentially.** Jackson W. Alsdorf*, Eric A. Jones, Diego J. Contreras, Charlie W. Cahoon, Ramon G. Leon, Wesley Everman; North Carolina State University, Raleigh, NC **(295)**

†**Effect of Increased Soil Temperatures on Weed Seed Mortality Vary Across Species.** Ashley N. Isaacson*[1], Carolyn J. Lowry[2]; [1]Penn State University, State College, PA, [2]The Pennsylvania State University, University Park, PA **(296)**

WEDNESDAY MORNING FEBRUARY 1

SYMPOSIUM - 03. Crop-Weed Management in a Rising CO2 and Warming World

LOCATION:	Salon II
TIME:	08:00 AM - 12:00 PM Eastern Time
CHAIR:	Martin Williams USDA-ARS Urbana, IL
MODERATOR:	Martin Williams USDA-ARS Urbana, IL

***SPEAKER**

08:00 AM **Our Changing Climate: Science, Impacts, and Vulnerabilities.** Donald J. Wuebbles*; University of Illinois, Urbana, IL **(297)**

08:20 AM **Modeling Invasive Weed Distributions Under Future Climate Scenarios.** Antonio DiTommaso*[1], Sophie Westbrook[1], Emma K. Nikkel[2], David R. Clements[3], Caroline Marschner[1]; [1]Cornell University, Ithaca, NY, [2]University of British Columbia, Vancouver, BC, Canada, [3]Trinity Western University, Langley, BC, Canada **(298)**

08:40 AM **Effect of CO_2 on Weedy Rice Growth, Seed Shattering, and Seed Dormancy.** Luis A. Avila*; Universidade Federal de Pelotas, Pelotas, Brazil **(299)**

09:00 AM **Impact of High Temperature Stress on Crops, Weeds, and Herbicide Efficacy.** Pv V. Prasad*; Kansas State University, Manhattan, KS **(300)**

09:20 AM **Climate Change and Pest Treatment Costs: What is Happening?** Bruce A. McCarl*; Texas A & M University, College Station, TX **(301)**

09:40 AM **Weeds and Weed Management in a Warmer, Wetter World.** Carolyn J. Lowry*[1], Ashley N. Isaacson[2], Alix Contosta[3], Erin Haramoto[4], Richard G. Smith[3], John M. Wallace[5]; [1]The Pennsylvania State University, University Park, PA, [2]Penn State University, State College, PA, [3]University of New Hampshire, Durham, NH, [4]University of Kentucky, Lexington, KY, [5]Penn State University, University Park, PA **(302)**

10:00 AM **Coffee Break**

10:15 AM **Does Environmental Change Increase the Impact of Weeds?** Montserrat Vilà*; Estación Biológica de Doñana (EBD-CSIC), Sevilla, Spain **(303)**

10:35 AM **Combined Effects of Climate Change and Intensification on Potential Yield.** Jonathan Storkey*; Rothamsted Research, Harpenden, United Kingdom **(304)**

10:55 AM **Crops and Weeds Running to Stand Still in the Red Queen's Climate Change.** David R. Clements*; Trinity Western University, Langley, BC, Canada **(305)**

11:15 AM **Climate Change, Weed Biology and Public Health.** Lewis H. Ziska*; Columbia University, Baltimore, MD **(306)**

WEDNESDAY MORNING FEBRUARY 1

Oral 01. Agronomic Crops - Day 1

LOCATION: Salon III
TIME: 09:00 AM - 05:00 AM Eastern Time
CHAIR: Daljit Singh
Bayer

Chesterfield, MO

MODERATOR: Daljit Singh
Bayer
Chesterfield, MO

CO-MODERATOR: Craig M. Alford
Corteva
Ankeny, IA

***SPEAKER**

09:00 AM **Diflufenican: a Tool for Managing *Amaranthus* Species in Corn and Soybean Cropping Systems.** John Buol*, Carl Coburn, Richard Leitz, Eric Riley, Ananda Datta; Bayer, St. Louis, MO **(307)**

09:15 AM **Depletion of Multiple-Herbicide-Resistant Waterhemp Seed in the Soil Seed Bank with Integrated Weed Management Strategies.** Peter H. Sikkema*, Nader Soltani; University of Guelph, Ridgetown, ON, Canada **(308)**

09:30 AM **How is Soybean Biomass Partitioning to Vegetation, Floral Development and Pod Formation Affected by Exposure to Simulated Drift Rates of Auxin Herbicides?** Koffi Badou-Jeremie Kouame*[1], Thomas R. Butts[1], Brad M. Davis[1], Leah M. Collie[2], Troy W. Dillon[1], Jeremy K. Sutterfield[1], Benjamin C. Thrash[1], Nick R. Bateman[3], John J. Adamcyzk[4], Gus M. Lorenz[2]; [1]University of Arkansas System Division of Agriculture, Lonoke, AR, [2]University of Arkansas, Lonoke, AR, [3]University of Arkansas System Division of Agriculture, Fayetteville, AR, [4]USDA, Poplarville, MS **(309)**

09:45 AM **Effectiveness of Tirexor® and Voraxor® on Glufosinate-Resistant Palmer Amaranth.** Pamela Carvalho-Moore*[1], Jason K. Norsworthy[1], Ty Smith[1], Casey H. Arnold[1], Matthew C. Woolard[1], Maria Leticia Zaccaro-Gruener[1], Tom Barber[2]; [1]University of Arkansas, Fayetteville, AR, [2]University of Arkansas System Division of Agriculture, Lonoke, AR **(310)**

10:00 AM **Coffee Break**

10:15 AM **Agriculturally Significant Weeds and Associated Soil Nematodes of South-Central New Jersey: the Good, the Bad, the Ugly.** Melissa A. Bravo*; Rutgers, The State University of New Jersey, Woodstown, NJ **(311)**

10:30 AM **A Survey of Twelve Common Rowcrop Weeds and Associated Soil Nematodes of Salem County, NJ.** Melissa A. Bravo*; Rutgers, The State University of New Jersey, Woodstown, NJ **(312)**

10:45 AM **The Potential of Inter-row Non-selective Herbicide Application to Manage Wild Oat (*Avena fatua* L.) in Wheat.** Eric N. Johnson*, Christian Willenborg, Steve Shirtliffe; University of Saskatchewan, Saskatoon, SK, Canada **(313)**

11:00 AM **Discovery of Tetflupyrolimet: A New Mode-of-Action Herbicide Interfering with De Novo Pyrimidine Biosynthesis for Effective Management of Herbicide-Resistant Grass Weeds Globally.** Atul Puri*, Thomas Selby, Steve Gutteridge, Mark Holliday, Adam Prestegord; FMC Corporation, Newark, DE **(314)**

11:15 AM **Control of Multiple-Herbicide-Resistant Waterhemp (*Amaranthus tuberculatus*) with Acetochlor-based Tank Mixtures in Soybean.** Hannah E. Symington*[1], Nader Soltani[2], Allan Kaastra[3], David C. Hooker[2], Darren E. Robinson[2], Peter H. Sikkema[2]; [1]University of Guelph, Rr#3 Watford, ON, Canada, [2]University of Guelph, Ridgetown, ON, Canada, [3]Bayer Crop Science, Guelph, ON, Canada **(315)**

11:45 AM **Spatiotemporal Changes in Glyphosate Efficacy on Key Weed Species in the US and Canada.** Christopher A. Landau*[1], Kevin W. Bradley[2], Erin E. Burns[3], Ryan P. DeWerff[4], Michael L. Flessner[5], Karla L. Gage[6], Aaron G. Hager[7], Joseph T. Ikley[8], Prashant Jha[9], William G. Johnson[10], Paul O. Johnson[11], Sarah Lancaster[12], Dwight Lingenfelter[13], Mark M. Loux[14], Eric J. Miller[15], Scott A. Nolte[16], Debalin Sarangi[17], Peter H. Sikkema[18], Christy L. Sprague[3], Mark VanGessel[19], Rodrigo Werle[20], Martin Williams[21], Bryan G. Young[22]; [1]University of Illinois, Crop Sciences, Urbana, IL, [2]University of Missouri, Columbia, MO, [3]Michigan State University, East Lansing, MI, [4]University of Wisconsin, Madison, WI, [5]Virginia Tech, Blacksburg, VA, [6]Southern Illinois University Carbondale, Carbondale, IL, [7]University of Illinois, Urbana, IL, [8]North Dakota State University, Fargo, ND, [9]Iowa State University, Ames, IA, [10]Purdue University, West Lafayette, IN, [11]South Dakota State University, Brookings, ND, [12]Kansas State University, Manhattan, KS, [13]Penn State University, University Park, PA, [14]The

Ohio State University, Columbus, OH, [15]Southern Illinois University, Carbondale, IL, [16]Texas A&M AgriLife Extension, College Station, TX, [17]University of Minnesota, St. Paul, MN, [18]University of Guelph, Ridgetown, ON, Canada, [19]University of Delaware, Georgetown, DE, [20]University of Wisconsin-Madison, Madison, WI, [21]USDA-ARS, Urbana, IL, [22]Purdue University, Brookston, IN **(317)**

12:00 PM **Lunch**

01:00 PM **Chaff Lining for Management of Weeds in Winter Wheat.** Matthew P. Spoth[1], Michael L. Flessner*[1], Kevin Bamber[1], Wykle C. Greene[1], Cynthia Sias[1], Eli C. Russell[1], Vipin Kumar[2], Vijay Singh[3]; [1]Virginia Tech, Blacksburg, VA, [2]Virginia Polytechnic Institute and State University, Painter, VA, [3]Virginia Tech, Painter, VA **(318)**

01:15 PM **Changes in Weed Population Density After Harvest with a Seed Impact Mill.** Eli C. Russell*, Michael L. Flessner, Matthew P. Spoth, Wykle C. Greene, Kevin Bamber; Virginia Tech, Blacksburg, VA **(319)**

01:30 PM **Reduction in Palmer Amaranth (Amaranthus Palmeri) Emergence and Growth Rate in Response to Cover Crop Termination Timing: Applications for Optimizing Herbicide Programs.** Cynthia Sias*, Michael L. Flessner, Kevin Bamber; Virginia Tech, Blacksburg, VA **(320)**

01:45 PM **The Cover Crop Evolution/Revolution- Past, Present, Future.** William S. Curran*; Penn State University, Bozeman, MT **(321)**

02:00 PM **Cross Resistance to Acetyl-CoA Carboxylase-Inhibiting Herbicides in a Johnsongrass [(*Sorghum halepense* (L.) Pers.] Accession Collected in Arkansas.** Fidel Gonzalez Torralva*, Jason K. Norsworthy; UNIVERSITY OF ARKANSAS, Fayetteville, AR **(322)**

02:15 PM **Georgia's Endangered Species Pilot Program: A Team Approach to Help Protect Species and the Farm.** Taylor M. Randell-Singleton*[1], Jenna C. Vance[1], William J. Chism[2], A Stanley Culpepper[1]; [1]University of Georgia, Tifton, GA, [2]US EPA, Point Of Rocks, MD **(323)**

02:30 PM **Survey of Herbicide Resistance in Pigweed (*Amaranthus* Spp.) Escapes in North Dakota Row Crops.** Joseph T. Ikley*[1], Stephanie DeSimini[1], Michael J. Christoffers[1], Megan O'Neil[2], Zack Bateson[2]; [1]North Dakota State University, Fargo, ND, [2]National Agricultural Genotyping Center, Fargo, ND **(324)**

02:45 PM **Does Interseeded Cover Crop Seed Size Influence Tolerance to Residual Herbicide Programs?** Tosh R. Mazzone*[1], John M. Wallace[1], Matthew R. Ryan[2], Chris J. Pelzer[2]; [1]Penn State University, University Park, PA, [2]Cornell University, Ithaca, NY **(325)**

03:00 PM **Coffee Break**

03:15 PM **A23980B: Corn Weed Control from Four Complimentary Active Ingredients.** Craig Austin*[1], Seth A. Strom[2], Mark Kitt[1], Thomas Beckett[1]; [1]Syngenta, Greensboro, NC, [2]Syngenta Crop Protection, Monticello, IL **(326)**

03:30 PM **Introducing Golden Pennycress (*Thlaspi arvense*) into the Corn-Soybean Rotation.** Mark L. Bernards*, Brent S. Heaton; Western Illinois University, Macomb, IL **(327)**

03:45 PM **Investigating the Genetic Control of Interspecific Hybridization Between *Sorghum bicolor* and *S. halepense* Using Recombinant Inbred Lines (RILs).** Usha R. Pedireddi*, Nithya K. Subramanian, George Hodnett, Patricia Klein, William Rooney, Muthukumar V. Bagavathiannan; Texas A&M University, College Station, TX **(328)**

04:00 PM **Classification of Herbicide Symptomology in Cotton Using Digital Images and Deep Learning.** Ubaldo Torres*[1], Joe Johnson[1], Shefali Sarang[1], Bishwa B. Sapkota[1], Scott A. Nolte[2], Muthukumar V. Bagavathiannan[1]; [1]Texas A&M University, College Station, TX, [2]Texas A&M AgriLife Extension, College Station, TX **(329)**

04:15 PM **Evaluation of Crop Herbicide Tolerance Using UAV Obtained Multispectral Imagery.** Ryan D. Langemeier*, Livia Pereira, Justin T. McCaghren, Steve Li; Auburn University, Auburn, AL **(330)**

WEDNESDAY MORNING FEBRUARY 1

ORAL - 07. Formulation, Adjuvant, & Application Technology

LOCATION:	Salon H
TIME:	09:30 AM - 02:00 PM Eastern Time
CHAIR:	Joseph T. Ikley North Dakota State University Fargo, ND
MODERATOR:	Joseph T. Ikley North Dakota State University Fargo, ND
CO-MODERATOR:	Gourav Sharma Virginia Tech Blacksburg, VA

***SPEAKER**

09:30 AM **Comparison of Weed Control Outcomes When Using Precision Application Versus Traditional Approaches in Corn (*Zea mays*).** Isaac H. Barnhart*[1], Greg Kruger[2], Kalvin Andrew Miller[3], Chris Proctor[4], Rodrigo Werle[5], Anita Dille[1]; [1]Kansas State University, Manhattan, KS, [2]University of Nebraska - Lincoln, North Platte, NE, [3]BASF, Seymour, IL, [4]University of Nebraska-Lincoln, Lincoln, NE, [5]University of Wisconsin-Madison, Madison, WI **(331)**

09:45 AM **Weed Control in Soybeans (Glycine Max) Using Precision Sprayers.** Isaac H. Barnhart*[1], Proctor Chris[2], Thiago H. Vitti[2], Kalvin Andrew Miller[3], Greg Kruger[4], Anita Dille[1]; [1]Kansas State University, Manhattan, KS, [2]University of Nebraska-Lincoln, Lincoln, NE, [3]BASF, Seymour, IL, [4]University of Nebraska - Lincoln, North Platte, NE **(332)**

10:00 AM **Coffee Break**

10:15 AM **Effects of PRE Herbicides on Precision Post Emergence Herbicide Application in Plasticulture Production.** Ana Claudia Buzanini*[1], Arnold W. Schumann[2], Nathan Boyd[3]; [1]University of Florida, Wimauma, FL, [2]University of Florida, Lake Alfred, FL, [3]University of Florida, Balm, FL **(333)**

10:30 AM **Herbicide Options in Industrial Hemp.** Wyatt J. Stutzman*, Kevin Bamber, Cynthia Sias, Eli C.

Russell, Michael L. Flessner; Virginia Tech, Blacksburg, VA **(334)**

10:45 AM **Assessing Deposition and Plant Responses Following Dicamba Applications in Cotton and Soybean Using Hooded and Open Boom Sprayers.** Devin Hammer*, Naresh Pai, Allen Olmstead; Bayer Crop Science, St. Louis, MO **(335)**

11:00 AM **A Novel Tool for Improving Selection of Nozzles and Adjuvants, CPDA's New *Application Enhancement Certification* Program.** James Reiss*; Precision Laboratories, LLC, Lake Geneva, WI **(336)**

11:15 AM **Dicamba Applications Using Drop Nozzles to Reduce Drift in Texas Cotton.** Hayden R. Taylor*[1], Scott A. Nolte[2], Peter A. Dotray[3], Joshua A. McGinty[4]; [1]Texas A&M University, College Station, TX, [2]Texas A&M AgriLife Extension, College Station, TX, [3]Texas Tech University, Texas A&MTexas Tech University, Texas A&M AgriLife Research, Texas A&M AgriLife Extension, Lubbock, TX, [4]Texas A&M AgriLife Extension, Corpus Christi, TX **(337)**

11:30 AM **Developer of Novel Adjuvant Systems for Herbicides.** Jim T. Daniel*[1], Chase T. Boman[2]; [1]Daniel Ag Consulting, Keenesburg, CO, [2]AgraSyst, Spokane, WA **(338)**

11:45 AM **Evaluation of Soybean and Cotton Injury to Low Volume Applications as Influenced by Application Type and GPA.** Livia Pereira*, Steve Li, Ryan D. Langemeier, Justin T. McCaghren; Auburn University, Auburn, AL **(339)**

12:00 PM **Lunch**

01:00 PM **What's New at the Winfield United Innovation Center?** Gregory K. Dahl*[1], Eric P. Spandl[2], Ryan J. Edwards[2], Joshua J. Skelton[3], Steven A. Fredericks[2], Aszhia K. Albrecht[4], Elizabeth R. Alonzi[2], Elizabeth Buescher[2], Cody Hoerning[2], Derick Jiwan[2]; [1]Winfield United, Eagan, MN, [2]Winfield United, River Falls, WI, [3]WinField United, Saint Paul, MN, [4]Land O' Lakes - Winfield United, River Falls, WI **(340)**

01:15 PM **Single Tank vs Dual Tank See and Spray™ Programs in Soybean.** Wesley Everman*[1], Diego J. Contreras[1], Jason K. Norsworthy[2], Tristen H.

Avent[2], Bryan G. Young[3], Marcelo Zimmer[4], Jasmine M. Mausbach[5], Michael M. Houston[2], Lauren M. Lazaro[6], William L. Patzoldt[6]; [1]North Carolina State University, Raleigh, NC, [2]University of Arkansas, Fayetteville, AR, [3]Purdue University, Brookston, IN, [4]Purdue University, West Lafayette, IN, [5]University of Nebraska-Lincoln, Lincoln, NE, [6]Blue River Technology, Sunnyvale, CA **(341)**

WEDNESDAY MORNING FEBRUARY 1

ORAL - 12. Integrated Weed Management

LOCATION:	Salon I
TIME:	09:00 AM - 05:00 PM Eastern Time
CHAIR:	Sarah E. Kezar Texas A&M University College Station, TX
MODERATOR:	Sarah E. Kezar Texas A&M University College Station, TX

***SPEAKER**

09:15 AM **Integrated Strategies of Weed Control in Semiarid Central Great Plains.** Vipan Kumar*[1], Rui Liu[2], Sachin Dhanda[1], Prashant Jha[3], Jason K. Norsworthy[4]; [1]Kansas State University, Hays, KS, [2]Washington State University, Prosser, WA, [3]Iowa State University, Ames, IA, [4]University of Arkansas, Fayetteville, AR **(342)**

09:30 AM **Development of a Novel Electrocution-Based Robotic Platform for Weed Control in Vegetables and Field Crops.** Ran Nisim Lati*[1], Avital Bechar[2]; [1]Department of Plant Pathology and Weed Research, Newe Yaar Research Center, Agricultural Research Organization (ARO) - Volcani Center, Ramat Yishay, Israel, [2]Institute of Agricultural Engineering, Agricultural Research Organization, The Volcani Center, Bet-Dagan, Israel, Bet Dagan, Israel **(343)**

09:45 AM **Investigating Novel Integrated Weed Management Systems on the Canadian Prairies.** Breanne D. Tidemann*[1], K. Neil Harker[1], Steve Shirtliffe[2], Christian Willenborg[2], Eric N. Johnson[2],

Robert Gulden[3], Newton Lupwayi[4], T. Kelly Turkington[1], Emma Stephens[4], Charles M. Geddes[4]; [1]Agriculture and Agri-Food Canada, Lacombe, AB, Canada, [2]University of Saskatchewan, Saskatoon, SK, Canada, [3]University of Manitoba, Winnipeg, MB, Canada, [4]Agriculture and Agri-Food Canada, Lethbridge, AB, Canada **(344)**

10:00 AM **Coffee Break**

10:15 AM **Soil Residual Herbicide Degradation as Influenced by Soil Microbial Activity in Cover Cropping Systems.** Lucas Oliveira Ribeiro Maia*[1], William G. Johnson[1], Bryan G. Young[2], Eileen J. Kladivko[1], Shalamar Armstrong[1]; [1]Purdue University, West Lafayette, IN, [2]Purdue University, Brookston, IN **(345)**

10:30 AM **Incorporating Winter Camelina or Canola into Rotational Cropping Systems of Colder Climates to Reduce Weed Pressure.** James V. Anderson*[1], Wun S. Chao[1], Xuehui Li[2], Russ W. Gesch[3], Marisol T. Berti[2], David P. Horvath[1]; [1]USDA-ARS, Fargo, ND, [2]North Dakota State University, Fargo, ND, [3]USDA-ARS, Morris, MN **(346)**

10:45 AM **Cereal Rye Cover Crop Effects on Fiber Hemp (*Cannabis sativa* L.) in Southeast Texas.** Jodie M. McVane*, Muthukumar V. Bagavathiannan; Texas A&M University, College Station, TX **(347)**

11:00 AM **Cereal Rye Cover Crop Significantly Delays Weed Emergence in Cotton.** Gustavo Camargo Silva*, Muthukumar V. Bagavathiannan; Texas A&M University, College Station, TX **(348)**

11:15 AM **Early Detection of Herbicide Resistance Using Commercial Weed Map Data.** Gregory E. Stewart*[1], Devin Kirk[2], Charles M. Geddes[3]; [1]Geco Engineering, North Vancouver, BC, Canada, [2]Geco Engineering, Vancouver, BC, Canada, [3]Agriculture and Agri-Food Canada, Lethbridge, AB, Canada **(349)**

11:30 AM **Effects of Cover Crop Management on Weed Emergence Across the Eastern United States.** Eugene P. Law*[1], Mark VanGessel[2], Steven Brian Mirsky[3], Victoria Ackroyd[4], Muthukumar V. Bagavathiannan[5], Kevin W. Bradley[6], William S. Curran[7], Adam S. Davis[8], Jeffrey Evans[9], Wesley Everman[10], Michael L. Flessner[11], Nicholas R.

Jordan[12], Lauren M. Lazaro[13], Ramon G. Leon[10], John Lindquist[14], Jason K. Norsworthy[15], Lovreet S. Shergill[16]; [1]University of Delaware, Beltsville, MD, [2]University of Delaware, Georgetown, DE, [3]USDA ARS, Beltsville, MD, [4]University of Maryland, College Park, MD, [5]Texas A&M University, College Station, TX, [6]University of Missouri, Columbia, MO, [7]Penn State University, Bozeman, MT, [8]University of Illinois, Urbana, IL, [9]Five Rivers Conservation Trust, Concord, NH, [10]North Carolina State University, Raleigh, NC, [11]Virginia Tech, Blacksburg, VA, [12]University of Minnesota, St. Paul, MN, [13]Blue River Technology, Sunnyvale, CA, [14]University of Nebraska, Lincoln, NE, [15]University of Arkansas, Fayetteville, AR, [16]Montana State University, Huntley, MT **(350)**

11:45 AM **Using "Stacked" Cultivation Implements to Improve Weed Control Selectivity.** Bryan Brown*[1], Daniel C. Brainard[2], Percy Marshall[3]; [1]New York State Integrated Pest Management Program, Cornell University, Geneva, NY, [2]Michigan State University, East Lansing, MI, [3]University of Arkansas, Fayetteville, AR **(351)**

12:00 PM **Lunch**

01:00 PM **Integrating Electrocution as a Supplemental Weed Control Tool in No-till Organic Soybean.** Madhav Dhakal*, Carolyn Garrity, Reza Afshar; Rodale Institute, Kutztown, PA **(352)**

01:15 PM **A New Enemy in Town: Resistant Ryegrass Heading North in Europe.** Alberto Collavo*, Bianca Assis Barbosa Martins, Bodo Peters; Bayer AG, Frankfurt Am Main, Germany **(353)**

01:30 PM **Effect of Winter *Brassica carinata* Production on Weed Control in Subsequent Peanut.** Pratap Devkota*, Sithin Mathew, Navjot Singh; University of Florida, Jay, FL **(354)**

01:45 PM **Automated Intra-row Weeding Informed by Sugar Beet Response to Hoeing and Flaming Across Early Growth Stages.** Margaret R. McCollough*[1], Frank Poulsen[2], Bo Melander[3]; [1]University of Maine, Orono, ME, [2]F. Poulsen Engineering Aps., Hvalsø, Denmark, [3]Aarhus University, Department of Agroecology, Slagelse, Denmark **(355)**

02:00 PM **HRAC Global Validation Criteria for Unique Herbicide Resistance Cases.** Harry J. Strek*[1], Mark A. Peterson[2], Marcel S C de Melo[3], Caio V S Rossi[4], Ian M. Heap[5]; [1]HRAC Global, Frankfurt Am Main, Germany, [2]Dow AgroSciences LLC (Ret.), Indianapolis, IN, [3]Bayer Brasil, Paulínia, Brazil, [4]Corteva, Mogi Mirim, Brazil, [5]The International Herbicide-Resistant Weed Database, Corvallis, OR **(356)**

02:15 PM **Pairing Nanoparticles with FANA Technology for Spray-on Gene Silencing in Weeds.** Katie Martin*[1], Todd A. Gaines[1], Veenu Aishwarya[2]; [1]Colorado State University, Fort Collins, CO, [2]AUM Lifetech, Philadelphia, PA **(357)**

02:30 PM **Species-Level Weed Biomass Estimation from Video Imagery Using 3D Point Clouds.** Daniel J. Ginn*[1], Paula J. Ramos-Giraldo[2], Michael A. Alcorn[3], Maria L. Cangiano[3], April M. Dobbs[2], Søren K. Skovsen[4], Matthew Kutugata[1], Ramon G. Leon[2], Chris Reberg-Horton[2], Muthukumar V. Bagavathiannan[1], Steven Brian Mirsky[5]; [1]Texas A&M University, College Station, TX, [2]North Carolina State University, Raleigh, NC, [3]USDA, Beltsville, MD, [4]Aarhus University, Aarhus, Denmark, [5]USDA ARS, Beltsville, MD **(358)**

02:45 PM **Hard Water Effect on Glyphosate Grass Control in the Enlist Weed System.** Justin T. McCaghren*, Steve Li, Ryan D. Langemeier, Livia Pereira; Auburn University, Auburn, AL **(359)**

03:00 PM **Multistate Assessment of Cowpea Cover Crop Germplasm, Seeding Rate, and Termination Timing Effects on Weed Suppression.** Carlene A. Chase*[1], Daniel A. Boakye[1], Parmeshwor Aryal[1], Nicholas T. Basinger[2], Lisa Fultz[3], Audrey V. Gamble[4], Erin Haramoto[5], Nithya Rajan[6], Vitalis Temu[7], Elide Valencia[8]; [1]University of Florida, Gainesville, FL, [2]The University of Georgia, Athens, GA, [3]Louisiana State University, Baton Rouge, LA, [4]Auburn University, Auburn, AL, [5]University of Kentucky, Lexington, KY, [6]Texas A&M University, College Station, TX, [7]Virginia State University, Petersburg, VA, [8]University of Puerto Rico, Mayaguez, Mayaguez, PR **(360)**

03:15 PM **Integrated Weed Management Practices to Control ALS and PPO-Inhibitor Resistant Palmer Amaranth.** Justin T. McCaghren*, Steve Li, Ryan D. Langemeier, Livia Pereira; Auburn University, Auburn, AL **(361)**

03:30 PM **Cultural Tactics Prevail in the War on Multiple Herbicide-Resistant Kochia (*Bassia scoparia*).** Charles M. Geddes*[1], Louis J. Molnar[1], Cindy A. Gampe[2], Prabhath Lokuruge[2]; [1]Agriculture and Agri-Food Canada, Lethbridge, AB, Canada, [2]Agriculture and Agri-Food Canada, Scott, SK, Canada **(362)**

03:45 PM **Cover Crop Monocultures and Mixtures Affect Weed Seedbank Diversity in the Southeast United States.** Aniruddha Maity*[1], Audrey V. Gamble[1], Annu Kumari[1], Anna Murry Johnson[1], Andrew J. Price[2]; [1]Auburn University, Auburn, AL, [2]USDA-ARS, Auburn, AL **(363)**

04:00 PM **Community Management of Herbicide Resistance: Insights from the Pacific Northwest.** Nick Bergmann*[1], Ian Burke[1], Chloe Wardropper[2]; [1]Washington State University, Pullman, WA, [2]University of Illinois, Champaign, IL **(364)**

WEDNESDAY AFTERNOON FEBRUARY 1

ORAL - 02. Horticultural Crops

LOCATION:	Salon K
TIME:	02:00 PM - 05:00 PM Eastern Time
CHAIR:	Kurt M. Vollmer University of Maryland Queenstown, MD
MODERATOR:	Kurt M. Vollmer University of Maryland Queenstown, MD
CO-MODERATOR:	Thierry E. Besancon Rutgers University CHATSWORTH, NJ

***SPEAKER**

02:00 PM **IR-4 Weed Science Update - Food Crops.** Roger B. Batts*, Jerry Baron, Venkat Pedibhotla; IR-4 Project HQ, NC State University, Raleigh, NC **(365)**

02:15 PM **Weed Control and Crop Tolerance as Influenced by Pendimethalin Rate and Timing of Application in Cranberry Production.** Thierry E. Besancon*[1], Marcelo L. Moretti[2], Katherine Ghantous[3], Jed Colquhoun[4], Cyrus A. Smith[5]; [1]Rutgers University, Chatsworth, NJ, [2]Oregon State University, Corvallis, OR, [3]UMass Amherst, UMass Cranberry Station, East Wareham, MA, [4]University of Wisconsin-Madison, Madison, WI, [5]Ocean Spray Cranberries, Inc., Bandon, OR **(366)**

02:30 PM **Reducing Herbicide Inputs in Plasticulture Peppers Using Spring-seeded Grass Cover Crops.** Dwayne D. Joseph[1], Thierry E. Besancon*[2], Cerruti Rr Hooks[3], Alan W. Leslie[3], Kurt M. Vollmer[4]; [1]University of Maryland, Chestertown, MD, [2]Rutgers University, Chatsworth, NJ, [3]University of Maryland, College Park, MD, [4]University of Maryland, Queenstown, MD **(367)**

02:45 PM **Can Vegetable Crop Seed Treatments Enhance Competitiveness with Weeds?** Jed Colquhoun*, Jacklyn Slivicke, Emma Schmidt; University of Wisconsin-Madison, Madison, WI **(368)**

03:00 PM **Coffee Break**

03:15 PM **Integrating Spring-seeded Grass Cover Crops for Weed Management in Cucurbit Production.** Kurt M. Vollmer*[1], Thierry E. Besancon[2], Alan W. Leslie[3], Dwayne D. Joseph[4], Cerruti Rr Hooks[3]; [1]University of Maryland, Queenstown, MD, [2]Rutgers University, Chatsworth, NJ, [3]University of Maryland, College Park, MD, [4]University of Maryland, Chestertown, MD **(369)**

03:30 PM **Flumioxazin Application Rate and Timing for Improved Crop Safety While Maintaining Efficacy in Potato.** Harlene M. Hatterman-Valenti*, Collin Auwarter, Avery Shikanai; North Dakota State University, Fargo, ND **(370)**

03:45 PM **An Update on NY Research in Snap Beans, Table Beets, and Carrots.** Lynn M. Sosnoskie*, Elizabeth C. Maloney; Cornell University, Geneva, NY **(371)**

04:00 PM **Evaluating In-season Management Strategies for Branched Broomrape Control in California Processing Tomatoes.** Matthew Fatino*, Bradley D. Hanson; University of California, Davis, Davis, CA **(372)**

04:15 PM **Weed Control and Hazelnut (*Corylus avellana* L.) Tolerance with Tiafenacil.** Joshua W. Miranda*, Marcelo L. Moretti; Oregon State University, Corvallis, OR **(373)**

04:30 PM **Speculations on Weed Management and Herbicide Use Under Agrivoltaic Panels in Cranberry Production.** Hilary A. Sandler*[1], Giverson Mupambi[2], Dwayne Breger[3], Samantha Glaze-Corcoran[4]; [1]UMass Cranberry Station, East Wareham, MA, [2]University of Massachusetts Amherst / Cranberry Station, East Wareham, MA, [3]UMass Clean Energy Extension, Amherst, MA, [4]University of Massachusetts, Amherst, MA **(374)**

04:45 PM **Development of Sustainable Weed Management Strategies for Michigan Cherry Growers by Identify Promising New Herbicides.** Sushila Chaudhari*, Christopher G. Galbraith, Monique Mose; Michigan State University, East Lansing, MI **(375)**

WEDNESDAY MORNING FEBRUARY 1

ORAL - 03. Turf and Ornamentals

LOCATION:	Salon C
TIME:	09:00 AM - 11:30 AM Eastern Time
CHAIR:	David E. Hillger Corteva Agriscience Thorntown, IN
MODERATOR:	David E. Hillger Corteva Agriscience Thorntown, IN

***SPEAKER**

09:00 AM **Weed Encroachment into a Fine Fescue Mixture Affected by Different Seeding Timings.** Ross C. Braun[1], Aaron J. Patton[1], Eric Watkins[2], Andrew B. Hollman[2], James A. Murphy[3], Bradley S. Park*[3], Alexander R. Kowalewski[4], Emily T. Braithwaite[4]; [1]Purdue University, West Lafayette, IN, [2]University of Minnesota, St. Paul, MN, [3]Rutgers University, New Brunswick, NJ, [4]Oregon State University, Corvallis, OR **(376)**

09:15 AM **Influence of Mowing Heights and Post-treatment Irrigation on Annual Bluegrass Control with Methiozolin.** John M. Peppers*, Shawn Askew; Virginia Tech, Blacksburg, VA **(377)**

09:30 AM **Seedbank Persistence of Annual Bluegrass (*Poa annua* L.) Across Multiple Climatic Zones in the United States.** Andrew W. Osburn*, Muthukumar V. Bagavathiannan; Texas A&M University, College Station, TX **(378)**

09:45 AM **Weed Control in Christmas Trees with Fluridone.** Marcelo L. Moretti*; Oregon State University, Corvallis, OR **(379)**

10:00 AM **Coffee Break**

10:15 AM **Factors Associated with the Exposure of Sub-canopy Stolons and Shoots to Herbicides in Dormant Zoysiagrass.** Shawn Askew*, Jordan M. Craft, Clebson G. Goncalves; Virginia Tech, Blacksburg, VA **(380)**

10:30 AM **Common Ragweed Control in Southern Appalachian Fraser Fir Christmas Tree Production.** Joe C. Neal*[1], Christopher D. Harlow[1], Jeff Owen[2], Brad Edwards[3]; [1]North Carolina State University, Raleigh, NC, [2]North Carolina State Univeristy, Raleigh, NC, [3]North Carolina Cooperative Extension, Jefferson, NC **(381)**

10:45 AM **Impact of Temperature on Germination and Growth of Japanese Stiltgrass [*Microstegium vimineum* (Trin.) A. Camus].** Jeffrey Derr*[1], Aman Rana[2]; [1]Virginia Tech, Virginia Beach, VA, [2]Virginia Tech HRAREC, Virginia Beach, VA **(382)**

11:00 AM **Investigating the Genetic Relatedness of Herbicide Resistant *Poa annua* Populations.** Claudia A. Rutland*[1], Joseph S. McElroy[2], Jinesh Patel[2]; [1]Auburn University Department of Crop, Soil, and Environmental Sciences, Auburn, AL, [2]Auburn University, Auburn, AL **(383)**

11:15 AM **Evaluation of Goosegrass (*Eleusine indica)* Populations Suspected Resistant to Revolver and Dismiss.** Bridgette C. Johnson*, Joseph S. McElroy; Auburn University, Auburn, AL **(384)**

11:30 AM **Pre-emergent Herbicides and Mulches in Propagation - Rooting and Growth After Transplant.** Anthony L. Witcher*[1], Isha Poudel[2]; [1]Tennessee State University, Mcminnville, TN, [2]University of Tennessee, Jackson, TN **(385)**

WEDNESDAY MORNING FEBRUARY 1

ORAL - 10. Physiology

LOCATION:	Salon B
TIME:	09:00 AM - 02:00 PM Eastern Time
CHAIR:	Chenxi Wu Bayer CropScience St Louis, MO
MODERATOR:	Chenxi Wu Bayer CropScience St Louis, MO

***SPEAKER**

09:00 AM **Florpyrauxifen-benzyl Resistance in Barnyardgrass (*Echinochloa crus-galli*) Evolved Prior to the Commercialization of Loyant^{TM}.** Hudson K. Takano*, Scott Greenwalt, David Ouse, Moriah Zielinski, Paul Schmitzer; Corteva Agriscience, Indianapolis, IN **(386)**

09:15 AM **Molecular Targets of Herbicides and Insecticides - Are There Useful Overlaps?** Stephen O. Duke*[1], Amar G. Chittiboyina[1], Thomas C. Sparks[2], Daniel R. Swale[3]; [1]University of Mississippi, Oxford, MS, [2]Agrilucent LLC, Greenfield, IN, [3]University of Florida, Gainesville, FL **(387)**

09:30 AM **Glyphosate Selection Increases *EPSPS* Copy Number and Glyphosate Resistance Within Individual Plants of *Hordeum glaucum*.** Christopher Preston*[1], Patricia Adu-Yeboah[1], Jenna M. Malone[1], Gurjeet Singh Gill[2]; [1]University of Adelaide, Glen Osmond, Australia, [2]University of Adelaide, Adelaide, Australia **(388)**

09:45 AM **Herbicides Pose a Unique Challenge: Insights from Omics Analyses of Plants Responding to Sublethal Stresses.** Jim Westwood*, Gourav

Sharma, Suzanne R. Laliberte, Shawn Askew, Jacob Barney, David C. Haak; Virginia Tech, Blacksburg, VA **(389)**

10:00 AM **Coffee Break**

10:15 AM **Chromosome-Scale Draft Genomes of Common and Giant Ragweed Reveal a Potential Mechanism of Glyphosate Resistance.** Eric Page*[1], Martin Laforest[2], Sara Martin[3], Katherine Bisaillon[4], Brahim Soufiane[4], Sydney Meloche[1], Francois Tardif[5]; [1]Agriculture and Agri-Food Canada, Harrow, ON, Canada, [2]Agriculture and Agri-Food Canada, Saint-jean-sur-richelieu, QC, [3]Agriculture and Agri-Food Canada, Ottawa, ON, [4]Agriculture and Agri-Food Canada, Saint-jean-sur-richelieu, QC, Canada, [5]University of Guelph, Guelph, ON, Canada **(390)**

10:30 AM **A Deeper Understanding of Resistance to Glufosinate in *Amaranthus palmeri*.** Matheus Machado Noguera*[1], Aimone Porri[2], Eduarda P. Mena Barreto[1], Juan Camilo Velásquez R[1], Martin Penkert[3], James W. Heiser[4], Jens Lerchl[5], Nilda Roma-Burgos[6]; [1]University of Arkansas, Fayetteville, AR, [2]BASF Global Research & Development, Limburgerhof, Germany, [3]BASF SE, Ludwigshafen, Germany, [4]University of Missouri, Portageville, MO, [5]BASF SE, Limburgerhof, Germany, [6]University of Arkansas, Fayetteville, Fayetteville, AR **(391)**

10:45 AM **Dose-Response Characteristics of Global Metabolome as a Potential Avenue to Capture the Secondary Toxicity of Herbicides in Weeds.** Shalini priya Etukuri*, Pawanjit Kaur Sandhu, Jisun Lee, Nishanth Tharayil; Clemson University, Clemson, SC **(392)**

11:00 AM **Identification and Characterization of GST Genes Involved in Group 15 Herbicide Detoxification in Resistant Grasses.** Evlampia Parcharidou[1], Rebecka Duecker[1], Roland S. Beffa*[2]; [1]Georg-August Universität Göttingen, Göttingen, Germany, [2]Senior Scientist Consultant, Frankfurt, Germany **(393)**

11:15 AM **Genetic Factors Influencing Snap Bean Tolerance to Preemergence Herbicides from Three Site-of-Action Groups.** Ana I. Saballos*[1], Nicholas Hausman[1], Ed Peachey[2], Patrick Tranel[3], Martin Williams[4]; [1]Global Change and Photosynthesis Research, US Department of

Agriculture, Agricultural Research Service, Urbana, IL, [2]Oregon State University, Corvallis, OR, [3]University of Illinois, Urbana, IL, [4]USDA-ARS, Urbana, IL **(394)**

11:30 AM **Genome-wide Characterization of MADS-box Transcription Factors Family in Dioecious Amaranth Species.** Lucas Kopecky Bobadilla*[1], Patrick Tranel[2]; [1]University of Illinois, Champaign, IL, [2]University of Illinois, Urbana, IL **(395)**

11:45 AM **Metabolic Resistance Profiles of Palmer Amaranth (*Amaranthus palmeri*) Populations to HPPD-inhibiting Herbicides.** Jeanaflor Crystal T. Concepcion*[1], Shiv S. Kaundun[2], James A. Morris[2], Devin Koester[1], Sarah-Jane Hutchings[2], Dean E. Riechers[3]; [1]University of Illinois at Urbana-Champaign, Urbana, IL, [2]Syngenta, Bracknell, United Kingdom, [3]University of Illinois-Urbana, Urbana, IL **(396)**

12:00 PM **Lunch**

01:00 PM **Comparative Analysis of *S*-Metolachlor Metabolism Among Palmer Amaranth (*Amaranthus palmeri*) and Waterhemp (*A. tuberculatus*) Populations from the U.S.** Jeanaflor Crystal T. Concepcion[1], Noeleen K. Brown[2], James A. Morris[3], Sarah-Jane Hutchings[3], Shiv S. Kaundun[3], Dean E. Riechers*[2]; [1]University of Illinois at Urbana-Champaign, Urbana, IL, [2]University of Illinois-Urbana, Urbana, IL, [3]Syngenta, Bracknell, United Kingdom **(397)**

01:15 PM **Overcoming Weed Resistance with Discovery of Novel Modes of Action.** Karthik Putta*, Paolo Castiglioni, Neville Anthony, David Lauffer, David Vierra, Balaji Aravindhan Pandian, Gary Schaefer, Peter Stchur, Brittany Baisch, Timothy Panosian, Jacqueline Heard, Thomas Meade; ENKO, Mystic, CT **(398)**

01:30 PM **Which Summer Annual Weeds May Emerge as Important Candidates as Model Species for Research to Combat Climate Change and Global Food Security?** Philip Westra*[1], Todd A. Gaines[1], Eric L. Patterson[2], Jacob S. Montgomery[1], John M. Lemas[1]; [1]Colorado State University, Fort Collins, CO, [2]Michigan State University, East Lansing, MI **(399)**

WEDNESDAY MORNING FEBRUARY 1

Graduate Student Luncheon

LOCATION: Salon J
TIME: 11:30 AM - 01:00 PM Eastern Time
MODERATOR: Sarah E. Kezar
Texas A&M University
College Station, TX

WEDNESDAY AFTERNOON FEBRUARY 1

SYMPOSIUM - 05. WSSA Research Priorities Survey Results: 13 Federal Agency-Perspectives and -Funding Opportunities

LOCATION: Salon II
TIME: 01:00 PM - 05:00 PM Eastern Time
CHAIR: James Kells
Michigan State University
East Lansing, MI
MODERATOR: James Kells
Michigan State University
East Lansing, MI
CO-MODERATOR: Steve Young
USDA ARS
Beltsville, MD

***SPEAKER**

01:00 PM **Introductions and Objectives.** James J. Kells*; Michigan State University, East Lansing, MI **(400)**

01:20 PM **Summary and Interpretation of WSSA Research Priorities Survey.** Daniel C. Brainard*[1], Erin Haramoto[2], Ramon G. Leon[3], James J. Kells[1], Lee Van Wychen[4], Pratap Devkota[5], Mithila Jugulam[6], Jacob Barney[7]; [1]Michigan State University, East Lansing, MI, [2]University of Kentucky, Lexington, KY, [3]North Carolina State University, Raleigh, NC,

[4]Weed Science Society of America, Alexandria, VA, [5]University of Florida, Jay, FL, [6]Kansas State University, Manhattan, KS, [7]Virginia Tech, Blacksburg, VA **(401)**

02:05 PM **Question and Answer Session on Survey Results.** Daniel C. Brainard*; Michigan State University, East Lansing, MI **(402)**

02:50 PM **Coffee Break**

03:05 PM **Research Panel Discussion.** Stephen O. Duke*; University of Mississippi, Oxford, MS **(403)**

03:50 PM **Management Panel Discussion.** Lee Van Wychen*; Weed Science Society of America, Alexandria, VA **(404)**

04:35 PM **Closing Comments and Next Steps.** James J. Kells*; Michigan State University, East Lansing, MI **(405)**

WEDNESDAY AFTERNOON FEBRUARY 1

ORAL - 09. Biocontrol of Weeds

LOCATION: Salon B
TIME: 02:00 PM - 02:30 PM Eastern Time

***SPEAKER**

02:00 PM **Development of the Brazilian Peppertree Biological Control Agent *Pseudophilothrips ichini* (Thysanoptera: Phlaeothripidae) in Florida.** Gregory Wheeler*; USDA, Davie, FL **(406)**

02:15 PM **Allelopathic Potential of *Parthenium hysterophorus* Against Selected Weeds of Wheat Crop.** Tauseef Anwar*[1], Huma Qureshi[2]; [1]Department of Botany, The Islamia University of Bahawalpur, Bahawalpur-63000, Punjab, Pakistan, Bahawalpur, Pakistan, [2]Institute of Biological Sciences, Gomal University, Dera Ismail Khan-29050, Khyber Pakhunkhwa,Pakistan, Dera Ismail Khan, Pakistan **(407)**

WEDNESDAY AFTERNOON FEBRUARY 1

SST - SST PhD - Finals

LOCATION:	Salon C
TIME:	02:00 PM - 03:00 PM Eastern Time
CHAIR:	Marty Schraer Syngenta Meridian, ID
MODERATOR:	Marty Schraer Syngenta Meridian, ID

***SPEAKER † STUDENT CONTEST**

WEDNESDAY AFTERNOON FEBRUARY 1

ORAL - 11. Soil and Environmental Aspects

LOCATION:	Salon B
TIME:	02:30 PM - 03:00 PM Eastern Time
CHAIR:	David P. Russell Auburn University Madison, AL
MODERATOR:	David P. Russell Auburn University Madison, AL

***SPEAKER**

02:30 PM **Dicamba Research Update.** Thomas C. Mueller*[1], Larry Steckel[2]; [1]University of Tennessee, Knoxville, TN, [2]University of Tennessee, Jackson, TN **(408)**

WEDNESDAY AFTERNOON FEBRUARY 1

Joint Student/ Photo Awards

LOCATION: Salon III
TIME: 05:00 PM - 07:00 PM Eastern Time
MODERATOR: Mandy Bish
University of Missouri
Columbia, MO

WEDNESDAY AFTERNOON FEBRUARY 1

Joint Dessert Social

LOCATION: Foyer
TIME: 09:00 PM - 11:00 PM Eastern Time

THURSDAY MORNING FEBRUARY 2

WSSA Business Meeting/Breakfast

LOCATION: Salon H
TIME: 06:30 AM - 08:00 AM Eastern Time
MODERATOR: A Stanley Culpepper
University of Georgia
Tifton, GA
CO-MODERATOR: Carroll Moseley
Syngenta
Brown Summit, NC

THURSDAY MORNING FEBRUARY 2

Registration

LOCATION: Arlington Registration
TIME: 08:00 AM - 10:00 AM Eastern Time

THURSDAY MORNING FEBRUARY 2

ORAL - 01. Agronomic Crops - Day 2

LOCATION:	Salon III
TIME:	08:00 AM - 12:00 PM Eastern Time
CHAIR:	Daljit Singh Bayer Chesterfield, MO
MODERATOR:	Daljit Singh Bayer Chesterfield, MO
CO-MODERATOR:	Craig M. Alford Corteva Ankeny, IA

***SPEAKER**

08:00 AM **Use of a Spray Drone for Corn Harvest Aid Herbicide Applications.** Ryan D. Langemeier*, Livia Pereira, Justin T. McCaghren, Steve Li; Auburn University, Auburn, AL **(409)**

08:15 AM **Investigations of Suspected Weed Resistance to XtendiMax® Herbicide with VaporGrip® Technology.** Aruna V. Varanasi*, Chandrashekar Aradhya, Jeffrey Herrmann, John Willis; Bayer CropScience, Chesterfield, MO **(410)**

08:30 AM **Evaluation of Physical Control Strategies for Kochia (*Bassia scoparia*) Patch Management.** Shaun M. Sharpe*, Julia Y. Leeson, Haben Asgedom Telda, Jennifer Town, Tim Dumonceaux, Krista Gill; Agriculture and Agri-Food Canada, Saskatoon, SK, Canada **(411)**

08:45 AM **Response of Soybean at Different Growth Stages to Sub-lethal Rates of Imazapyr.** Michael W. Marshall*; Clemson University, Blackville, SC **(412)**

09:00 AM **Detection and Site-Specific Treatment of Late-Season Weed Escapes in Rice Using Unmanned Aerial Systems and Deep Learning.** Bholuram

Gurjar*[1], Bishwa B. Sapkota[1], Isidor Ceperkovic[1], Ubaldo Torres[2], Matthew Kutugata[1], Daniel E. Martin[3], Muthukumar V. Bagavathiannan[1]; [1]Texas A&M university, College Station, TX, [2]Texas Tech University, College Station, TX, [3]United States Department of Agriculture, College Station, TX **(413)**

09:15 AM **Effects of Delaying Rye Cover Crop Termination on Cover Crop Biomass, Weed Suppression, and Corn Yield.** Daniel A. Boakye*[1], Parmeshwor Aryal[1], Carlene A. Chase[1], Muthukumar V. Bagavathiannan[2], Andrea Basche[3], Heather Darby[4], Michael L. Flessner[5], Erin Haramoto[6], Ramon G. Leon[7], Steven Brian Mirsky[8], Alison Robertson[9], Matt Ruark[10], Matthew R. Ryan[11], Nicholas J. Seiter[12], Kelley Tilmon[13], Peter Tomlinson[14], John F. Tooker[15], Mark VanGessel[16], John M. Wallace[17], Jared Adam[15], Tauana Ferreira de Almeida[18], Ashley Decker[12], Kristen Loria[11], Rashelle Matthiessen-Anderson[9], Jodie M. McVane[2], Fernando H. Oreja[7], DeAnn Presley[14], Amy Raudenbush[19], Lindsey Ruhl[20], Cynthia Sias[5], Barbara Scott[16], Ethan Sweep[21], Ashley Waggoner[22]; [1]University of Florida, Gainesville, FL, [2]Texas A&M University, College Station, TX, [3]University of Nebraka, Lincoln, NE, [4]The University of Vermont, St. Albans, VT, [5]Virginia Tech, Blacksburg, VA, [6]University of Kentucky, Lexington, KY, [7]North Carolina State University, Raleigh, NC, [8]USDA ARS, Beltsville, MD, [9]Iowa State University, Ames, IA, [10]University of Wisconsin, Madison, WI, [11]Cornell University, Ithaca, NY, [12]University of Illinois, Urbana, IL, [13]Ohio State University, Wooster, OH, [14]Kansas State University, Manhattan, KS, [15]The Pennsylvania State University, University Park, PA, [16]University of Delaware, Georgetown, DE, [17]Penn State University, University Park, PA, [18]University of Nebraska, Lincoln, NE, [19]The Ohio State University, Wooster, OH, [20]University of Vermont, St. Albans, VT, [21]USDA Agricultural Research Service, Beltsville, MD, [22]University of Wisconsin, Mas, WI **(414)**

09:30 AM **Maverick™ Corn Herbicide: Application Flexibility with Broad-Spectrum Control.** Jonathon Kohrt*[1], Garrison J. Gundy[2], Trevor D. Israel[3], Chad L. Smith[4], John Pawlak[5], Eric Ott[5]; [1]Valent USA LLC, West Des Moines, IA, [2]Valent U.S.A LLC, Mcpherson, KS, [3]Valent USA,

Harrisburg, SD, [4]Valent USA, Hallsville, MO, [5]Valent USA LLC, San Ramon, CA **(415)**

09:45 AM **Response of North Carolina Palmer Amaranth Accessions to Very-Long-Chain-Fatty-Acid Inhibiting Herbicides.** Amanda M. Foderaro*; Syngenta, Raleigh, NC **(416)**

10:00 AM **Coffee Break**

10:15 AM **Survey and Screening to Evaluate Herbicide Efficacy in Triple Stack Sugarbeet Cropping Systems in SE Montana.** Het Samir Desai*[1], Fabian Menalled[1], Lovreet S. Shergill[2]; [1]Montana State University, Bozeman, MT, [2]Montana State University, Huntley, MT **(417)**

10:30 AM **Multi-state Survey of Waterhemp (Amaranthus Tuberculatus) and Palmer Amaranth (Amaranthus Palmeri) in the USA on Sensitivity to Multiple Herbicides.** Daljit Singh*[1], Alejandro Perez-Jones[2], Drew Tyre[3], Chandrashekar Aradhya[2]; [1]Bayer, Chesterfield, MO, [2]Bayer Crop Science, Chesterfield, MO, [3]Bayer Crop Science, Lincoln, NE **(418)**

10:45 AM **Evaluation of See & Spray Programs for Weed Management in Cotton Production.** William L. Patzoldt*[1], Lauren M. Lazaro[1], Michael M. Houston[1], Diego J. Contreras[2], Wesley Everman[2], Tristen H. Avent[3], Jason K. Norsworthy[3]; [1]Blue River Technology, Sunnyvale, CA, [2]North Carolina State University, Raleigh, NC, [3]University of Arkansas, Fayetteville, AR **(419)**

11:00 AM **A23980B: A Step Change for Residual Weed Control in Corn.** Sudeep A. Mathew*[1], Scott E. Cully[2], Mark Kitt[3]; [1]Syngenta, Germantown, MD, [2]Syngenta Crop Protection, Marion, IL, [3]Syngenta, Greensboro, NC **(420)**

11:15 AM **What's New in Industry.** Sudeep A. Mathew*[1], Marshall Hay[2], Mark Kitt[3]; [1]Syngenta, Germantown, MD, [2]Syngenta Crop Protection, Vero Beach, FL, [3]Syngenta, Greensboro, NC **(421)**

11:30 AM **Estimating Dry-down Performance of Desiccants in Lentils Using Hyperspectral Imaging.** Charles M. Geddes[1], Keshav Dev Singh[2], Prabahar Ravichandran*[3]; [1]Agriculture and Agri-Food Canada, Lethbridge, AB, Canada, [2]University of Saskatchewan, Lethbridge, AB, Canada,

[3]Agriculture and Agri-Food Canada, Truro, NS, Canada **(422)**

THURSDAY MORNING FEBRUARY 2

SYMPOSIUM - 02. Novel Technology for Weed Management: An Update on WSSA Member Activities from Around the US

LOCATION:	Salon I
TIME:	08:00 AM - 12:00 PM Eastern Time
CHAIR:	Lynn M. Sosnoskie Cornell University Geneva, NY
MODERATOR:	Lynn M. Sosnoskie Cornell University Geneva, NY
CO-MODERATOR:	Marcelo L. Moretti Oregon State University Corvallis, OR

***SPEAKER**

08:00 AM **Novel Technology for Weed Management - Symposium Introduction.** Lynn M. Sosnoskie*[1], Marcelo L. Moretti[2]; [1]Cornell University, Geneva, NY, [2]Oregon State University, Corvallis, OR **(423)**

08:09 AM **Identifying and Overcoming Barriers of Adoption for Robotic Weed Control Implementation in Specialty Crops.** Matthew A. Cutulle*[1], Lynn M. Sosnoskie[2], Thierry E. Besancon[3], Nathan Boyd[4], Sushila Chaudhari[5], Matthew B. Bertucci[6], Joe Maja[7], Tara Wade[8], Felipe Silva[9], Laura Pineda-Bermudez[2]; [1]Clemson University, Charleston, SC, [2]Cornell University, Geneva, NY, [3]Rutgers University, Chatsworth, NJ, [4]University of Florida, Balm, FL, [5]Michigan State University, East Lansing, MI, [6]University of Arkansas, Fayetteville, AR, [7]Clemson University, Blackville, SC, [8]University of Florida, Gainesville, FL, [9]Clemson University, Clemson, SC **(424)**

08:27 AM **Smart Weed Control in Vegetables with Band Steaming.** Steve Fennimore*; University of California - Davis, Salinas, CA **(425)**

08:45 AM **Electrical Weed Control in Perennial Cropping Systems.** Marcelo L. Moretti*; Oregon State University, Corvallis, OR **(426)**

09:03 AM **Electrical Weeding and Inter-row Mowing in Annual Crops.** Annika Rowland*[1], Lynn M. Sosnoskie[2], Chris J. Pelzer[1], Matthew R. Ryan[1]; [1]Cornell University, Ithaca, NY, [2]Cornell University, Geneva, NY **(427)**

09:21 AM **The Future of Precision Weed Management in Specialty Crops.** Nathan Boyd*[1], Arnold W. Schumann[2], Ana Claudia Buzanini[3]; [1]University of Florida, Balm, FL, [2]University of Florida, Lake Alfred, FL, [3]University of Florida, Wimauma, FL **(428)**

09:39 AM **Weed Control Efficacy and Perennial Crop Tolerance Using an Herbicide Sprayer Based on Chlorophyll Fluorescence Sensing in Perennial Crops.** Thierry E. Besancon*[1], Lynn M. Sosnoskie[2], Wesley M. Bouchelle[1], Elizabeth C. Maloney[2], Bethany M. Schulteis[3]; [1]Rutgers University, Chatsworth, NJ, [2]Cornell University, Geneva, NY, [3]Cornell University, Ithaca, NY **(429)**

09:57 AM **Coffee Break**

10:12 AM **See & Spray Ultimate - A New Tool for Weed Management in Agronomic Crops.** Lauren M. Lazaro*[1], William L. Patzoldt[1], Scott Brown[1], Jesaelen Gizotti de Moraes[1], Michael M. Houston[2], Gaylia Ostermier[3], Chad Yagow[3]; [1]Blue River Technology, Sunnyvale, CA, [2]Blue River Technology, Stoneville, MS, [3]John Deere, Urbandale, IA **(430)**

10:30 AM **Advances in Unmanned Aerial Systems-based Spray Applications, 2023.** Vijay Singh*[1], Dhiraj Srivastava[2], Vipin Kumar[2], Daniel E. Martin[3], William Reynolds[4]; [1]Virginia Tech, Painter, VA, [2]Virginia Polytechnic Institute and State University, Painter, VA, [3]United States Department of Agriculture, College Station, TX, [4]LeadingEdge Aerial Technologies, New Smyrna Beach, FL **(431)**

10:48 AM **Current Status of Harvest Weed Seed Control Research in the Eastern United States.** Michael L. Flessner*[1], Prashant Jha[2], Lovreet S. Shergill[3], Lynn M. Sosnoskie[4], Muthukumar V. Bagavathiannan[5], Wesley Everman[6], Eugene P. Law[7], Lauren M. Lazaro[8], Steven Brian Mirsky[9],

Mark VanGessel[10]; [1]Virginia Tech, Blacksburg, VA, [2]Iowa State University, Ames, IA, [3]Montana State University, Huntley, MT, [4]Cornell University, Geneva, NY, [5]Texas A&M University, College Station, TX, [6]North Carolina State University, Raleigh, NC, [7]University of Delaware, Beltsville, MD, [8]Blue River Technology, Sunnyvale, CA, [9]USDA ARS, Beltsville, MD, [10]University of Delaware, Georgetown, DE **(432)**

11:06 AM **Implementing Harvest Weed Seed Control Strategies in the Midwestern and Southern U.S.** Prashant Jha*[1], Jason K. Norsworthy[2], Vipan Kumar[3], Karla L. Gage[4], Steven Brian Mirsky[5], Muthukumar V. Bagavathiannan[6], Michael L. Flessner[7], Michael Walsh[8], Avery J. Bennett[1], Alexis L. Meadows[1], Austin H. Schleich[1]; [1]Iowa State University, Ames, IA, [2]University of Arkansas, Fayetteville, AR, [3]Kansas State University, Hays, KS, [4]Southern Illinois University Carbondale, Carbondale, IL, [5]USDA ARS, Beltsville, MD, [6]Texas A&M University, College Station, TX, [7]Virginia Tech, Blacksburg, VA, [8]University of Sydney, Narrabri, Australia **(433)**

11:24 AM **Current Status of Harvest Weed Seed Control Research in the Northern Great Plains and Pacific Northwest Region of the United States.** Lovreet S. Shergill*[1], Todd A. Gaines[2], Ian Burke[3], Drew J. Lyon[3], Judit Barroso[4], Eric P. Westra[2]; [1]Montana State University, Huntley, MT, [2]Colorado State University, Fort Collins, CO, [3]Washington State University, Pullman, WA, [4]Oregon State University, Adams, OR **(434)**

THURSDAY MORNING FEBRUARY 2

ORAL - 04. Pasture, Range, Forest, & Rights of ways, Wildland, and Aquatic Invasive plants

LOCATION:	Salon II
TIME:	09:00 AM - 11:30 AM Eastern Time
CHAIR:	Scott A. Nolte Texas A&M AgriLife Extension College Station, TX
MODERATOR:	Scott A. Nolte

	Texas A&M AgriLife Extension College Station, TX
CO-MODERATOR:	Byron B. Sleugh Corteva Agriscience Carmel, IN

***SPEAKER**

09:00 AM **Will it Dye or Die? Effects of Application Speed on Droplet Distribution and Swath Width of Boomless Nozzles.** Kayla L. Broster*, Thomas H. Duncan, John D. Byrd, Jr.; Mississippi State University, Starkville, MS **(435)**

09:15 AM **Chemical Control of Terrestrial Alligatorweed in Mississippi.** Gray Turnage*; Mississippi State University, Starkville, MS **(436)**

09:30 AM **Assessment of Reverse Cut Stem Treatments for Difficult to Control Tropical Vines.** Stephen F. Enloe*; University of Florida, Gainesville, FL **(437)**

09:45 AM **Aquatic Weed Control Research Update.** Benjamin P. Sperry*; US Army Corps of Engineers, Gainesville, FL **(438)**

10:00 AM **Coffee Break**

10:15 AM **Smutgrass (Sporobolus Indicus) Identification and Management with Unmanned Aerial Systems.** Zachary S. Howard*[1], Scott A. Nolte[2], Daniel E. Martin[3], Chenghai Yang[3]; [1]Texas A&M University, College Station, TX, [2]Texas A&M AgriLife Extension, College Station, TX, [3]United States Department of Agriculture, College Station, TX **(439)**

10:30 AM **How Much Pre-plant Weed Suppression is Practical for Riparian Forest Buffer Establishment?** Arthur E. Gover*, Emily N. Rojik; Penn State, University Park, PA **(440)**

10:45 AM **Evaluation of Factors That Influence Herbicide Spray Retention on Floating and Emergent Aquatic Vegetation.** Christopher R. Mudge*[1], Benjamin P. Sperry[2], Michael W. Durham[3], Kurt D. Getsinger[4]; [1]U.S. Army Engineer Research & Development Center, Baton Rouge, LA, [2]US Army Corps of Engineers, Gainesville, FL, [3]University of Florida, Gainesville, FL, [4]U.S. Army Engineer

Research and Development Center, Vicksburg, MS **(441)**

11:00 AM **Control of Wolftail Sedge (*Carex cherokeensis*) in Forages.** David P. Russell*; Auburn University, Madison, AL **(442)**

THURSDAY MORNING FEBRUARY 2

ORAL - 06. Teaching and Extension

LOCATION: Salon J
TIME: 09:00 AM - 12:00 PM Eastern Time
CO-MODERATOR: Debalin Sarangi
University of Minnesota
St. Paul, MN

***SPEAKER**

09:00 AM **Preventing Preventable Problems on the Farm.** Eric P. Prostko*; University of Georgia, Tifton, GA **(443)**

09:15 AM **And You Think Pigweed Control is a Challenge!!** John D. Byrd, Jr.*, Kayla L. Broster, Thomas H. Duncan; Mississippi State University, Starkville, MS **(444)**

09:30 AM **US-Herbicide Resistance Action Committee (HRAC): Who We Are and What We Do?** Zahoor A. Ganie[1], Greg Elmore[2], Lowell Sandell[3], Chandrashekar Aradhya*[4]; [1]FMC, Newark, DE, [2]Bayer Crop Science, Saint Louis, MO, [3]Corteva, Ashland, NE, [4]Bayer Crop Science, Chesterfield, MO **(445)**

09:45 AM **Public Parks, Family Farms, and Cover Crops: Implementing Ecological Weed Management with Diverse Stakeholders in Oregon Hazelnuts.** Steven Haring*, Marissa Lane-Massee, Cal Penkauskas, Alejandro Brambila, Ari Brown, Lauren Hallett; University of Oregon, Eugene, OR **(446)**

10:00 AM **Coffee Break**

10:15 AM **Challenges with Co-existence of Multiple Herbicide-Resistant Soybean Traits.** Amit J. Jhala*; University of Nebraska Lincoln, Lincoln, NE **(447)**

10:30 AM **Climate Change and Weed Management: Synthesizing Information for New York Farmers.** Caroline Marschner*[1], Sharon Bachman[2], Catalina Ferreria-Dias[3], Isabella Colucci[1], Rebecca Stup[1], Claire Liu[1], Antonio DiTommaso[1]; [1]Cornell University, Ithaca, NY, [2]Cornell Cooperative Extension, East Aurora, NY, [3]Cornell University, Santa Ana, CA **(448)**

10:45 AM **Advancing Physical Weed Control: Farmers Love Field Days, But On-farm Research is Very Challenging.** Eric R. Gallandt*[1], Daniel C. Brainard[2]; [1]University of Maine, Orono, ME, [2]Michigan State University, East Lansing, MI **(449)**

11:00 AM **A Story of Success and Challenges in Eradicating Palmer Amaranth (*Amaranthus palmeri*) in Minnesota.** Debalin Sarangi*[1], Eric Y. Yu[1], Anthony Brusa[2], Shane Blair[3], Monika Chandler[3], Denise Thiede[3], Anthony Cortilet[3], Melissa Wilson[1], Jeffrey Gunsolus[1], Roger Becker[1], Thomas J. Peters[4]; [1]University of Minnesota, St. Paul, MN, [2]University of Minnesota, Saint Paul, MN, [3]Minnesota Department of Agriculture, St. Paul, MN, [4]North Dakota State University, Fargo, ND **(450)**

THURSDAY AFTERNOON FEBRUARY 2

US-HRAC Meeting

LOCATION: Salon K
TIME: 12:00 PM - 04:00 PM Eastern Time

THURSDAY AFTERNOON FEBRUARY 2

WSSA Board of Directors Meeting

LOCATION: Alexandria
TIME: 02:00 PM - 04:00 PM Eastern Time
MODERATOR: Carroll Moseley
Syngenta
Brown Summit, NC
CO-MODERATOR: Gregory K. Dahl
Winfield United
Eagan, MN

THURSDAY AFTERNOON FEBRUARY 2

NEWSS Board of Directors Meeting

LOCATION: Crystal Boardroom
TIME: 02:00 PM - 04:00 PM Eastern Time
MODERATOR: Wesley Everman
North Carolina State University
Raleigh, NC

Crystal Gateway Marriott

2022-2023 Weed Science Society of America Board of Directors

President, Stanley Culpeper

President-Elect, Carroll Moseley

Vice-President, Greg Dahl

Past-President, Anita Dille

Secretary, Lauren Lazaro

Treasurer, Greg Elmore

Director of Publications, Chris Willenborg

Chair, Constitution and Operating Procedures, John Lindquist

Member-at-Large, Michael Flessner

Member-at-Large, Lynn Sosnoskie

Graduate Student Member, Sarah Kezar

Aquatic Plant Management Society, Ben Sperry

Canadian Weed Science Society, Darren Robinson

North Central Weed Science Society, Brett Miller

Northeastern Weed Science Society, Steve Pyle

Southern Weed Science Society, John Byrd

Western Society of Weed Science, Alan Helm

Executive Director of Science Policy, Lee VanWychen

Executive Secretary, Eric Gustafson

2022-2023 Northeastern Weed Science Society Officers and Executive Committee

President, Jacob Barney

President-Elect, Wes Everman

Vice-President, Erin Hitchner

Treasurer, Larissa Smith

Past President, Caren Schmidt

Membership Chair, Matt Elmore

Editor, Lane Heimer

Graduate Student Rep., Eric Jones

Graduate Student Rep. Elect, Katie Diehl

Research & Education, Carolyn Lowry

Sustaining Membership, John Wallace

Public Relations, John Kaminski

WSSA Representative, Steve Pyle

CAST Representative, Anthony Witcher

Meeting Conduct Reminder

The WSSA Code of Ethics defines professional conduct binding on all members of the Society. Members should recognize that this code of ethics and conduct signifies a voluntary assumption of the obligation of self-discipline and members should strive to uphold and maintain the honor and dignity of the Society.
The following rules and standards of conduct have been developed for the safe and efficient operation of the WSSA and for the benefit and protection of the rights and safety of all. WSSA members are expected to observe the highest standards of professional conduct at all times, while at work or engaged in Society business.

1. **Obey all laws, rules and regulations governing our business.** The WSSA is subject to applicable federal laws in the state or country of the meeting or event. It is the policy of the WSSA that all laws, rules and regulations are complied with fully and completely. If it is unclear whether an action or activity is a legal or ethical violation, contact the WSSA President and/or Executive Secretary immediately for advice. Any incident or situation that violates the law or this policy should be immediately reported to the President and/or Executive Secretary; in person or via this contact information: info@wssa.net or 720-445-4789.

2. **Be honest, truthful, fair and trustworthy in all WSSA activities and relationships.** The WSSA expects each member to treat other members with respect and honesty. This includes providing information that is accurate, complete, objective, timely, relevant, and understandable.

3. **Respect and protect WSSA assets.** Assets are anything of value owned by the WSSA. All WSSA members are expected to be custodians of those assets. Members are responsible to maintain Society assets in good condition and to protect them from loss. This includes, but is not limited to, real assets and equipment as well as "soft assets" such as intellectual property, member lists, and other confidential information owned by the Society. WSSA assets of any kind should not be used for personal benefit.

4. **Avoid all conflicts of interest between Society business and personal affairs.** All WSSA members are expected to act with total objectivity regarding WSSA business. Accordingly, it is improper for a WSSA member to be in a position where their personal interests' conflict, or appears to conflict, with WSSA interests. WSSA members should not use their position within the WSSA to influence WSSA members or others for their personal benefit. If a member believes that a conflict of interest has developed or may develop, it should be promptly reported to the WSSA President and/or Executive Secretary. The Society's Conflict of Interest Policy is outlined in 5.8.

5. **Promote a culture of respect for all WSSA members.** The WSSA supports and adheres to laws and regulations dealing with fair member practices. Membership discrimination in our Society based on sex, race, age, religion, national origin, or sexual preference will not be tolerated. Sexual harassment will not be tolerated. Sexual harassment is defined as unwelcome sexual advances, requests for sexual favors, and other verbal or physical conduct of a sexual nature that explicitly or implicitly affects an individual's membership in the society, unreasonably interferes with an individual's activities within the society, or creates an intimidating, hostile or offensive society environment. This also includes inappropriate use of nudity and/or sexual images in public spaces (including within visual presentations, Twitter, and other online media); deliberate intimidation, stalking, or unwelcome following; harassing photography or recording; and sustained disruption of talks or other events.

6. **Use your best efforts to maintain a safe environment and protect the Society.** The WSSA believes in and supports the laws designated to keep our Society safe and designed to protect the environment. If you believe that an unsafe condition exists in our Society, bring it immediately to the attention of the President, a member of the Board, or Executive Secretary. If an accident takes place, report it pursuant to policy and immediately take action to address the problem. Any incident or situation that violates the law or this policy should be immediately reported to the President and/or Executive Secretary; in person or via this contact information: info@wssa.net or 720-445-4789.

7. **Promote an ethical culture for all WSSA members.** The WSSA always expects all of its members to conduct themselves ethically and to encourage and support that behavior in their fellow members. Members exercise integrity in scientific research activities and in the reporting of results.

8. **Relation of professionals to the public.** They shall not knowingly permit the publication of reports or other documents for any unsound or illegitimate undertaking.

9. **Respect fellow WSSA members.** Members shall freely give credit for work done by others to whom the credit is due and shall refrain from plagiarism in oral and written communication, and not knowingly accept credit rightfully due to another person.

10. **Reporting alleged violations.** Violation of the WSSA Code of Ethics and Conduct may subject a member to disciplinary action up to and including membership revocation. At the annual conference, when "Safe WSSA" members become aware of a complaint, they will seek out the alleged victim(s) and offender(s) separately and gather facts. Immediate responses at the conference may range from warning a harasser to cease his or her behavior immediately, to ending a speaker's talk early if the speaker uses inappropriate language or images, to requiring a harasser to leave the conference immediately with no refund, to banning a harasser from future events "either indefinitely or for a certain time period." Law enforcement may be engaged if circumstances require.

PERSONAL TIME SCHEDULE

Time	Monday		Tuesday	Wednesday	Thursday
6:30					
7:00					
7:30					
8:00					
8:30					
9:00			Poster Session	Poster Session	
9:15					
9:30					
9:45					
10:00					
10:15					
10:30					
10:45					
11:00					
11:15					
11:30					
11:45					
Noon					
1:00					
1:15					
1:30					
1:45					
2:00					
2:15					
2:30					
2:45					
3:00					
3:15					
3:30					
3:45					
4:00					
4:15					
4:30					
4:45	General Session and Awards Presentation				
5:00					
5:30					
6:00					
6:30					
7:00					
7:30		Awards Reception			
8.00					
8:30					
9:00					

3MT

Subtelomeric Rearrangements Cause Glyphosate Resistance in *Eleusine indica*. Nicholas A. Johnson*[1], Nathan D. Hall[2], Qin Yu[3], Chun Zhang[4], Eric L. Patterson[1]; [1]Michigan State University, East Lansing, MI, [2]Michigan State, East Lansing, MI, [3]University of Western Australia, Perth, Australia, [4]Guangdong Academy of Agricultural Sciences, Guangzhou, China (129)

Genomic structural variation (SV) can have profound effects on an organism's evolution. Copy number variation (CNV), a source of SV, is associated with adaptive evolution in eukaryotes. Resistance to the most widely used herbicide, glyphosate, has evolved through CNV in many weedy plant species, including the economically important cosmopolitan grass, *Eleusine indica* (goosegrass). We obtained high-quality reference genomes for both glyphosate-susceptible and -resistant goosegrass, fine assembled the duplication of glyphosate's target site gene enolpyruvylshikimate-3-phosphate synthase (EPSPS), and revealed a novel instance of subtelomeric CNV of EPSPS leading to herbicide resistance evolution. Therefore, the association of subtelomeric CNV with adaptive evolution is evident in *E. indica*, adding to limited knowledge of the importance of subtelomeres as rearrangement hotspots and novel variation generators. We hypothesize glyphosate resistance is conferred by the translocation of EPSPS from its native location on chromosome three into the subtelomeric region and subsequent duplication in this region.

Confirmation of Paraquat Resistance in Horseweed (*Erigeron canadensis*) in New York. Bethany M. Schulteis*[1], Elizabeth C. Maloney[2], Caio A. Brunharo[3], Lynn M. Sosnoskie[2]; [1]Cornell University, Ithaca, NY, [2]Cornell University, Geneva, NY, [3]Penn State University, University Park, PA (245)

Herbicides are the most popular form of weed control across cropping systems because of their low cost and high efficacy. In grapes and apples, glyphosate has been a frequently used chemical although there is significant concern about the potential for crop injury. Consequently, paraquat has become a popular alternative for many producers. In 2020, a vineyard manager (Ontario County, NY) and an apple grower (Wayne County, NY) reported poor control of horseweed following paraquat applications and inquired about resistance screening. Seed from both populations, as well as two paraquat-sensitive populations from upstate NY, were surface-planted in 7.6 cm diameter pots filled with a commercial potting mix and placed in a greenhouse set to a constant temperature of 25°C with a 16:8 light:dark cycle. Seedlings were hand-thinned to one plant per pot once they had produced 3 to 4 true leaves. Small (2.5 to 3.5 cm in diameter) horseweed plants were treated with paraquat at doses of 0 (untreated check), 0.14, 0.28, 0.56, 1.12 (1X), 2.24, 4.48 and 8.96 kg ai ha^{-1} of paraquat (as Gramoxone 2.0 SL, Syngenta Crop Protection). Applications were made using a cabinet sprayer with a single nozzle (8002E, Teejet Technologies) boom set to deliver 187 L ha^{-1}. Each herbicide by rate combination was replicated 12 times for each of the four horseweed populations and the study was repeated in time. Adjuvants were included according to label recommendations. At 21 days after treatment, surviving plants were harvested and weighed to assess biomass accumulation. Data was analyzed with mixed models ANOVA and log-logistic regression using R statistical software. Results demonstrated significant ($P<0.05$) differences among the populations with respect to herbicide injury responses. Both sensitive horseweed samples were almost completely controlled by paraquat applied at 0.14 kg ai ha^{-1}, the lowest dose evaluated in the study, except for the untreated check. The dose required to reduce the biomass of the putative resistant populations was 0.56 kg ai ha-1. The lethal dose required to kill 50% of the treated plants was roughly equal to the 1X rate for the trial of 1.12 kg ai ha^{-1}. The putative paraquat-resistant horseweed populations had plants surviving treatments of 2.24 and 4.48 kg ai ha^{-1}, although biomass was reduced 70 to 75% relative to the untreated checks. Results from this study show that the suspected resistant horseweed populations collected from a vineyard and an apple orchard in the Finger Lakes region of New York survived labeled doses of paraquat applied under greenhouse conditions, while susceptible populations were controlled at substantially lower product rates. Current studies are evaluating horseweed responses to alternate herbicide chemistries and surveying perennial crop growers to determine if paraquat resistance is widespread throughout the state.

Preliminary Responses of Weeds, the Weed Soil Seedbank, and Microarthropod Communities to Zasso Electrical Weeding. Aleah L. Butler-Jones*[1], Elizabeth C. Maloney[2], Charlee R. Smith[3], Emily J. Sausser[4], Lynn M. Sosnoskie[2]; [1]Cornell University, Ithaca, NY, [2]Cornell University, Geneva, NY, [3]The Ohio State University, Columbus, OH, [4]Lebanon Valley College, Annville, PA (246)

Improved Herbicide Selectivity in Tomato by Safening Action of Benoxacor, Fenclorim, Melatonin, and 2,4,6- Trichlorophenoxyacetic Acid. Tabata Raissa De Oliveira*, Antonio Augusto Tavares, Josiane C. Argenta, Varsha Varsha, Te-Ming (Paul) Tseng; Mississippi State University, Starkville, MS (247)

Sulfentrazone Efficacy and Safety in Cabbage and Broccoli. Laura Pineda-Bermudez*[1], Elizabeth C. Maloney[1], Baylee L. Carr[2], Wesley M. Bouchelle[3], Thierry E. Besancon[3], Lynn M. Sosnoskie[1]; [1]Cornell University, Geneva, NY, [2]Rutgers University - Weed Science, Chatsworth, NJ, [3]Rutgers University, Chatsworth, NJ (248)

Specialty crops are significant contributors to the agricultural economy of the U.S. Weeds threaten yield quantity and quality in agricultural systems through direct crop competition, by complicating and contaminating harvest, and by harboring pests and pathogens. However, there is a limited number of herbicides available for specialty crops, especially cole crops like cabbage (Brassica oleracea var. capitata L.) and broccoli (Brassica oleracea var. italica L.). In 2021 and 2022, researchers at Cornell University and

Rutgers University initiated small-plot research trials to evaluate the efficacy and safety of sulfentrazone in transplanted broccoli and cabbage. Five treatments were applied in May (Cornell) or August (Rutgers) and were replicated 3-4 times in randomized crop block design. Treatments included oxyfluorfen (as GoalTender 1.17 l/ha PRE-T (pre-transplant)), sulfentrazone (as Spartan Charge at 0.28 and 0.56 l/ha PRE-T), and S-metolachlor (as Dual Magnum at 0.75 l/ha POST-T (immediately post-transplant)) followed by (fb) oxyfluorfen (as GoalTender 0.47 l/ha POST (14 days later)). An untreated check was included for comparison. Weed control (e.g., weed cover and density counts) and crop injury ratings were collected, weekly to bi-weekly, for 28 days after transplanting (DAT) and at 42 DAT. Weed biomass was collected and weighed at the time of crop harvest. Averaged across years and crops, mean percent weed cover was significantly ($P<0.05$) reduced by all herbicide treatments, relative to the untreated check, at 14, 21, 28, and 42 DAT in the Cornell trials. Differences were observed among herbicides with respect to cover with the lowest sulfentrazone PRE-T rate providing significantly ($P<0.05$) less weed suppression compared to the other treatments at 28 and 42 DAT. All treatments, except for the lowest sulfentrazone PRE-T rate, statistically ($P<0.05$) reduced common ragweed, common lambsquarters, smartweed and knotweed, and other broadleaf weed densities relative to the untreated check at 28 DAT. At harvest, weed biomass was 627 g/m^2 in the untreated checks; biomass was significantly reduced only in the oxyfluorfen (69 g/m^2) and S-metolachlor POST-T fb oxyfluorfen POST treatments (11 g/m^2). When averaged across years, the greatest amount of stunting was observed in highest rate of sulfentazone PRE-T and S-metolachlor POST-T fb oxyfluorfen POST treatments (15-21% at 21 and 28 DAT). Broccoli was more sensitive than cabbage and all injury was transient. Similar trends in weed cover and density counts were observed in the Rutgers trial.In the Cornell trials, mean head weight was not impacted by treatment ($P>0.05$); significant rainfall in NY in 2021 and drought conditions in NY in 2022, may have negatively affected crop yields. In NJ, all herbicide treatments significantly ($P<0.05$) increased cabbage and broccoli head weight. Results were intended to support possible label expansion into NY but the registrant has since decided not to proceed; data derived from these studies will be used to enhance NJ recommendations and guidelines with respect to sulfentrazone use in cabbage and support future labeling efforts in other cole crops.

Weed Impacts on Alfalfa Forage Accumulation and Nutritive Value. Chandra L. Montgomery*[1], Albert T. Adjesiwor[2]; [1]University of Idaho, Moscow, ID, [2]University of Idaho, Kimberly, ID (249)

Annual weeds can impact the economics of alfalfa production by reducing forage yield, and nutritive value or by contaminating hay. Field studies were conducted in Idaho in 2021 and 2022 to evaluate the effect of weed control treatments on alfalfa forage accumulation, weed biomass, and nutritive value. In addition, the relationship between the proportion of individual weed species biomass and alfalfa nutritive value was assessed. There were eight treatments including the untreated check. Treatments comprised pre-emergence incorporated, early postemergence (after 80% alfalfa emergence), and postemergence (third trifoliate alfalfa) herbicide applications. Data collection included weed control efficacy, weed and alfalfa biomass, and alfalfa nutritive value. Additional samples were collected and combined in these alfalfa (%) to weed biomass (%) proportions: 0:100, 20:80, 40:60, 60:40, 80:20, and 100:0, for wet chemistry analysis of forage nutritive value to evaluate the relationship between the proportion of individual weed species biomass and alfalfa nutritive value. The acetochlor only treatment provided less than 50% weed control while the EPTC only treatment provided 54 to 81% weed control. The control provided by acetochlor and EPTC were generally less than treatments containing imazamox and imazamox plus bromoxynil. Weed biomass in forage (23 to 55% of total biomass) due to poor or no weed control reduced crude protein and increased fiber concentrations. This in turn reduced the relative feed value. The relationship between the proportion of individual weed species biomass and alfalfa nutritive value was linear for all weed species evaluated.

GST Inhibition - Influence on Quizalofop P-ethyl Metabolism in Winter Wheat. Amber Pelon*[1], Franck E. Dayan[2], Todd A. Gaines[2]; [1]Colorado State University, Berthoud, CO, [2]Colorado State University, Fort Collins, CO (250)

Weed competition has the most significant negative impact on wheat grain yield, relative to other pests. Understanding the contribution of metabolism in overall resistance to herbicides can lead to new methods for controlling weeds in wheat. There is a lot of information about GSTs' role in the detoxification of herbicides. The present study is assessing the role of Phase 2 plant cell metabolism by testing glutathione S-transferase (GST) inhibition to see if it influences the metabolism of quizalofop p-ethyl (QPE) in winter wheat (*Triticum aestivum*). Previous literature reported increased resistance to herbicides with higher GST activity in black grass (*Alopecurus myosuroides*) and Asia minor bluegrass (*Polypogon fugax*). Resistance could be reversed by inhibiting GST activity. This research aims to determine if repeated applications of 4-chloro-7-nitrobenzoxadiazole (NBD), a GST inhibitor, on winter wheat sprayed with QPE cause the wheat to be more sensitive to QPE. Preliminary data has shown higher QPE-acid in plants sprayed with repeated NBD applications. Understanding the impact of GST inhibitors on plant metabolism will bring new knowledge to wheat breeders, herbicide producers, and growers that can lead to sustainable weed management practices in wheat and other important staple crops. In particular, identifying GST enzymes involved in herbicide metabolism would lead to important molecular markers to insure that these traits are included in breeding selection programs.

Surface Ponds as a Source of Weed Contamination in Container Nurseries. Alisha O. Ray*[1], Anthony LeBude[1], James Altland[2], Christopher D. Harlow[1], Joe C. Neal[1]; [1]North Carolina State University, Raleigh, NC, [2]USDA-ARS, Wooster, OH (252)

Nursery crop producers of the Southeastern U.S. use open ponds of captured water for irrigating container-grown plants, often without filtration. Many growers perceive irrigation water as a significant source of weed seeds, but data on the presence of weed seeds in nursery irrigation ponds are lacking. In the present study we evaluated irrigation pond water as a source of weed seeds in container nurseries. The presence and diversity of viable weed seeds in irrigation water samples from six commercial container nurseries in central and eastern NC was documented for two consecutive years. Irrigation pond water was sampled using a custom-fabricated filtration system. Six 20,000-gal samples were filtered at each location in autumn, spring, and mid-summer. The 20,000-gal sample size was selected to approximate a single day of irrigation on one acre of container nursery stock. Irrigation filtrates were collected in 355 µm-mesh sieves. Filtrate samples were spread onto the surface of a peat/bark/perlite substrate in flats and placed in a covered hoop house at the Horticulture Field Lab in Raleigh, NC. Seedling emergence was counted every 7 days for 12 weeks. Seedlings were identified to species and removed after counting. Unknown species were transplanted and grown to maturity for identification. Irrigation samples from all locations and seasons contained viable weed seeds. Differences between seasons were observed in each evaluated year. The average weed seed emergence across locations in the first recorded year was 16.4, 9.6, and 10.3 seeds per 20K gallons in autumn, spring and summer, respectively. The average weed seed emergence across locations in the second recorded year was 19.3, 15.2, and 90.2 seeds per 20K gallons in autumn, spring, and summer, respectively. Many species present in samples coincided with the season in which those weeds were mature and actively shedding seeds in the nursery. However, *Euphorbia maculata* emerged in all seasons, even when no mature plants were present at the nurseries. The population of weed seeds in irrigation was also associated with temporal seed-shed for some weed species, such as *Salix nigra*. An intense rain event also increased the number of seeds distributed via irrigation for species with buoyant seeds. Irrigation could be a source of seed spread in container nurseries for both new and endemic weed species.

Investigating New Peanut Cultivar Responses to Postemergence Applications of Chlorimuron. Nicholas J. Shay*[1], Eric P. Prostko[2]; [1]The University of Georgia, Tifton, GA, [2]University of Georgia, Tifton, GA (253)

Peanuts are a long-season crop that takes approximately 140-150 days to mature making it challenging to provide season-long weed control. Competition from weeds is costly and can cause severe economic losses. Ranked as one of the most troublesome weeds in the southeastern peanut-producing states, Florida beggarweed [*Desmodium tortuosum* (Sw.) DC.] can reduce peanut yields by up to 19% through direct competition for resources as well as negatively impacting fungicide deposition and harvest efficiency. Postemergence (POST) applications of herbicides such as chlorimuron are an excellent tool to help reduce economic losses from late-season weeds such as Florida beggarweed. However, prior research has shown that POST applications of chlorimuron can cause significant yield losses to sensitive cultivars such as Georgia-06G and Tifguard. Additionally, previous studies have also shown that chlorimuron can increase the incidence of tomato spotted wilt virus (TWSV) (*Tospovirus bunyaviridae*). Therefore, small-plot field trials were established in 2021-2022 to evaluate the response of new peanut cultivars to POST applications of chlorimuron. The experimental design was a randomized complete block with 3-4 replications and a 4 X 5 factorial treatment arrangement. Four chlorimuron timings (None,

65, 75, and 90 DAP) and 5 peanut cultivars (AUNPL-17, Georgia-18RU, Georgia-20VHO, TifNV High O/L, and FloRun™ 331) were evaluated. Chlorimuron was applied at 9 g ai ha^{-1} in combination with NIS at 0.25% v/v. All data were subjected to ANOVA using PROC GLIMMIX and means separated using the Tukey-Kramer method (P=0.10). Results indicated that there was no interaction between chlorimuron timing and peanut cultivar. When averaged over cultivar, chlorimuron applied 90 DAP resulted in a 7% increase in TSWV when compared to the non-treated control. Peanut yield was not reduced by any application of chlorimuron with yields ranging from 6,411 to 6,786 kg ha^{-1} (P=0.1243). Despite the slight increase in TSWV, these results suggest that the newer peanut cultivars are sufficiently tolerant to POST applications of chlorimuron

Weed Community Shift in Nitrogen Immobilized Soils. Maria A. Gannett*, Antonio DiTommaso, Jenny Kao-Kniffin; Cornell University, Ithaca, NY (254)

As more weed species develop resistance to herbicides, it is critical to develop new non-herbicide-based weed management strategies. A potential method of weed management is through soil carbon (C) amendments, which stimulate soil microbial growth and immobilize nitrogen (N). Agricultural weeds, many of which grow well in high N soils, may have reduced growth in high C, low available N soils. In a two-year field experiment, we implemented five amendment treatments increasing in C: an unamended weed-free control, an unamended weedy control, rye hay for a total of 3,500 kg C $ha^{-1}yr^{-1}$, sawdust for a total of 5,000 kg C $ha^{-1}yr^{-1}$, and rye hay and sawdust combined for a total of 8,500 kg C $ha^{-1}yr^{-1}$. We planted corn and soybean and each treatment was replicated 5 and 6 times, respectively. Throughout each season we measured available nitrate with ion exchange resin strips to monitor nitrogen immobilization. We also measured soil respiration to monitor soil microbial activity. Aboveground weed and crop biomass were measured to monitor crop-weed competition. Weed species were identified before being dried and weighed to measure changes in weed community composition. Soil microbial activity increased with increasing total C added to the soil, but N availability was lowest in plots treated with the highest C:N ratio amendment. Weed biomass was also lowest in plots treated with the highest C:N ratio amendment. Amendment treatments selected for specific weed species and weed species traits were correlated with environmental variables. Although corn and soy crop biomass in all treatments was lower than in the weed-free control, targeted N immobilization, may be able to help manage certain weed species in organic cropping systems, especially for leguminous crops.

Effect of Fall- or Spring-Planted Cover Crops on Weed Suppression in High Plains. Sachin Dhanda*[1], Vipan Kumar[1], Augustine Obour[1], Elizabeth Yeager[2], John Holman[3], Anita Dille[2]; [1]Kansas State University, Hays, KS, [2]Kansas State University, Manhattan, KS, [3]Kansas State University, Garden City, KS (255)

The widespread evolution of herbicide resistance in kochia (*Bassia scoparia* L.) and Palmer amaranth (*Amaranthus palmeri* L.) warrants the development of alternative, ecological-based integrated weed management strategies in the High Plains. A field study was conducted at Kansas State University Agricultural Research Center near Hays, KS during 2020-21 and repeated in 2021-22 to determine the impact of fall- or spring-planted cover crops (CC) on weed suppression in no-till dryland wheat-sorghum-fallow rotation. The field site had a natural seedbank of glyphosate and dicamba-resistant kochia and glyphosate-resistant Palmer amaranth. A four-way CC mixture (triticale/winter peas/radish/rapeseed) was planted during fall after wheat harvest and terminated at triticale heading stage in spring with glyphosate and/or glyphosate + (acetochlor + atrazine). After termination of fall-planted CC, grain sorghum hybrid was planted. In contrast, a three-way CC mixture (oats/spring peas/barley) was planted in sorghum stubble during early spring and terminated at oats heading stage with glyphosate and/or glyphosate + (flumioxazin + pyroxasulfone). Chemical-fallow and nontreated weedy check were also included in both fall- or spring-planted CC treatments. In 2021 growing season, fall-planted CC terminated with glyphosate + (acetochlor + atrazine) reduced total weed density by 50% and weed biomass by 68% at sorghum harvest compared to chemical-fallow. In 2022 growing season, same treatment reduced total weed density by 88% and weed biomass by 96% at sorghum harvest as compared to weedy check treatment. Sorghum yield was not significantly different in chemical-fallow and both fall-planted CC treatments in 2021, however the yield was significantly higher (1130 kg ha^{-1}) in fall-planted CC terminated with glyphosate + (acetochlor + atrazine) followed by chemical-fallow in 2022. Compared to chemical-fallow, spring-planted CC terminated with glyphosate + (flumioxazin + pyroxasulfone) reduced total weed density by 24 to 72% and weed biomass by 14 to 23% at 90 days after termination across both years. Overall, these results suggest that both fall- and spring-planted CC mixtures in combination with effective residual herbicides at CC termination can provide effective suppression of herbicide-resistant kochia and Palmer amaranth in the High Plains region.

Utility of John Deere's See &Amp; Spray™ Ulimate in Southern U.S. Row Crops. Tristen H. Avent*[1], Jason K. Norsworthy[1], William L. Patzoldt[2], Lauren M. Lazaro[2]; [1]University of Arkansas, Fayetteville, AR, [2]Blue River Technology, Sunnyvale, CA (256)

Increasing costs of inputs and the need for greater environmental stewardship have driven the development of precision application technologies. Recently, John Deere commercially launched See & Spray™ Ultimate for use in the 2023 growing season. See & Spray can apply targeted broadcast applications while simultaneously broadcasting with split-tank and separate plumbing systems. Currently, this technology is supported in select crops and fallow systems. The University of Arkansas Systems Division of Agriculture (UADA) has conducted more than 15 trials over two years evaluating the utility of this technology in soybean and cotton. Preliminary results indicate the See & Spray Ultimate can provide comparable weed control to a broadcast standard program and reduce herbicide inputs. However, data still needs to be evaluated to determine the likelihood of success with varying weed species and sizes, machine settings, and environments. The UADA is conducting trials to determine 1) the effect of weed size, species, location, and machine settings on weed mortality and accuracy; 2) the performance of See & Spray when using cover crops in cotton; 3) the long-term effect of low and high sensitivity settings in single-tank programs on weed seedbank management compared to a standard broadcast program; and 4) the impact of See & Spray programs on crop response, weed control, and yield in cotton and soybean. Many of these trials will be conducted in collaboration with other universities and Blue River Technology to collect data across the U.S. and determine the fit of this new technology for producers across the country.

Transgenerational Effect of Weed Competition on Wheat. Albert O. Kwarteng*[1], Albert T. Adjesiwor[2]; [1]University of Idaho, Moscow, ID, [2]University of Idaho, Kimberly, ID (257)

Plants are subjected to several forms of stress and have evolved mechanisms to cope with these stress conditions. Studies have shown that plants can store and recollect memory of a previous stress exposure which affects their reponse to future stresss. This research is aimed at understanding how multigenerational weed exposure affects phenotypic plasticity, hormonal and gene expression, and DNA methylation in wheat (*Triticum aestivum*). Wheat was planted in the center of 3L plastic pots surrounded by either 8 kochia (*Bassia scoparia*), 8 Italian ryegrass (*Lolium multiflorum*), 8 wheat, or no surrounding plants. Treatments were arranged in a completely randomized design with 15 replications. Seeds harvested from the first generation were used to plant the second generation, and the process was repeated under the same conditions to obtain the second, third and fourth generations. Relative to wheat with no surrounding plants, wheat-kochia, wheat-ryegrass, and wheat-wheat treatments reduced seed yield by 7, 26, and 43%, respectively, in the first generation; 90, 93, and 89%, respectively, in generation two; 63, 85, and 84%, respectively, in generation three, and 80, 82, and 80%, respectively in the fourth generation. The number of seed heads per plant in the wheat only treatment increased from 2 heads in generation one to 12 heads per plant in the fourth generation. These results suggest a potential maladaptive impact of transgenerational memory of weed stress on wheat. The fifth generation, together with the biochemical, transcriptomic, and epigenetic data would provide a better understanding of the mechanisms involved in these observations.

Assessing the Efficacy of a Biculture Living and Dead Cover Crop Mixture for Weed Suppression in Sweet Corn. Veronica Yurchak*, Alan W. Leslie, Bill Phillips, Cerruti Rr Hooks; University of Maryland, College Park, MD (258)

Modern vegetable production systems are often characterized by monoculture fields and intensive use of tillage and/or synthetic agrochemicals for managing weeds. A growing public interest in more sustainable and eco-friendly production practices has resulted in increased demand that crops be

produced with lower inputs. Field studies were conducted over three field seasons to investigate the use of conservation tillage in concert with an interplanted living mulch and/or cover crop residue for managing weeds in sweet corn and compared with the traditional practice of using conventional tillage and a pre- and post-emergent herbicide. Whole plot treatments included: (1) conventional till, (2) no-till with cover crop residue, (3) living mulch + cover crop residue, and (4) living mulch + winter killed residue. The split-plot factor consisted of herbicide treatments: (1) at-planting application of residual herbicides or (2) no herbicide. The cover crop systems suppressed weeds as well as the traditional practice throughout the cropping cycle in all three years. In addition, there was no significant improvement in weed suppression with the application of herbicides within the cover crop treatments.
vjohnso4@umd.edu

A Deep Learning Approach for Nutsedge (*Cyperus* Spp.) Detection in Strawberry Plasticulture Production. Alex G. Rodriguez*[1], Arnold W. Schumann[2], Nathan Boyd[3]; [1]University of Florida, Wimauma, FL, [2]University of Florida, Lake Alfred, FL, [3]University of Florida, Balm, FL (259)

In Florida strawberry plasticulture systems, non-uniform fumigation is the only effective management option to control nutsedge, which typically occurs in patches. A precision fumigation system has the potential to greatly reduce fumigation and labor costs, and building a rapid and accurate detection model for nutsedge is key for the development of such a system. This study presents a deep learning approach for detecting nutsedge in strawberry plasticulture production using a state-of-the-art real-time object detection system, You Only Look Once (YOLO). In this work, YOLOv5 and YOLOv7 algorithms were trained on a dataset of 4,780 pre-processed images and 10,050 bounding box annotations of nutsedge in plastic at different stages of growth and under different lightning conditions. Among the models evaluated, YOLOv5x outperformed all other algorithms in terms of mAP@50, precision, and accuracy. It achieved a mAP@50 of 91.2%, a precision of 91.4%, and a recall of 88.5%. In contrast, the YOLOv7x model was the least accurate among the models evaluated, with a mAP@50 of 69.6%, a precision of 78.1%, and a recall of 61.8%. These results demonstrate that object detection models such as YOLOv5x have the potential to be used for detecting nutsedge and can provide the framework for developing a precision fumigation system for strawberry plasticulture production.

Weed Seed Fate in the Combine During Wheat and Soybean Harvest. Eli C. Russell*, Michael L. Flessner, Matthew P. Spoth, Kevin Bamber; Virginia Tech, Blacksburg, VA (260)

Harvest weed seed control (HWSC) targets weed seeds as they pass through the combine by concentrating, removing, or destroying them. Most methods of HWSC target the chaff fraction as it exits the combine. If the combine is properly tuned, most of the weed seeds should be exiting the combine in the chaff fraction and therefore subjected to HWSC. However, if those weed seeds exit in the straw fraction, they will not be subjected to HWSC and will be returned to the soil seed bank. So, the purpose of this research was to determine what percentage of weed seeds for economically important weeds in wheat and soybean escaped HWSC by exiting the combine in the straw fraction. For this initial research, combine settings were adjusted to minimize grain loss while effectively cleaning the grain. In wheat, the percentage of seeds escaping HWSC in the straw fraction for Italian ryegrass (*Lolium perenne* L. ssp. *multiflorum* (Lam.) Husnot), canola (*Brassica napus* L.), and hairy vetch (*Vicia villosa* Roth) were 1.15%, 0.05%, and 0.00%, respectively. In soybean, the percentage of seeds lost in the straw fraction for redroot pigweed (*Amaranthus retroflexus* L.), common ragweed (*Ambrosia artemisiifolia* L.), and barnyardgrass (*Echinochloa crus-galli* (L.) P. Beauv.) were 4.28%, 0.57%, and 1.02%, respectively. Our first-year results indicate that a low percentage of weed seeds are escaping HWSC in the straw fraction. These data indicate that very few weed seeds are exiting in the straw fraction.

Evaluating the Effect of Cover Crop Termination Management on Palmer Amaranth (Amaranthus Palmeri) Suppression in Soybean. Cynthia Sias*, Michael L. Flessner, Kevin Bamber; Virginia Tech, Blacksburg, VA (261)

Palmer amaranth (Amaranthus palmeri S. Wats.) is an exemplary weed species displaying the ability to adapt and evolve resistance to multiple herbicide modes of action. There is a need for additional weed suppression tactics to control such species. Growing interest in the use of cover crops (CC) has led to questions regarding the most appropriate forms of CC management prior to cash crop planting in order to maximize weed suppression benefits. Experiments were conducted in 2021 and 2022 to test: 1) cover crop termination timing (i.e. green or brown), 2) CC biomass (i.e. low, medium, high, or no cover), and 3) CC termination method (i.e. rolled versus standing) on Palmer amaranth suppression. Treatments included planting brown (cereal rye terminated two weeks before soybean planting), planting green (cereal rye terminated at soybean planting), and a no cover (winter fallow) check. Palmer amaranth emergence counts were evaluated at 4 and 6 weeks after soybean planting, and yield was calculated at harvest. To explore the ideal combination of termination tactics to increase weed suppression, data were analyzed using SAS to conduct ANOVA using a=0.05. Among CC options, CC plots reduced Palmer amaranth germination when compared to the no CC check plots. Plots that were left standing also displayed greater weed suppression capability when compared to the rolled plots. This result is potentially due to a difference in light penetration reaching the soil surface. More data are necessary to make final conclusions, however, this information will provide better recommendations for farmers interested in using cover crops for weed suppression.

Using Black Soldier Fly Larvae Composting to Reduce Weed Seed Germination. Jeanine Arana*[1], Stephen L. Meyers[2], Luz Aide Cardona Giraldo[2], Nathaly Belen Vargas Arroyo[2], Laura D. Rodriguez Baquero[3]; [1]Purdue University/Horticulture, West Lafayette, IN, [2]Purdue University, West Lafayette, IN, [3]University of Wisconsin-Madison, Madison, WI (262)

Black soldier fly (*Hermetia illucens*) larvae (BSFL) composting is biotechnology used for organic waste management. However, there is little information available on how BSFL composting affects weed seed germination. We designed an experiment to evaluate the effect of BSFL on weed seed germination of six weed species, including barnyardgrass (*Echinochloa crus-galli),* common ragweed (*Ambrosia artemisiifolia*), giant foxtail (*Setaria faberi*), ivyleaf morningglory (*Ipomoea hederacea*), redroot pigweed (*Amaranthus retroflexus*), and velvetleaf (*Abutilon theophrasti*). The experiment had 5 treatments; inside a transparent container, we placed (1) 100 seeds each of six common weed species alone or (2) 100 weed seeds with a standard Gainesville diet or 100 seeds plus ~2,000 BSFL fed with either (3) standard Gainesville diet, (4) vegetable waste, or (5) food leftovers. After six weeks, the larvae started pupating, and we extracted the pupae from the bins. Once pupae were removed from the compost, the compost was placed on a tray with potting soil to evaluate weed seed germination. The weeds that emerged were identified, counted, and removed for four weeks. Then we subjected the trays to cold stratification for 3 months. After this, weed seed germination data were collected again for four more weeks. Total germination in the untreated control was 28% for barnyardgrass, 37% for common ragweed, 25% for giant foxtail, 65% for morningglory, 2% for redroot pigweed, and 64% for velvetleaf. Germination in the Gainesville diet control treatment was 13% for barnyardgrass, 18% for common ragweed, 5% for giant foxtail, 17% for morningglory, 0% for redroot pigweed, and 32% for velvetleaf. With the exception of velvetleaf in the vegetable waste substrate and giant foxtail in the food leftovers substrate, treatments with BSFL reduced weed seed germination more than the Gainesville diet control. All treatments with BSFL greatly reduced the germination of all weeds to = 1%, except for velvetleaf. Potential causes of the high weed seed germination reduction include high moisture developed by the BSFL composting because of the type of feed used and carbon dioxide and ammonia generated during BSFL composting. We believe that the seeds' characteristics, such as a thick seed coat, may interfere with the potential of the BSFL composting process.

Weed Genetic Control: Potential Application for Controlling Dioecious Amaranths. Lucas Kopecky Bobadilla*[1], Patrick Tranel[2]; [1]University of Illinois, Champaign, IL, [2]University of Illinois, Urbana, IL (263)

Dioecious amaranths (*Amaranthus* spp.) are a group of highly invasive and economically important weeds that pose a significant threat to global agriculture. Effective control of dioecious amaranths is challenging due to their ability to rapidly adapt to various environments and the evolution of herbicide resistance to multiple sites of action. As a result, there is a need to develop alternative control strategies that are more sustainable and specific to dioecious amaranths. One potential approach is the use of genetic control methods, involved whereby the modification of the weed's own genome could reduce its competitiveness or fertility. There are several types of genetic control methods that have been proposed for use against dioecious amaranths. One approach is the use of gene drive systems, which involve the insertion of a modified gene that is able to spread rapidly through the population due to its high transmission rate. The first step to developing a gene drive system is to identify the target gene for manipulation. Using genomics and transcriptomics tools, we are working towards the identification of the sex-determination genes in *Amaranthus* spp. With the identified candidates, the next step would be the development of a transformation protocol for *Amaranthus* spp. for functional validation. While the potential of genetic control methods for controlling dioecious amaranths is promising, there are also several challenges that need to be addressed. One concern is the potential for negative impacts on non-target species and the possibility of unintended consequences from releasing genetically modified organisms into the environment. In addition, there are regulatory and social barriers to adopting these technologies, which will need to be addressed to bring them to fruition. Despite these challenges, genetic control methods can provide a more sustainable and targeted approach for controlling dioecious amaranths and other invasive weeds.

Determining the Critical Period of Weed Control in Organic Carrots. Emma Kubinski*[1], Mac Burgess[2], Fabian Menalled[3]; [1]Montana State University/Land Resources and Environmental Science, Bozeman, MT, [2]Montana State University/ Plant Sciences and Plant Pathology, Bozeman, MT, [3]Montana State University, Bozeman, MT (264)

An essential aspect of developing an integrated weed management plan is understanding the appropriate timing of weed removal that minimizes labor needs while optimizing yields, crop quality, and economic returns. This information is of special importance in carrot (*Daucus carota*) production given the crop's poor competitive ability. This challenge is further compounded in organic agriculture where synthetic chemical control cannot be relied upon, often limiting weed management to labor-intensive methods. To determine the critical period of weed control, it is required to assess the maximum amount of time early-season weed competition (i.e., the critical timing of weed removal) and the minimum weed-free period required to prevent unacceptable yield reductions (i.e., the critical weed-free period). To evaluate this, we conducted a field study in Bozeman, MT during the 2021 and 2022 growing seasons. The experiment followed a randomized block design with six weed removal treatments applied based on the carrot leaf stage, plus a season-long weed-free and a season-long treatment. In addition to yields, we also measured carrot marketability and sugar content. Preliminary findings suggest that the critical period of weed control ranges between weed-free to the 2-leaf stage and weedy to the 4-leaf stage. To generalize our results, the carrot leaf stage will be correlated to growing-degree days. A four-parameter log-logistic regression analysis is being used to assess the length of the critical period of weed control and to graphically represent the impact of the length of crop-weed competition on yield, sugar content, and marketability. Results from this study will enable organic carrot growers to design more efficient and affordable integrated weed management programs.

Allelopathic Weed-suppressive Effect of Cotton Chromosome Substitution Lines Under Field Conditions. Worlanyo Segbefia*[1], Varsha Singh[1], Ncomiwe A. Maphalala[2], Tabata Raissa De Oliveira[1], Ziming Yue[1], Te-Ming (Paul) Tseng[1]; [1]Mississippi State University, Starkville, MS, [2]Mississippi State University-Department of Plant and Soil Sciences, Starkville, MS (265)

Weed competition with crops is a significant concern for farmers worldwide. Allelopathy enables certain plants to produce compounds (allelochemicals) that wade off competition from weeds. Allelochemicals can be released in various ways and are affected by environmental circumstances. Because greenhouse and lab testing are restricted in their ability to imitate natural ecological conditions and interactions, field trials are required to test the efficiency of weed-suppressing cotton chromosome substitution lines. Eight chromosomal substitution lines were planted in a 20-foot two-row plot at Pontotoc, MS, for this study. After three weeks of planting (WAP), seeds of four weed species were put between the planted rows: redroot pigweed *(Amaranthus retroflexus),* morning-glories (*Ipomoea spp.*), and common lambs quarters *(Chenopodium album).* Cotton plant heights were measured weekly for six weeks. Weed density data were collected for all weeds in the field, including native weeds. Weed reduction percentages were lowest for Enlist (32%), followed by UA48 (35%), and highest for T26lo (52%), followed by BNTN 16-15 (60%). Allelopathy is a complex process with significant agricultural potential in a world where pesticide resistance is developing. In addition, these weed-suppressive cotton chromosome substitution lines could be used to make organic fertilizers, minimizing farmers' reliance on synthetic herbicides for cotton weed suppression.Keywords: allelopathy, weed density reduction, Gossypium hirsutum, chromosome substitution

Prediction of Weed Emergence Patterns in a Rice-Wheat Cropping System Using Population-Based Threshold Models. Faryal Ali*[1], Irfan Afzal[2]; [1]University of agriculture Faisalabad Pakistan, Mardan, Pakistan, [2]University of agriculture Faisalabad Pakistan, Faisalabad, Pakistan (267)

Soybean Weed Management with Residual Herbicide Programs. Zachary R. Treadway*, Jennifer Dudak, Todd A. Baughman; Oklahoma State University, Ardmore, OK (268)

SOYBEAN WEED MANAGEMENT WITH RESIDUAL HERBICIDE PROGRAMS. Z.R. Treadway*, T.A. Baughman, J.L. Dudak; Oklahoma State University, Ardmore, OK ABSTRACT Weed management is crucial to maximizing soybean yield. Producers are facing the ever-growing issue of herbicide resistance and how to combat it. Resistance to acetolactate synthase (ALS), glyphosate, and protoporphyrinogen oxidase inhibiting (PPO) herbicides along with reports of glufosinate and auxin resistance has only increased this dilemma. One useful method for combatting herbicide resistance is through the use of PRE residual herbicide combinations with multiple modes of action. Experiments were conducted from 2020-2022 to evaluate four modes of action for PRE residual weed control in dicamba-tolerant soybean. Treatments included chloransulam-methyl (36 g ai ha^{-1}), metribuzin (417 g ai ha^{-1}), pyroxasulfone (95 g ai ha^{-1}), or sulfentrazone (163 g ai ha^{-1}). Herbicides were applied alone, as well as in two and three way combinations. Treatments were followed by a POST application of dicamba (567 g ae ha^{-1}) + glyphosate (329 g ae ha^{-1}). Soybean injury never exceeded 10% with any treatment at any point during the growing season. Palmer amaranth (*Amaranthus palmeri* S. Watson) was present in five of seven site years. Control of Palmer amaranth 2 WAP was 93% or greater with all three-way combinations across all locations. In three of the five site years where Palmer amaranth was present, all PRE combinations provided at least 94% control. At 4 WAP, control of Palmer amaranth was 90% or greater with all three-way combinations in four of the five site years. Control was at least 98% at three of the five site years, with pyroxasulfone in a two or three-way combination with chloransulam-methyl and metribuzin; and chloransulam-methyl + metribuzin + sulfentrazone. Following the POST application of dicamba + glyphosate, control of Palmer amaranth was at least 97% with all three-way combinations across all site years, as well as pyroxasulfone + chloransulam-methyl or metribuzin. Large crabgrass (*Digitaria sanguinalis* (L.) Scop.) was present in five of seven site years. Control 2 WAP was at least 90% at all locations with any treatment that included metribuzin except in combination with sulfetrazone. Large crabgrass control 4 WAP was at least 91% in four of five site years with all two- and three-way combinations that included metribuzin. Following the POST application, control of large crabgrass was 90% or greater with all treatments across all site years. All treatments increased yield when compared to the untreated check, but there was no consistent yield increase among treatment combinations. However, it was noted that chloransulam-methyl applied alone was the only treatment that yielded equal to or below the herbicide treatment trial average at all locations. This is most likely due the prevalence of ALS-resistant Palmer amaranth in Oklahoma soybean. Additionally, only treatments that included pyroxasulfone yielded equal to or greater than the trial average. These experiments highlight the importance of a residual herbicide program to combat troublesome weeds across varying geographical and environmental conditions. The use of residual herbicides also aids in protecting new POST herbicide technologies that Oklahoma soybean producers are implementing as part of their weed management program.

Evaluation of Pre-emergence Herbicides for Weed Management in Florida's Strawberry Plasticulture Production. Ruby Tiwari*[1], Nathan Boyd[2], Pamela Roberts[1], Samira Daroub[3], Ramdas Kanissery[1]; [1]University of Florida, Immokalee, FL, [2]University of Florida, Balm, FL, [3]University of Florida, Everglades, FL (269)

Johnsongrass (*Sorghum halepense*) Promotes Biological Nitrification Inhibition in its Rhizosphere. Megan L. Schill*, Dinesh Phuyal, Nithya Rajan, Nithya K. Subramanian, Muthukumar V. Bagavathiannan; Texas A&M University, College Station, TX (270)

ADMIN

Introduction and Announcements. Carroll Moseley*[1], Wesley Everman[2]; [1]Syngenta, Brown Summit, Nc, NC, [2]North Carolina State University, Raleigh, NC

IWSC 2024 - Jerusalem and IWSS Update. Nilda Roma-Burgos*; University of Arkansas, Fayetteville, Fayetteville, AR

Keynote Presentation - Jake Li - Deputy Assistant Administrator for Pesticide Programs - Protecting Endangered Species from Pesticides: Why it Matters to Your Work. Jake Li*; US EPA - Deputy Assistant Administrator for Pesticide Programs, Washington, DC

Presidential Addresses. A Stanley Culpepper*[1], Jacob Barney[2]; [1]University of Georgia, Tifton, GA, [2]Virginia Tech, Blacksburg, VA

Awards Presentation. Mandy Bish*[1], Caren A. Schmidt[2]; [1]University of Missouri, Columbia, MO, [2]BASF, Durham, NC

Presentation of Honorary Members and Fellows. Carroll Moseley*[1], Wesley Everman[2]; [1]Syngenta, Brown Summit, Nc, NC, [2]North Carolina State University, Raleigh, NC

ORAL

Directional Selection of Competitive Ability in *Setaria faberi* Confirmed with GWAS. Sandra R. Ethridge[1], Saket Chandra[1], Wesley Everman[1], David L. Jordan[1], Anna M. Locke[2], Micheal D. Owen[3], Ramon G. Leon*[1]; [1]North Carolina State University, Raleigh, NC, [2]USDA-ARS, Raleigh, NC, [3]Iowa State University, Ames, IA (195)

The evolution of plant species usually occurs over thousands or millions of years, however it has been proposed that rapid evolution in certain fitness traits could occur on a scale of decades, such as herbicide resistance. This rapid evolution of has yet to be proven in weed species for complex polygenic traits. For this study, we hypothesized that weed growth and competitive ability shifts can occur in just a few years due to highly selective forces in agroecosystems. A resurrection approach was taken in order to characterize life-history traits of a population of *Setaria faberi* under competitive and non-competitive situations. Seeds were collected from a single population of *Setaria faberi* during a 34-year time span, 1983 to 2017. A replacement series was conducted in which the oldest year-line, 1983, was put in competition with the newer year-lines. This study showed oscillating selection overtime, with plant competitive ability decreasing before 1995 and increased progressively thereafter. The competitive ability was due to the increase in leaf area and biomass of the newer year-lines when in competition. In addition, a genome-wide association study (GWAS) was conducted and identified four loci associated with increased competitive ability over time. Results from this study confirmed the ability of weeds to exhibit rapid evolution of complex life-history traits that could increase weediness in agricultural systems.

Impact of Combined Temperature and Moisture Stress on Palmer Amaranth (*Amaranthus palmeri*) Growth and Fecundity: Adaptations and Constraints. Sarah E. Kezar*, Aniruddha Maity, Muthukumar V. Bagavathiannan; Texas A&M University, College Station, TX (196)

The response of crops to anticipated changes in temperature and available water has been well documented, but the counterpart of weed species has not. Palmer amaranth (*Amaranthus palmeri*) is inherently a high-fecundity specie in the current climate, but its competitive ability level is unknown under elevated temperatures and moisture stress. This study was conducted to understand the impact of temperature and soil moisture stresses on growth of Palmer amaranth. The effects of four temperature levels of 25/20°C, 30/25°C, 35/30°C, and 40/35°C day/night temperature with a 14/10 hours photoperiod and four soil moisture stress levels as the sub-treatment, i.e., field capacity, 75% field capacity, 50% field capacity, and 25% field capacity were of interest. The experiment was conducted in the controlled environment growth chambers (Conviron MTPS) located at the Norman E. Borlaug Center for Southern Crop Improvement Greenhouse Facility, Texas A&M University, College Station, TX. Palmer amaranth accumulated vegetative biomass with the highest recorded in plants at field capacity whereas biomass production was reduced with increasing degrees of water stress, regardless of temperature regime. Root growth responded by producing more biomass in the field capacity and 75% field capacity treatments. The impact of all treatment levels on photosynthetic output, such as stomatal conductance and transpiration rate, is also of interest in this study (LI-6800 Portable Photosynthesis System). Results reflected how Palmer amaranth and other C4 species can maintain a high photosynthetic capacity under moisture and temperature stress using osmotic adjustment. The threats to croplands experiencing resistant weed pressures under the face of environmental changes are increasingly important as crop productivity and profitability are directly impacted. Agriculturalists must produce at an unprecedented pace to provide for a growing global population, begging the critical need to understand critical yield limiting factors such as resistant weeds like Palmer amaranth.

The International Weed Genomics Consortium: a Review of Current Reference Genomes and Future Directions. Sarah Morran*[1], Todd A. Gaines[1], Dana R. MacGregor[2], Eric L. Patterson[3], Mithila Jugulam[4], Roland S. Beffa[5]; [1]Colorado State University, Fort Collins, CO, [2]Rothamsted Research, Harpenden, United Kingdom, [3]Michigan State University, East Lansing, MI, [4]Kansas State University, Manhattan, KS, [5]Senior Scientist Consultant, Frankfurt, Germany (197)

New Insights in Amaranth PPO Target Site Resistance and PPO1 and PPO2 Biochemistry. Jens Lerchl*[1], Aimone Porri[2], Inessa De[1], Lucie Meyer[1], Nilda Roma-Burgos[3], Matheus Machado Noguera[4], Bernhard Grimm[5], Boris Hedtke[5], Daniel Wittmann[6], Patrick Tranel[7], Kathryn Seebruck[8]; [1]BASF SE, Limburgerhof, Germany, [2]BASF Global Research & Development, Limburgerhof, Germany, [3]University of Arkansas, Fayetteville, Fayetteville, AR, [4]Univ Arkansas, Fayetteville, AR, [5]Humboldt University, Berlin, Germany, [6]BASF SE, Berlin, Germany, [7]University of Illinois, Urbana, IL, [8]University Illinois, Urbana, IL (198)

Group 14 PPO herbicides have become an important tool to control resistant weeds. With the widespread use of PPO inhibitors target site mutations have evolved in weeds, especially in waterhemp and Palmer amaranth. We have aimed at a better understanding of the herbicide resistance risk in amaranth for existing and newly developed PPO herbicides. Benchmark compounds have been compared with the new active trifludimoxazin specifically designed to combat target site mutations and combinations thereof. Results include a summary on the in vitro efficacy of PPO group 14 herbicides on mutant PPO enzyme variants as well as in planta data. Results show the importance of PPO2 enzyme in Palmer amaranth versus Arabidopsis thaliana and new insights in protoporphyrinogen IX oxidase isoforms and its implications for herbicide resistance and herbicide tolerance traits.

Modeling the Effect of Density, Emergence Time, Inter- and Intra-Specific Competition of Barnyard Grass (*Echinochloa Crus-Galli)* on Growth and Yield of Dry-Seeded Rice: Simulations with INTERCOM. Tahir Hussain Awan*[1], Bhagirath Singh Chauhan[2], John Lindquist[3]; [1]Rice Research Institute, Kala Shah Kaku, Punjab, Pakistan, Lahore, Pakistan, [2]bThe Centre for Plant Science, Queensland Alliance for Agriculture and Food Innovation (QAAFI), The University of Queensland, Gatton 4343, Queensland, Australia, Queensland, Australia, [3]University of Nebraska, Lincoln, NE (199)

Dry-seeded rice (DSR) is an emerging resource-conserving technology in many Asian countries including Pakistan, but weeds remain the major threat to the production of DSR systems. DSR crop yields on-farm rarely approach their production potential, partly as a result of growth reduction due to weed interference. Among the weeds, *Echinochloa crus-galli* mimics rice and is a stronger competitor of rice for resources than other weeds. This weed has developed resistance to herbicides and poses a serious threat to rice production in the world. An adequate knowledge of the biology, ecology and critical period of competition of *E. crus-galli* with DSR is fundamental in designing effective, sustainable and integrated weed management programs for DSR. Cost-effective weed management requires accurate estimates of yield and the potential yield loss resulting from weed infestations. However, crop yield and the effects of weeds are highly variable across weed density and their time of emergence. This may be accomplished through early crop vigor or delaying the weed emergence. An Eco physiological model of rice-barnyard grass competition (INTERCOM) may be useful for predicting the effects of weed density and their time of emergence on crop and weed growth and competitive ability. To evaluate the ability of the model to predict the dry-seeded rice (DSR) growth and yield, the effect of barnyardgrass interference on rice yield loss was evaluated using two season rice data sets collected from the field experiments. Predicted and observed monoculture leaf-area-index, leaf, stem, aboveground, and panicle biomass of both rice and barnyardgrass were in close agreement. For inter-competition between species, height is very important, and the model captured the height very well. The normalized deviation for all values was near zero, which means the model calibration efficiency is good. Model-simulated results were very much similar to our field-observed results. Predicted and observed weed-free rice yields were ranged from 6.42-7.47 t ha^{-1} The simulated results depicted that rice panicle yield was affected by the increasing weed density at early emergence. As weed emergence was delayed, there were no observed effects of different densities as the weed was usually unable to survive and did not cause much reduction in panicle yield. Percentage yield reduction decreased with the decrease in weed density and the delay in their emergence. Barnyardgrass at 70 plants m^{-2}, emerging 2 days after rice emergence (DARE), reduced the grain yield by 65-70%. When the weed emergence was delayed to 45 DARE, the reduction in grain yield was only 2-5%. This reduction in yield was not so much different from the reduction caused by the weed density of 10 plants m^{-2} (0.7-3.2%). having the same emergence time as with rice. The model predicted that barnyard grass emergence with any densities at 45 DARE had negligible effect on rice growth and yield. The model suggests that the use of competitive rice cultivars or delaying the weed emergence may reduce the need for chemical weed control. Nomenclature: Barnyardgrass *Echinochloac crus-galli* (L.) Beauv., ECHCG; rice, *Oryza sativa* L. Keywords: Economic threshold, integrated weed management, weed ecology, IPM, weed-crop interference. Rice-weed competition.

A Comparative Genomics Approach to Investigate the Parallel Evolution of Weediness in *Alopecurus myosuroides* (Black Grass) and *Alopecurus aequalis* (Orange Foxtail). Dana R. MacGregor*[1], Jon Wright[2], Hannah Rees[2], Christian Harrison[1], Jun Li[3], Paul Neve[4], Seanna McTaggart[2], Anthony Hall[2]; [1]Rothamsted Research, Harpenden, United Kingdom, [2]Earlham Institute, Norwich, United Kingdom, [3]Nanjing Agricultural University, Nanjing, China, [4]University of Copenhagen, Copenhagen, Denmark (200)

Alopecurus aequalis (common name orange foxtail), is an annual to short-lived perennial bunchgrass native to much of the temperate Northern Hemisphere. It can be found growing wild in wet, heavy soil from Eurasia to North America. Despite this incredibly widespread distribution, only China and Japan it has become a serious and problematic weed in winter wheat and barley causing considerable yield losses and economic consequences. Like many other weeds, *A. aequalis* has evolved both target-site-based resistance (TSR) and non-target-site-based resistance (NTSR). Interestingly and despite extreme geographic separation, a closely related species *Alopecurus myosuroides* (common name black-grass) has evolved to occupy similar niches. Like *A. aequalis, A. myosuroides* has a widespread distribution across the Northern Hemisphere but has become the predominant agricultural weed in Western European winter wheat and barley. Again like *A. aequalis. A. myosuroides* exhibits TSR in similar locations of homologous sequences and NTSR using similar mechanisms e.g. induction of cytochrome P450 mono-oxygenases and glutathione s-transferases. The mechanisms underpinning TSR are clear, but the mechanistic or genetic causes of NTSR are not well defined. These two closely related species present a model to investigate the genetic mechanisms underpinning the parallel evolution towards a weediness phenotype driven by NTSR. To explore this, we are generating a high-quality genome and annotation for *A. aequalis*. We will use this new *A. aequalis* resource in comparative genomic analysis between existing genomes for *A. myosuroides* to identify structural rearrangement, SNP variation and expansion and contraction of gene families between these two species to identify variation driving the evolution of weediness. Once annotated, we have resources for comparative transcriptomics analysis using sensitive and resistant members of each species to determine if UK and Chinese *Alopecurus* species have evolved herbicide resistance via altered regulation of a common suite of genes. Together these analyses help to explore the genetic basis of weediness in these problematic species.

Does the Emergence Pattern of *Echinochloa crus-gali* (Barnyardgrass) Differ with EPOST Timings in Corn? Theresa Piskackova*; Czech University of Life Sciences Prague, Prague 6 - Suchdol, Czechia (Czech Republic) (201)

Predictive models for weed emergence can be an important part of precision weed management to minimize weed intervention while maximizing the benefits to the crop. Weed emergence models are often based on data from bare ground studies, which usually represent the worst-case scenario; however, weed emergence in the crop represents many factors which could affect the pattern of emergence. In this study we examine the emergence pattern of *Echinochloa crus-gali*, barnyardgrass, in corn treated with pre-emergence herbicides at five timings. The study was conducted in Suchdol, Czech Republic in 2021 and 2022 and corn was planted on April 22 and 28, respectively. Treatments were assigned in a randomized complete block design with 3 replications to small plots 2.25 m wide by 6.5 m long. Four herbicides were applied at each of five timings and two plots were left as nontreated controls in each replication (dimathinamid-P at 1008 g ha^{-1}, pethoxamid at 1200 g ha^{-1}, isoxaflutole at 96 g ha^{-1}, and mesotrione at 480 g ha^{-1}; applications at 18, 22, 27, 32, and 36 days after planting). *Echinochloa crus-gali* were enumerated in 3 0.25 m^2 quadrats in every plot from first emergence after planting to full canopy closure (mid-May through mid-July). Total emergence and density at canopy closure was affected by herbicide and application timing. Emergence pattern was most affected by year even when accounting for thermal time.

A Single GST Identified Via an Association Mapping and Transcriptomics Approach Endows High Levels of Resistance to Atrazine in *Amaranthus palmeri*. Shiv S. Kaundun*; Syngenta, Bracknell, United Kingdom (202)

A Single GST Identified via an Association Mapping and Transcriptomics Approach Endows High Level of Resistance to Atrazine*A. palmeri* has become one of the most troublesome weeds in the Americas due in part to resistance evolution to multiple herbicide modes of action. Whilst target site resistance to a number of herbicides is well characterized in the species, the genes involved in non-target site resistance remain elusive so far due to the lack of quality genomes and mapping populations from carefully conducted crosses between susceptible and resistant individuals. In this study, we have employed an alternative transcriptomics and association mapping strategy to determine the non-target-site gene(s) linked to atrazine resistance in *Amaranthus palmeri*. RNA-seq experiments were conducted on nine sensitive and 10 atrazine-resistant US populations before and after atrazine application. Candidate genes from the transcriptome analysis were further validated via independent QPCR assays and heterologous expression in *E. coli* and tobacco. Only a single glutathione-*S*-transferase (GST) gene belonging the Phi-class, denoted ApGSTF1, was constitutively over-expressed in resistant (several 100-folds) as compared to sensitive populations. The single GST was further confirmed to confer very high levels of resistance to atrazine in both the GST-transformed *E. coli* and tobacco experiments. ApGSTF1 also endowed resistance to two other chlorotriazine herbicides but not to the other PSII herbicides tested.This study identified the first fully validated NTSR gene in the increasingly problematic *Amaranthus palmeri* species. It also demonstrates the power of association mapping to determine the most commonly present gene involved in NTSR among several resistant populations as opposed to gene identification in single resistant parents using a classical linkage mapping tactic. deepak.kaundun@syngenta.com

Utilizing Thermal Requirements of California Weedy Rice (*Oryza sativa spontanea*) to Predict Seedling Emergence Timing Under Field Conditions. Liberty B. Galvin*[1], Whitney Brim-DeForest[2], Kassim Al-Khatib[1]; [1]University of California, Davis, Davis, CA, [2]University of California, Davis, Yuba City, CA (203)

Post-emergent control applications are being developed for weedy rice (*Oryza* spp.) in California rice cropping systems. Identification of this pest is difficult due to the conspecificity with cultivated rice. Therefore, species-specific biological knowledge should be utilized to optimize application timing to ensure maximum efficacy. Rice fields west of the University of California Davis campus were hand-seeded in rows with California weedy rice accessions 1, 2, 3, and 5 in both 2019 and 2020 to simulate an infested weed seed bank. Fields were flooded the same day as seeding; seeds were not primed beforehand. Seedlings were removed from the soil upon emergence and burial depth was noted daily for 28 days after flooding. Soil temperature was used to calculate thermal time to emergence for all accessions. Accessions 1 and 5 required twice as much thermal accumulation (growing degree days, Celsius days) to initiate emergence compared with accessions 2 and 3 in 2019. In the same year, accession 3 required only 84 C days to reach 90% emergence, while accessions 1 and 2 required ~123 C days, and accession 5, 157 C days. In 2020, accession 5 required 3 times more thermal accumulation to initiate emergence compared with accessions 1, 2, and 3. Accession 2 and 3 required ~120 C days to reach 90% emergence in 2020, while accession 1 and 5 required 1481 and 222 C days, respectively. Four-parameter, log-logistic models were utilized, but had significant amounts of error and were not well-fit to the data collected (RMSE=4.264, AIC=4251 in 2019; RMSE=2.018, AIC=3152 in 2020). Incorporation of the flooding depth variable increased error, and consequently was not included in the analysis. Of the seeds counted, 98% emerged from 1 cm within the soil surface, so this is unlikely the source of variation in the data. Additional studies are needed to determine other environmental sources of influence, e.g., cloud cover at the time of flooding, over California weedy rice.

***In-vitro* Screening to Phenotype Cereal Rye (*Secale cereale*) Lines for Allelopathic Activity.** Democrito B. Rebong*[1], Ramon G. Leon[1], Chris Reberg-Horton[1], J. Paul Murphy[1], Virginia Moore[2], Steven Brian Mirsky[3], Lisa Kucek[4], Nancy Ehlke[5]; [1]North Carolina State University, Raleigh, NC, [2]Cornell University, Ithaca, NY, [3]USDA ARS, Beltsville, MD, [4]USDA Agricultural Research Service, Madison, WI, [5]University of Minnesota, St. Paul, MN (204)

Weeds are among the most important challenges in organic production due to limited effective control tools, the need for frequent control, and the high cost of fuel and labor. The use of cover crops in organic systems can improve weed control through competition with weeds and reducing their germination and emergence. Various research programs have conducted screenings for allelopathy and competition in different cereal crops with successful integration of allelopathy and competition research in rice (*Oryza sativa* L.), common wheat (*Triticum aestivum* L.), and barley (*Hordeum vulgare* L.). For rye (*Secale cereale* L.), old and new cultivars were developed for grain yield, forage, human food, bioethanol production, and biogas production. There have been no cultivars developed specifically for allelopathic potential and weed suppression. In this study, we screened 100 lines derived from crosses with commonly planted rye varieties 'ND Gardner' and 'Aroostook', using equal-compartment-agar method (ECAM) at the laboratory. Results showed that 19 lines from ND Gardner crosses and 3 lines from Aroostook crosses demonstrated high allelopathic activity with good to excellent weed control based on visual observations of symptoms (chlorosis, stunting and necrosis) on a percent scale compared to cereal rye-free control. These selected materials will be further tested in the field during the 2023 planting season. The ECAM is a simple and rapid procedure that can be very useful in evaluating allelopathy in rye genotypes under laboratory conditions which can be applied in a large screening program.

Demographic Implications of Self-fertilization and Migration for the Maintenance of Genetic Variation and Successful Adaptation to New Environments in Annual Weeds. Samuel R. Revolinski*, Ian Burke; Washington State University, Pullman, WA (205)

High rates of self-fertilization have long been associated with weediness in plants, especially downy brome (*Bromus tectorum*). However, the recent emergence of extremely problematic cross-fertilizing weeds indicates that there may be situations where cross-fertilization is advantageous for weedy species. Theoretically, complete self-fertilization reduces effective population sizes by one half, thus reducing the genetic variation present in populations that can be used to adapt. However, near-obligate self-fertilizing plants have been successful in adapting to, and subsequently invading many environments, often having substantial amounts of standing genetic variation within a single locale of the invaded range. However, it is not understood how near-obligate or obligate self-fertilizing species can maintain genetic variation within locales or why self-fertilizing species have been so successful in invading many environments, especially marginal or disturbed ecosystems. A single locus, self-fertilizing, two-island Fisher-Wright forward genetic simulation with migration and a polygenic black-hole sink simulation with varying rates of self-fertilization was used to answer fundamental questions about the implications of self-fertilization on invasiveness of annual weeds. The Fisher-Wright simulation demonstrated that, with even low levels of migration between locales and differential selection on the allelic state within locales, genetic variation could be maintained indefinitely within locales. The black-hole sink simulation indicated that self-fertilization does not impede adaptation compared to cross-fertilization in marginal environments, but that in rich environments where adult plants produced many off-spring, cross fertilization increased the likelihood of adaption by weedy species to the new environment. This study supports in cooperating methods to prevent seed spread in best management practices and informing biosecurity monitoring using self-fertilization frequencies.

Mapping of Metabolic Resistance to Tembotrione in Palmer Amaranth. Carlos Alberto Gonsiorkiewicz Rigon*[1], Jacob S. Montgomery[1], Anita Küpper[2], Roland S. Beffa[3], Franck E. Dayan[1], Todd A. Gaines[1]; [1]Colorado State University, Fort Collins, CO, [2]Bayer AG, Frankfurt, Germany, [3]Senior Scientist Consultant, Frankfurt, Germany (206)

Amaranthus palmeri is one of the most problematic weeds in the United States. 4-hydroxyphenylpyruvate dioxygenase (HPPD) inhibitors as tembotrione are widely used to control this species in corn systems. A tembotrione-resistant population of Palmer amaranth from Nebraska (NER) was characterized to have a faster hydroxylation than a susceptible population (NES). The objective of the study was to identify responsible genes and map the resistant trait in Pseudo-F2 populations. The hypothesis was that a cytochrome P450 gene is involved in the resistant trait and that a *trans*-regulatory element is involved in the increased metabolism mechanism. Two crosses of susceptible and resistant plants were performed to generate F1 and subsequent pseudo-F2 populations. Dose-response curve using increasing doses of tembotrione was performed with parental resistant NER, susceptible NES, and F1 plants to determine a delimiting rate that best separate S from R. Segregation in the pseudo-F2 population was measured in response to tembotrione at 77 g a.i. ha^{-1}, in cross A and B. RNAseq and genotype-by-sequencing (GBS) were used on individuals from these segregating populations to identify candidate genes and quantitative trait loci (QTL), respectively. For QTL mapping, 382 plants were selected, 181 most susceptible, 150 most resistant plants, and parental plants. Plant tissue was collected before herbicide application and had DNA extraction using CTAB method. GBS libraries were made with ApeKI sequenced using NovaSeq S4 with 150 bp paired-end reads (Illumina). The sequence was performed at the University of Minnesota Genomics Center. Whole-genome sequencing reads were aligned to the male reference genome for Amaranthus palmeri (id55760) using BWA. Variant calling was performed using GATK 4.0 and filtered. Based on RNAseq analysis and subsequent validation in yeast as a heterologous system, *CYP72A219-like* gene was identified as responsible for the tembotrione metabolism. The segregation results demonstrate that the trait is multigenic. Future work will focus on functionally validating new candidate genes that are involved in *CYP72A219-like* regulation based on QTL.

Coupling Seedbank Management and Cultivation Tool Stacking for Improved Weed Control in Organic Vegetables. Margaret R. McCollough*[1], Rebecca Champagne[2], Eric R. Gallandt[1], Daniel C. Brainard[3]; [1]University of Maine, Orono, ME, [2]The University of Maine, Orono, ME, [3]Michigan State University, East Lansing, MI (207)

The combined effects of reducing the germinable weed seedbank and increasing physical weed control efficacy has the potential to improve weed management on organic farms. Physical weed control (PWC) remains a principal management strategy for organic farmers; however, efficacy is often low and variable. In an organic vegetable weed management systems trial, two levels of seedbank management were assessed in a factorial design with two PWC treatments. Seedbank management was either omitted or implemented using silage tarps followed by flame weeding, while cultivation treatments consisted of finger weeders as a single tool or stacked with hilling discs, torsion weeders, or hoe ridgers. Experiments were replicated over three years in Old Town, Maine, and Holt, Michigan, and used bush bean, table beet, and sweet corn as test crops. Results from the Maine site show that stacking finger weeders and hoe ridgers consistently reduced intra-row weed stands by 91% compared to 70% when finger weeders were used alone. Tool stacking with hoe ridgers also reduced intra-row weed recruitment 14 days after the last cultivation, resulting in 10 weeds m^{-2} compared to 45 weeds m^{-2} when hoe ridgers were not used. When finger weeders were stacked with torsion weeders, intra-row efficacy increased from 74% to 96% in one out of two cultivation events performed. Improved efficacy was not observed when tool stacking with hilling discs. Tarping for six weeks prior to planting the crop and pre-emergence flame weeding reduced early-season weed densities from 90 weeds m^{-2} to fewer than 20 weeds m^{-2}. In plots receiving seedbank management, the germinable weed seedbank decreased from 4,100 seeds m^{-2} in year one to 2,000 seeds m^{-2} in year three. In plots receiving no seedbank management, the seedbank increased from 3,900 seeds m^{-2} in year one to 11,200 seeds m^{-2} in year two. However, these trends were not sustained; regardless of seedbank management treatment, final seedbank assays collected in year four, following three years of treatment, possessed seedbank densities equivalent to those observed at the start of the trial. Partial budget analysis showed that the addition of seedbank management can result in higher variable costs; however, they were offset by greater crop yields, leading to higher net returns. Data from the Michigan site awaits analysis; however, results from Maine show that coupling tool stacking for increased efficacy and targeted weed seedbank to encourage fatal weed seed germination has the potential to both reduce weed densities and increase profitability in organic vegetable systems.

Phenotypic Variation Among Populations of the Invasive Weed *Parthenium hysterophorus* L. (Asteraceae) as a Tool to Assess Multiple Introductions. Maor Matzrafi*[1], Sahar Malka[2], Hanan Eizenberg[3]; [1]039904412, Ramat-yishay, Israel, [2]The Robert H. Smith Faculty of Agriculture, Food and Environment, The Hebrew University of Jerusalem, Jerusalem, Israel, [3]Department of Plant Pathology and Weed Research, Agricultural Research Organization – Volcani Institute, Newe-Ya'ar Research Center, Ramat-yishay, Israel (208)

Parthenium hysterophorus L. (Asteraceae) is a noxious weed that has invaded more than 50 countries worldwide. In Israel, *P. hysterophorus* was first detected in 1980 at Tirat-Zvi, in the north-eastern region. In recent years, there has been increasing concern about the spread of this weed in agricultural and non-agricultural habitats across the country. The different locations where *P. hysterophorus* was found might suggest multiple introductions into the Israeli flora from foreign sources. Differences in response to environmental conditions among *P. hysterophorus* populations and transgenerational variation may reinforce the hypothesis of multiple introductions. Therefore, phenotypic variation in germination and phenology of field and progeny *P. hysterophorus* populations was studied. Seeds were collected from field populations in five locations with varying climatic and environmental conditions, and pollen-exclusion cages were used to produce the progeny populations. Seed germination of field and progeny populations was examined at constant temperatures (10, 15, 20, 25, 30 and 35 °C) and water potentials (0, -0.2, -0.4, -0.6, -0.8 and -1 MPa at 20°C optimal temperature). Plants were grown in a greenhouse with controlled environmental conditions, and development was evaluated over a period of three months. According to the temperature model developed, seed germination occurred from 10 °C to 30 °C, with variation in optimal germination temperatures for field (19.5-21 °C) and progeny (20-22.5 °C) populations. Germination was recorded for all water potentials except -1MPa. Differences in germination capabilities under negative water potentials were observed among generations and populations. Modelling life cycle traits revealed significant differences in phenological characteristics among field and progeny populations. Phenotypic variation within and among generations and populations of the invasive weed *P. hysterophorus* can be used to assess the number of local introductions from foreign countries. By understanding the biology and phenology of different populations of *P. hysterophorus*, weed management efforts can be improved to control this invasive plant.

EPA's Plant Assessment Tool: A Non-Target Plant Exposure Model Used in Endangered Species Assessments. Melissa Bridges*[1], Elizabeth Donovan[1], Gretchen E. Dykes[1], William P. Eckel[1], Frank T. Farruggia[1], Jerrett Fowler[1], Kristina Garber[1], Brian Kiernan[1], Stephan Miller[2], Chuck Peck[1], Dana Spatz[1], Holly Summers[1], Peter Tellez[1], Dirk Young[1]; [1]US Environmental Protection Agency, Washington, DC, [2]US Environmental Protection Agency, Washington, VA (209)

The U.S. Environmental Protection Agency (EPA), Office of Pesticide Programs (OPP), Environmental Fate and Effects Division (EFED) evaluates the potential risk to non-target plants and animals from use of conventional pesticides, including the potential impact on federally endangered or threatened (listed) species. EFED developed the Plant Assessment Tool (PAT) to estimate pesticide exposure to non-target plants in three ecological systems representing upland terrestrial, wetland, and aquatic habitats. PAT is a tool used to support the evaluation of whether pesticide exposure could affect listed plants, their critical habitats, and habitat for taxa that depend on plants. PAT is used in conjunction with the Pesticide in Water Calculator (PWC) to incorporate pesticide-specific environmental fate and transport data to derive estimated environmental concentrations (EECs) in each of the three habitat types (aquatic, wetland, and terrestrial). EECs from each habitat type can be compared to multiple plant toxicity endpoints, allowing for a more extensive characterization of a pesticide's potential impact to non-listed and listed plants species as compared to EFED's older plant exposure model, TerrPlant. Toxicity endpoints used by PAT can be from individual test species or species sensitivity distributions. PAT compares terrestrial and wetland habitat EECs (expressed as lbs a.i./A) to endpoints derived from terrestrial plant seedling emergence (OCSPP guideline 850.4100) and vegetative vigor (OCSPP guideline 850.4150) studies or other suitable studies. PAT compares wetland water column and aquatic habitat EECs (expressed as µg a.i./L) to aquatic plant toxicity endpoints, including *Lemna* spp. (OCSPP guideline 850.4400), algae (OCSPP guideline 850.4500), and cyanobacteria (OCSPP guideline 850.4550). This presentation will provide a summary of the conceptual models for the three habitats, modeling approaches, and plant toxicity considerations in PAT.

EPA's Consideration of Pesticide Use, Mitigating Environmental Exposure, and Updating Labels in Endangered Species Evaluations. Frank T. Farruggia*[1], Amy Adams[2], Jennifer Connolly[3], Kristina Garber[1], Stephen Muela[4], Tracy Perry[3]; [1]US Environmental Protection Agency, Washington, DC, [2]United States Environmental Protection Agency, Washigton, DC, [3]United States Environmental Protection Agency, Washington, DC, [4]United States Environmental Protection Agency, Washtington, DC (210)

The U.S. Environmental Protection Agency (EPA), Office of Pesticide Programs (OPP), Environmental Fate and Effects Division (EFED) evaluates the potential risk to non-target plants and animals from use of conventional pesticides, including the potential effects on federally endangered or threatened (listed) species and designated critical habitats (CH). In the assessment process, EPA considers labelled pesticide use site locations, usage information from surveys and censuses, chemical properties, and species life history information to characterize the potential exposure to listed species and CH. These refinements inform EPA's individual-level effects determinations and predictions of likelihood of potential jeopardy to populations or adverse modification (J/AM) to CHs. Depending on the results of the assessment, mitigations may need to be identified. Potential mitigations vary by chemical, use site, application method, and primary exposure routes (e.g., spray drift, runoff), and mitigations can vary in scale (geographically localized or nation-wide). This presentation will provide a summary of the elements of EPA's evaluation when considering pesticides and their potential effects to listed species and CHs.

Public Comment Periods: How Can WSSA Help EPA? Part 1. Kelly V. Tindall*, Bradley Kells; US Environmental Protection Agency - Office of Pesticide Programs, Washington, DC (211)

This presentation will help stakeholders understand the kinds of information they can provide in public comments that will inform pesticide regulatory decisions, and how to provide that information. Presenters will provide an overview of the regulatory decision processes for pesticides highlighting opportunities for public participation, including when comment periods occur and how to submit comments. The presentation will explain the type of data or information that are used in human health and ecological risk assessments (including guideline studies, which are essentially protocols for required studies used to inform risk assessments), the type of data or information useful in assessing the benefits of a pesticide, and the data quality considerations that determine if and how the data are used.. Finally, this presentation will demonstrate how to determine where a pesticide is in the Registration Review process and how to navigate Regulations.gov. Due to the amount of content presented, the material will be divided in two presentations.

Public Comment Periods: How Can WSSA Help EPA? Part 2. Bradley Kells*, Kelly V. Tindall; US Environmental Protection Agency - Office of Pesticide Programs, Washington, DC (212)

This presentation will help stakeholders understand the kinds of information they can provide in public comments that will inform pesticide regulatory decisions, and how to provide that information. Presenters will provide an overview of the regulatory decision processes for pesticides highlighting opportunities for public participation, including when comment periods occur and how to submit comments. The presentation will explain the type of data or information that are used in human health and ecological risk assessments (including guideline studies, which are essentially protocols for required studies used to inform risk assessments), the type of data or information useful in assessing the benefits of a pesticide, and the data quality considerations that determine if and how the data are used.. Finally, this presentation will demonstrate how to determine where a pesticide is in the Registration Review process and how to navigate Regulations.gov. Due to the amount of content presented, the material will be divided in two presentations.

The Importance of Herbicide Resistance Information for Regulatory Agencies. Cameron H. Douglass[1], John Orlowski*[2]; [1]US Department of Agriculture, Office of the Chief Economist, Office of Pest Management Policy, Washington, DC, [2]US Environmental Protection Agency, Office of Pesticide Programs, Biological and Economic Analysis Division, Washington, DC (213)

In the United States, pesticides, such as herbicides, are regulated by the Environmental Protection Agency (EPA), using guidelines and policies stipulated by applicable federal laws and codified in the Code of Federal Regulations. The Agency relies on a variety of information to inform regulatory decisions. One consideration is the benefit of an herbicide taking into account existing weed resistance to the herbicide or to alternative herbicides. EPA's ability to appropriately account for the resistance-related benefits and/or risks of a given herbicide depends on the availability of relevant data and information from registrants and the scientific community alike. Important sources of information on herbicide resistance include peer-reviewed published research, online databases, Extension weed control guides, and registrant-provided reports, when available. Recently EPA has also considered real-time, in-season reports on herbicide resistance, including information on Extension crop blogs and social media outlets, to better understand the benefits and risks of herbicide in relation to herbicide-resistant weeds. These up-to-date Extension communications are becoming increasingly critical sources of information for EPA on the status of herbicide resistance across the US, and help to inform regulatory decisions.

Climate Impacts of Herbicide Regulations: Case Studies with Cover Cropping and No-till. Laura Dodson[1], Fengxia Dong[2], Cameron H. Douglass*[3], Rebecca Nemec[4], Ryan Olver[2], Michelle Ranville[3]; [1]US Department of Agriculture, Economic Research Service, Washington, DC, [2]US Department of Agriculture, Economic Research Service, Kansas City, MO, [3]US Department of Agriculture, Office of the Chief Economist, Office of Pest Management Policy, Washington, DC, [4]US Department of Agriculture, Office of the Chief Economist, Washington, DC (214)

Herbicides are important tools used by US growers that help to facilitate the adoption of conservation practices such as no- and reduced-tillage, and cover cropping. There has been particular interest recently in understanding the potential impacts - including with regards to carbon sequestration benefits - of regulatory changes and restrictions on grower adoption and implementation of these conservation practices. Current data describing interactions between cover cropping and no- and reduced-tillage adoption amongst US field corn and soybean growers and the use by those growers of 2,4-D, atrazine and/or glyphosate will be discussed in this talk. A preliminary modeling approach aiming to quantify the impacts on conservation practice adoption and estimated carbon sequestration benefits of possible future regulatory changes for these herbicides will also be presented.

Pesticide Task Forces: Role, Function, and Benefits. Montague U. Dixon*; Syngenta Crop Protection, Greensboro, NC (215)

ESA And FIFRA: A Regulatory Perspective on Successes and Challenges. Reuben D. Baris*; Corteva Agriscience, Indianapolis, IN (216)

On January 11, 2022 the US EPA approved amended registrations for Enlist One and Enlist Duo herbicides set to expire in January 2022, providing certainty for growers in the 2022 growing season. With EPA's decision to grant a seven-year amended registrations they announced a new process for registering new actives setting a new standard for pesticide registrations in the U.S. while ensuring the continued protection of endangered species. There has been a steady shift over the last few decades by the US EPA to evolve regulatory standards for evaluating pesticide registrations in the face of increasing litigation pressure. New policies implemented by US EPA shift the paradigm for pesticide evaluations. The Enlist registration decisions, and more recent evaluations can be viewed as case studies for collective renewed focus on the process and to chart an efficient path for endangered species protection, agriculture, and pesticide regulations.

Types of Registrations Under FIFRA - Section 3, Special Local Needs, Section 18 Emergency Use: What They Are and How They Are Utilized. Rachel Hardie*; Syngenta Crop Protection, Greensboro, NC (217)

The Purpose and Benefits of Regulatory Task Forces EPA uses the best available science and data to build models and conduct risk assessments to determine if a pesticide and its application will cause any unreasonable adverse effects to human health and the environment. Supplying EPA with the best available data falls on the shoulders of the pesticide registration owners. Task forces have been an excellent tool for industry to work together and provide EPA with a wide range of data – from toxicity studies for one specific active ingredient to exposure endpoints that are applicable to all active ingredients. Regulatory task forces have numerous benefits, both for industry and EPA, including reduced financial costs, more efficient review process, and uniform decision making.

Paraquat Dichloride Interim Registration Review Decision Update. Jordan Moseley*; Syngenta Crop Protection, LLC, Greensboro, NC (218)

The US Environmental Protection Agency (EPA) published the final interim decision for paraquat dichloride registration review on August 2nd, 2021. EPA reviews each registered pesticide every 15 years to ensure that the pesticide can carry out its intended function while not causing unreasonable adverse affects to human health and the environment. The registration review process considers the best available science and latest data to make informed decisions regarding the use and registration status of each pesticide. For paraquat dichloride, the public docket for registration review was opened in December 2011. The August 2021 final interim decision marked the culmination of nearly ten years of work assessing the registered uses, risks, and benefits of paraquat. In the final interim decision, EPA identified potential human health risks of concern to occupational handlers mixing, loading, and applying paraquat for various use scenarios. Potential post-application risks to workers and risks to bystanders from spray drift were also identified. In addition, the Agency noted potential ecological risks to mammals, birds, terrestrial invertebrates, terrestrial plants, as well as some aquatic invertebrates and some aquatic plants. As a result, several changes have been required to product labels to address the potential risks to human health and the environment and to implement more generic label requirements for herbicide products.

Regulatory Update on Atrazine. Mark White*; Syngenta, Macon, MO (219)

Atrazine is the foundation of many weed control programs in field corn, sorghum, sugarcane, sweet corn, and certain specialty crops. Since 1958, atrazine been essential for high-yield agriculture, and it is critical to Integrated Pest Management and sustainable practices such as conservation tillage and no-till. Atrazine is used in more than 60 countries, and in the United States it is critical for controlling devastating pest weeds in the Midwestern, Southern and South-eastern growing regions. Atrazine is contained in more than 70 products, many of them mixtures with multiple herbicides, to control the most difficult weeds that infest farmlands. As we continue to see the increased development of weeds with resistance to certain herbicide modes of action, it is even more important than ever to maintain atrazine as a tank mixture partner for many other herbicides.Atrazine is currently undergoing a registration review re-evaluation by EPA as part of the 15 year registration review process. The agency has proposed significant changes to the atrazine label based in part on the revised (draft) Conservation Equivalent Level of Concern (CE-LOC) of 3.4 ppb over a 60 day rolling average. The proposed label changes are also being driven, in part, by the Endangered Species Act (ESA) and the current ESA consultation process with the Fish and Wildlife Service (FWS) and the National Marine Fisheries Service (NMFS). Proposed changes to the label include watershed based rate restrictions, a pick list of 12 mitigation practices, and additional restrictions based on environmental conditions. These changes could have significant impacts on effective weed management in the four major labeled crops (field corn, sweet corn, grain sorghum, and sugar cane). EPA has decided to convene a Scientific Advisory Panel (SAP) in 2023 to consider the appropriateness of the proposed CE-LOC and the derivation process surrounding it. The Revised Interim Decision will not be finalized until the conclusion of the SAP.

Persistent Herbicides in Compost: A Regulatory Pathway. Travis Bui*; Corteva Agriscience, Indianapolis, IN (220)

Composting is the biologic process that converts organic materials into nutrient-rich soil amendments through natural decomposition. Raw ingredients for compost include a variety of organic materials such as grass, hay, plant debris and/or animal manures. Herbicides, including synthetic auxins, have been safely used by farmers, ranchers and lawn care operators and homeowners for decades; however, some herbicides are not quickly degraded by normal, commercialized composting processes. Unintended exposures may arise from off-label herbicide use, which may result in residues entering compost feedstuffs, or from composting of treated plant materials when composters are unaware of previous herbicide use. Current guideline non-target plant (NTP) studies (vegetative vigor and seedling emergence), which are used to support herbicide product registrations, do not represent the route of exposure from contaminated compost. Following EPA guidance, Corteva developed non-target, compost route-of-exposure dose-response study designs to generate toxicity endpoints that are appropriate for use in EPA regulatory risk assessments.

Introduction to the Center for Excellence in Regulatory Science in Agriculture at North Carolina State University. Keith L. Edmisten*, Danesha Seth Carley; North Carolina State University, Raleigh, NC (221)

The Economic, Nutritional, and Environmental Benefits of International Cooperation for the Establishment of MRLs on Specialty Crops. Dirk C. Drost*; Minor Use Foundation, Inc, Raleigh, NC (222)

This presentation will shine a light on the economic, environmental, and nutritional benefits of international research cooperation for the establishment of MRLs on specialty crops. Specialty crops are often high-value crops that have cultural importance, or are important for trade and economic growth in developing countries. They are also often overlooked in R&D investments in agriculture since they represent less of a percentage of total tonnage and acreage. Through innovative partnerships with pesticide manufacturers, universities, international organizations, and governments around the world, the Minor Use Foundation conducts research to enable access to crop protection solutions for these important commodities. More broadly, the Foundation promotes international cooperation and information-sharing to reduce redundancy in the pesticide-registration process. (i.e., joint international reviews, global zoning work, aligned trade standards) and builds a network of pre-eminent experts in pesticide science to create an inclusive, robust global dialogue on pesticide risk assessment, risk management, and risk communication. This presentation will explore the work of the Minor Use Foundation and its global impact.

Influence of *Sorghum halepense* (Johnsongrass) Density on Early Growth of *Zea mays* (Corn). Ryan P. O'Briant*[1], Antonio DiTommaso[2], Lynn M. Sosnoskie[3]; [1]Cornell University - Soil and Crop Sciences, Brooktondale, NY, [2]Cornell University, Ithaca, NY, [3]Cornell University, Geneva, NY (223)

Competitive and Hormetic Responses of Velvetleaf (*Abutilon theophrasti*) Historical Seed Collection. Sandra R. Ethridge*[1], Saket Chandra[1], Wesley Everman[1], Anna M. Locke[2], David L. Jordan[1], Micheal D. Owen[3], Ramon G. Leon[1]; [1]North Carolina State University, Raleigh, NC, [2]USDA-ARS, Raleigh, NC, [3]Iowa State University, Ames, IA (224)

Abutilon theophrasti Medik, or velvetleaf, is a weed of major economic importance in row crops, such as corn, cotton, and soybean. When in competition with row crops, *A. theophrasti* has negative impacts on crop height, leaf area index, biomass, and yield. Beginning in the 1980s and 1990s, producers used acetolactate synthase (ALS)-inhibiting herbicides and glyphosate to control *A. theophrasti* in agronomic systems. There have not been studies that seek to understand the multigenerational changes that are occurring in *A. theophrasti* populations that tracks the changes in competitive ability and increased tolerance to sublethal rates of glyphosate and ALS-inhibiting herbicides. The objective of this study was to understand morphological changes, competitive ability, and herbicide tolerance of a population of *A. theophrasti* from a seed collection spanning 28 years, from 1988 to 2016. A common garden experiment, a competition study, and a herbicide dose-response study were conducted with the historical seed collection. The results indicated that plant architecture and life-history traits of the studied *A.*

theophrasti population did not change over time, however, when in competition in biomass and number of leaves for each plant over time. Also, when newer year-lines exhibited higher competitive ability. Furthermore, each year-line after 1988 had a hormetic response to the lowest evaluated dose of glyphosate, and this hormetic response increased progressively from 1995 to 2016. This study shows that it is critical to examine the evolutionary potential of weed species to be able to anticipate the rate of change and also mitigate this process for the future.

Soybean and Palmer Amaranth Identification Using Machine Learning. Dhiraj Srivastava*[1], Vijay Singh[2]; [1]Virginia Polytechnic Institute and State University, Painter, VA, [2]Virginia Tech, Painter, VA (225)

Palmer amaranth (Amaranthus palmeri) is one of the most troublesome weeds in the US which interferes with soybean crop. Integrated weed management using spot spraying or selective placement practices is an effective approach for reducing herbicide usage and increasing weed control efficiency. Recently, emphasis has been placed on eco-friendly artificial intelligence (AI) -based weed robots and unmanned aerial spot sprayers (UAS). However, lack of quality data on crop and weed identification is the biggest obstacle to create reliable AI technology robust enough for mapping under different weather conditions and locations. Cost of image-labeling scales linearly with data size with only marginal increase in model performance. Therefore, image labelling process is not cost-efficient, while gathering unlabeled data is generally inexpensive. Semi-supervised learning offers to solve this problem by only requiring a partially labeled dataset and being label-efficient by utilizing the unlabeled images as well for learning process. The focus of this research is to present a new architecture based on self-supervised contrastive learning to detect Palmer amaranth in soybean. This research was conducted in 2022 at Painter, VA, where Unmanned Aerial Systems (UAS)-based red, green and blue (RGB) images were acquired at different growth stages of soybean and Palmer amaranth using DJI M-300 flown at an altitude of 12 meters. Image data for object detection models were annotated using 'LabelImg'. Considering the tradeoff between speed and accuracy, object detection method 'You Only Look Once' (YOLOv6) was found to be a suitable model for real-time deployment with precision and recall of Palmer amaranth greater than 80% at mean average precision (mAP) of 0.5 with an average inference speed of 6.75 milliseconds tested on the device NVIDIA GeForce RTX 2080 GPU. These findings have the potential to advance weed management through robots and UAS, reducing herbicide usage and associated environmental impacts, leading to enhanced economic viability of farming operations.

Do Cover Crops Affect Weed Community Structure? Uriel D. Menalled*[1], Stephane Cordeau[2], Antonio DiTommaso[1], Richard G. Smith[3], Matthew R. Ryan[1]; [1]Cornell University, Ithaca, NY, [2]INRAE, Lyon, France, [3]University of New Hampshire, Durham, NH (226)

Cover crops are multifunctional tools that can suppress weeds while providing other important ecosystem services. Two field experiments were conducted across four site-years in New York to test the effect of cover crops on weed community structure. The first experiment compared four fall-planted cover crop treatments (canola, cereal rye, hairy vetch, and a hairy vetch × cereal rye mix) to a control treatment that received tillage. The second experiment was identical, except it compared four spring-planted cover crop treatments (buckwheat, sorghum sudangrass, sunn hemp, and a sunn hemp × sorghum sudangrass mix) to a tilled control. To test if crop-weed competition is affected by the relatedness of cover crops to weeds, the phylogenetic distance between species in each treatment was compared. Results show that cover crop competition can reduce the overall phylogenetic distance of weed communities, most strongly suppressing phylogenetically related weeds. However, planting timing affected the ability of cover crops to filter weed phylogenies because while all cover crops in this experiment led to meaningful reductions in weed biomass, only over-wintering species changed the phylogenetic spread of weed communities relative to the controls. Cover crops that were planted in the fall and overwintered were well established by the spring when summer annual weeds emerged, suggesting that there is a temporal component to phylogenetic filtering from crop-weed competition. Thus, our results demonstrate that asymmetric crop competition during weed establishment can influence weed community structure, where closely related species are most strongly filtered out of the community. This research has implications for the targeted management of problematic weed species, including herbicide resistant biotypes, and it contributes to a larger body of research on using cover crops for sustainable weed management.

Varietal Tolerance of Cucurbits Crops with Postemergence Applications of *S*-Metolachlor. Carrie J. Mansue*[1], Thierry E. Besancon[2], Lynn M. Sosnoskie[3], Kurt M. Vollmer[4]; [1]Rutgers University, Mays Landing, NJ, [2]Rutgers University, Chatsworth, NJ, [3]Cornell University, Geneva, NY, [4]University of Maryland, Queenstown, MD (227)

Cucurbits are economically important vegetable crops in the mid-Atlantic and Northeast regions of the United States. Weeds can cause significant cucumber and summer squash yield reduction when allowed to compete with these crops. Season-long interference by various broadleaf and grassy weed species can reduce cucumber and summer squash yield depending on annual weather conditions, weed species, and planting density. S-metolachlor is a very long chain fatty acid (VLCFA) biosynthesis inhibiting herbicide (WSSA Group 15) that provides selective control of many annual grassy weeds as well as some small seeded broadleaf weeds and is effective on suppressing yellow nutsedge. Considering the limited and sometimes contradictory information on summer squash, cucumber and pumpkin tolerance to S-metolachlor applied PRE or early POST as well as the limited number of effective residual herbicides in cucurbits, greenhouse studies were conducted in Delaware, New Jersey and New York to evaluate crop tolerance on five cultivars on each species in response to four rates of S-metolachlor applied POST after plants have reached the two-leaf stage. The experimental design was a two-factor factorial consisting of cucurbit variety and S-metolachlor rate with four replications. S-metolachlor rates consisted of 800

(1/2X), 1,600 (1X), 3,200 (2X) and 6,400 (4X) g ha-1 applied postemergence when plants reached the two-leaf stage. A nontreated control was also included. Crop injury was rated by scoring crop canopy for leaf injury (necrosis, chlorosis, and leaf curl) and general stunting on a scale of 0% to 100%. For squash and cucumber studies, the surface area of the third-emerged leaf was analyzed using Leaf-IT and vegetation was collected for dry weight measurements. In pumpkin trials, there was no significant interaction with variety and rates of S-metolachlor at 1x or 2x. In comparing the cucumber and squash trials when apply rates of S-metolachlor at 1x, 2x, and 4x, cucumber varieties showed the most sensitivity in stunting, leaf area and dry weight measurements. Cucumbers ranged from 22% to 66% stunting as compared to squash ranging between 7% to 9%. Leaf area and dry weight data of cucumber varieties ranged from 25% to 73% leaf area and 46% to 87% dry weight compared to squash ranging from 110% to 113% leaf area and 106% to 120% dry weight, all comparable to the untreated control. The least sensitive variety of cucumbers was Diamondback. Diamondback at all rates of S-metolachlor showed lower injury compared to the other four varieties used in the trials.

Neutral Weed Communities: the Junction Point Between Crop Productivity, Biodiversity, and Ecosystem Services. Marco Esposito*[1], Valerio Cirillo[2], Albino Maggio[2], Antonio DiTommaso[3]; [1]Cornell University & University of Napoli Federico II, Ithaca, NY, [2]University of Naples Federico II, Naples, Italy, [3]Cornell University, Ithaca, NY (228)

Weeds are a fundamental component of agroecosystems and, if not appropriately managed, can cause severe crop yield losses. Considering the increase in environmental and agronomic concerns associated with herbicide application and soil tillage, it is necessary to identify new weed management solutions. We propose the promotion of *neutral weed communities*, which are weed communities that coexist with crops and do not negatively affect their yield and quality. Establishing neutral weed communities can enable reduced herbicide use and soil tillage while enhancing ecosystem services and biodiversity. We report scientific evidence of neutral weed communities and survey ecological explanations for why different weed communities have different effects on crop production. We also propose two weed management approaches to obtain neutral weed communities. The first approach aims to maximize weed biodiversity using traditional approaches such as cropping system diversification and integrated weed management. Higher weed biodiversity is associated with lower dominance of competitive weed species that reduce crop yield. The second approach relies on modern tools such as robots and biotechnology to manipulate the density of specific weed species. This approach can remove highly problematic species and minimize niche overlap between the weeds and crop. Given the complexity of interactions between crops, weeds, and other components of the agroecosystem, we highlight the need for multidisciplinary research to illuminate mechanisms determining the neutrality of weed communities.

Optimizing Impregnated Pyroxasulfone for Cotton. Brock A. Dean*, Charlie W. Cahoon, Zackary R. Taylor, Jose H. de Sanctis, Jacob Forehand, Hunter Lee; North Carolina State University, Raleigh, NC (229)

Palmer amaranth (*Amaranthus palmeri*) is a troublesome weed now confirmed resistant to 9 herbicide modes of action. Moving forward, cotton farmers will need all available tools to successfully manage herbicide-resistant Palmer amaranth. One such tool is the use of granular fertilizers impregnated with residual herbicides. During 2022, two separate experiments were established near Rocky Mount and Clayton to determine the optimal rate of granular ammonium sulfate (AMS) impregnated with pyroxasulfone and the optimal timing for use in cotton. Cotton cultivar 'DP2115 B3XF' was planted on 11 May and 12 May at Rocky Mount and Clayton, respectively. For the rate study, AMS rates included 161, 214, 292, 321, 374, 428, and 482 kg ha^{-1}, equivalent to 34, 45, 56, 67, 79, 90, and 101 kg N ha^{-1}, respectively. Pyroxasulfone was impregnated on granular AMS to achieve a final herbicide rate of 118 g ai ha^{-1} across all AMS rates. For the timing study, 118 g ai ha^{-1} pyroxasulfone was impregnated on granular AMS applied at a rate of 321 kg ha^{-1} (67 kg N ha^{-1}) to 5- to 7-leaf, 9- to 11-leaf, and 1st bloom cotton. For both studies, weed control and cotton tolerance to impregnated pyroxasulfone was compared to pyroxasulfone (118 g ai ha^{-1}) applied postemergence over-the-top (POST) and postemergence-directed (POST-DIR). A nontreated was also included for comparison. Prior to pyroxasulfone applications, all plots (including the nontreated) were kept weed free using multiple applications of glyphosate + glufosinate; no residuals were used prior to pyroxasulfone applications. Visual estimates of Palmer amaranth control and cotton injury were collected bi-weekly until 70 days after application (DAA) and Palmer amaranth density and biomass were collected prior to cotton defoliation. At the conclusion of the season, cotton was machine harvested and weighed to determine yield. For both studies, cotton injury was minimal outside of some foliar necrosis in response to some AMS sticking to damp foliage. For the rate study, Palmer amaranth was controlled 58 to 89% and 63 to 86% 70 DAA at Clayton and Rocky Mount, respectively. Furthermore, there were no differences in late season Palmer amaranth control across AMS rates and all pyroxasulfone impregnated AMS treatments performed similarly to pyroxasulfone applied POST and POST-DIR. For the timing study, 70 DAA, Palmer amaranth was controlled 73 to 90% and 82 to 99% at Clayton and Rocky Mount, respectively. Like the rate study, there were no differences in late season Palmer amaranth control across timings nor when comparing pyroxasulfone impregnated AMS to pyroxasulfone applied POST or POST-DIR. For both studies, cotton yield among plots receiving pyroxasulfone impregnated on AMS, POST, or POST-DIR were similar regardless of AMS rate or timing.

Anaerobic Soil Disinfestation: A Catalyst for Organic Watermelon Production in South Carolina? Sohaib Chattha*, Matthew A. Cutulle; Clemson University, Charleston, SC (230)

Managing weeds and diseases in organic crop production is difficult. Palmer amaranth (*Ameranthus palmeri* S. Watts) and yellow nutsedge (*Cyperus esculentus* L.) are two problematic weed species in southeastern U.S. organic watermelon (*Citrullus lanatus*) production. Weed control options in organic watermelon production are limited and require non-chemical tactics. Anaerobic soil disinfestation (ASD) is an ecological alternative to chemicals and is an effective method to suppress soil-borne diseases and weeds in various crops. Two experiments were conducted jointly at Clemson University Coastal Research and Education Center and United States Department of Agriculture Vegetable Laboratory, Charleston, South Carolina to study the impact of ASD for controlling weeds. The objective of first study was to evaluate the potential of standard carbon amendments to control weeds such as yellow nutsedge and palmer amaranth and to check reactivity of 20 commercial and/or USVL watermelon cultivars and rootstock in ASD in a hoophouse. The standard carbon amendments were chicken manure (CM) + molasses (M). Weed control percentage, weed count, plant vigor, plant height and plant biomass data were recorded. CM+M significantly improved soil anaerobic conditions compared to control. The carbon amendments treatment effectively controlled weeds by 90-95% compared to control only 10-20%. Four watermelon cultivars: Extazy, powerhouse, exclamation and sangaria showed highest plant vigor and plant biomass. The results suggested that ASD incorporated with CM+M treatments can effectively control weeds in organic watermelon production. Future research will focus on evaluating the performance of these treatments in field studies.

Weed Potential of Roll-crimped Cereal Rye Differs Across Cultural Practices. Laurel Wellman*[1], John M. Wallace[2]; [1]Penn State University, State College, PA, [2]Penn State University, University Park, PA (231)

Incomplete cereal rye (*Secale cereale* L.) termination with the roller crimper can generate a weed management problem for reduced tillage organic growers. Cereal rye is commonly used for weed suppression post-termination as a surface mulch. Organic growers rely on the roller crimper to effectively terminate rye while leaving the shoots and roots intact to promote mulch longevity and maximize weed suppression throughout the subsequent cash crop. Rye termination using the roller crimper is phenology-based, and roll crimping is most effective at anthesis. Cultural practices, such as adjusting seeding rate, sowing arrangement, and fall fertility can influence groundcover and biomass potential of cereal rye, but little is known about how these practices affect cereal rye termination efficacy when using a roll-crimper. Specifically, these practices may alter the evenness of the cereal rye stand, compromising the grower's ability to target roll crimping at anthesis. Although the trigger for roll crimping is phenology-based, other characteristics of the rye at the time of roll-crimping may also influence termination efficacy, such as the height and diameter of tillers. Two complementary field experiments were initiated in the fall of 2021 to assess cultural practices for improving cereal rye performance within CCORNT (cover crop-based, organic rotational no-till) soybean production. The first experiment is a two-factor randomized complete block design with two rye sowing arrangements (grid sown vs. standard 19-cm rows), four seeding rates with an unseeded control (0, 31, 63, 126, 188 kg ha^{-1}), and three replications. The second experiment is a randomized complete block design with a split-plot treatment structure and five replications. Fall fertility treatments are imposed in the main plot (0, 3.4, 6.7 Mg ha^{-1} fall-applied poultry litter), and soybean planting strategy (planting green at boot stage followed by roll-crimping and planting at anthesis vs. roll-crimping and planting at anthesis) is imposed at the split-plot level. At the time of soybean planting, the distribution of the rye stand was assessed by measuring the height, diameter, and Zadoks growth stage of individual tillers. The evenness of the stand was assessed using Gini coefficients, a measure of inequality along a frequency distribution. In late June, rye regrowth was assessed by counting tillers, and in late July, the same rye seedheads were hand harvested and the number of seeds counted to determine seed rain potential. A subset of seeds was germinated in the lab to determine the viability of seeds and ensuing weed potential. We found that fall applied poultry litter increased the height, diameter and growth stage of rye at the time of roll crimping, while increasing rye seeding rate was negatively correlated with rye height and diameter, but positively correlated with growth stage. Increasing seeding rate increased inequality in the distribution of rye stem diameter, but trended towards decreased inequality in rye growth stage. Poultry litter decreased inequality in the distribution of rye height and growth stage. Poultry litter minimized the number of tillers of regrowth after roll crimping and both poultry litter and increased seeding rate minimized the number of viable volunteer rye seeds, effectively diminishing the impact of volunteer rye. Preliminary results suggest increasing seeding rate and fall applied poultry litter may improve the evenness of the rye stand and reduce the weed potential of cereal rye in subsequent crops.

False-Green Kyllinga (*Kyllinga gracillima*) Establishment from Seed is Influenced by Turfgrass Mangement Regime. Trevor S. Watson*, Matthew T. Elmore, Daniel P. Tuck; Rutgers University, New Brunswick, NJ (232)

Replicate field research experiments were conducted from 2021-2022 in North Brunswick, NJ on Kentucky bluegrass (*Poa pratensis*) to evaluate false-green kyllinga (*Kyllinga gracillima*) establishment from seed. Two turfgrass maintenance regimens and three seeding treatments were evaluated in a split-plot randomized complete block design with four replications. Maintenance regimens formed whole plots. The high maintenance regimen consisted of 200 kg N ha^{-1} annually and preventative fungicide applications during summer. The low maintenance regimen was fertilized once annually at 25 kg N ha^{-1} and no fungicides were applied. Sub-plot treatments were either seeded with false-green kyllinga seeded at 5 kg or 500 kg ha^{-1}, and smooth crabgrass (*Digitaria ischaemum*) at 500 kg ha^{-1}. Smooth crabgrass was used as a standard for comparison. Weed cover was determined visually and using a grid intersect count. Turfgrass cover was determined using digital image analysis (Turf Analyzer software) in May at kyllinga emergence. Data were subjected to ANOVA in SAS 9.4 using Fisher's protected LSD ($P=0.05$) was used to separate means. Main effect interactions were detected ($P < 0.05$) in August and September intersect count data in both years. In August of both years, false-green kyllinga cover was greater under low maintenance (81% in 2021 and 48% in 2022) than high maintenance (24% in 2021 and

18% in 2022) when seeded at 500 kg ha^{-1}. In August and September of both years, crabgrass cover was not affected by maintenance. Under high maintenance, crabgrass cover was 88 and 93% in August 2021 and 2022, respectively. When seeded at 5 kg ha^{-1}, false-green kyllinga cover was greater under low maintenance (16%) than high maintenance (6%) in August 2021, but maintenance did not affect cover in 2022. Kentucky bluegrass cover was 94 and 98% under low and high maintenance, respectively in May 2022, compared to 87 and 97% cover in May 2021. Greater low maintenance turfgrass cover in 2022 could explain why maintenance regimen had less of an impact on weed cover. False-green kyllinga establishment from seed was influenced more by turf management regimen than smooth crabgrass. This research also indicates that false-green kyllinga establishment can occur from seed in turfgrass. Future studies will focus on false-green kyllinga control with preemergence herbicides.

Soybean Response to Simulated Drift of Herbicides Commonly Used to Manage Roadside Weeds.

Estefania Gomiero Polli*, Travis Gannon, Ronald R. Rogers, Mathieu LeCompte, Khalied Ahmed; North Carolina State University, Raleigh, NC (233)

Management of roadside vegetation is essential to provide safe travel routes for drivers and passengers, and to preserve the integrity of road system infrastructure. Herbicides are an effective and economical way to manage vegetation along roadsides. However, the high occurrence of herbicide off-target movement cases in recent years has raised concerns among agencies responsible for managing roadsides. The objective of this research was to investigate the response of soybean to simulated drift of five herbicides commonly used along roadsides. A field study was conducted at Sandhills Research Station in Jackson Springs, NC from January to October of 2022. The study was arranged as a strip-plot design on a randomized complete design replicated 3 times. The whole plots were a two-level factorial of 5 herbicides (triclopyr, triclopyr + clopyralid, 2,4-D + dichlorprop, sulfometuron, indaziflan) and 4 herbicides rates (1%, 5%, 10% and 100% of field dose), and the strip-plot was application timing (18, 12, 6, or 0 weeks before planting and 4 or 8 weeks after planting). An untreated control was included for each block within application timing. Soybean was planted on May 18th and harvested on October 28th. Applications were performed using a 3-nozzle boom CO_2 backpack with a carrier volume of 140 L ha^{-1} (triclopyr + clopyralid, sulfometuron, indaziflan) and 935 L ha^{-1} (triclopyr, 2,4-D + dichlorprop) to simulate real roadside application parameters. Yield data were recorded and converted into a percentage of yield loss (YL) as compared to the untreated control. YL data were subjected to analysis of variance in SAS software and treatment means were computed using Fisher's least significant procedure (a=0.05). Herbicide treatments applied post-planting presented higher YL (44%) than treatments applied pre-planting (10%). While YL by indaziflan and sulfometuron was observed throughout all application timings, YL by triclopyr, triclopyr + clopyralid, and 2-4-D + dichlorprop was higher when applied post-planting. YL was observed for all herbicides even at the lowest rates. Triclopyr applied at 1% and 5% of the field rate caused 7% and 25% YL, respectively. The findings of this study suggest that drift of triclopyr, triclopyr + clopyralid, and 2,4-D + dichlorprop is less likely to decrease yield when it occurs before soybean planting. Additionally, drift of 2,4-D + dichlorprop, indaziflan, and sulfometuron at very low rates had little impact (<10%) on soybean YL.

Multi-Seed *Zea* Pellets (MSZP) for Increasing Agroecosystem Biodiversity. Sophie Westbrook*[1], Masoume Amirkhani[2], Alan G. Taylor[2], Michael T. Loos[2], John E. Losey[1], Antonio DiTommaso[1]; [1]Cornell University, Ithaca, NY, [2]Cornell University, Geneva, NY (234)

Intensive agricultural crop production is typically associated with low biodiversity. Low biodiversity is associated with a deficit of ecosystem services, which may limit crop yield (e.g., low pollination of insect-pollinated crops) at the individual field level or exacerbate the landscape-level impacts of intensive agriculture. To increase biodiversity and enhance ecosystem services with minimal loss of crop production area, farmers can plant desirable non-crop species near crop fields. Adoption of this practice is limited by inefficiencies in existing establishment methods. We have developed a novel seed molding method allowing non-crop species to be planted with a conventional *Zea mays* (corn) planter, reducing labor and capital costs associated with native species establishment. *Asclepias syriaca* (common milkweed) was selected as a model native species because *Asclepias* plants are the sole food source for *Danaus plexippus* (monarch butterfly) larvae. Stratified *A. syriaca* seeds were added to a mixture of binder (maltodextrin) and filler (diatomaceous earth and wood flour) materials in a 3D-printed mold with the dimensions of a *Z. mays* seed. The resulting Multi-Seed *Zea* Pellets (MSZP), shaped like *Z. mays* seeds, were tested against non-pelleted *A. syriaca* seeds in several indoor and outdoor pot experiments. Molding into MSZP did not affect percentage emergence or time to emergence from a 2 cm planting depth. Intraspecific competition among seedlings emerged from an MSZP did not differ from competition among seedlings emerged from a cluster of non-pelleted seeds. These findings demonstrate the potential of MSZP technology as a precise and efficient method for increasing agroecosystem biodiversity.

Using 3-D Image Reconstruction to Characterize Canopy Architecture and Predict Biomass. April M. Dobbs*[1], Daniel J. Ginn[2], Søren K. Skovsen[3], Muthukumar V. Bagavathiannan[2], Steven Brian Mirsky[4], Chris Reberg-Horton[1], Prashant Jha[5], Ramawatar Yadav[5], Ramon G. Leon[1]; [1]North Carolina State University, Raleigh, NC, [2]Texas A&M University, College Station, TX, [3]Aarhus University, Aarhus, Denmark, [4]USDA ARS, Beltsville, MD, [5]Iowa State University, Ames, IA (235)

Variability in biomass production poses a challenge for growers when using cover crops for weed control. However, most methods for assessing cover crop biomass are laborious and impractical for measuring within-field variability in real-time. This study used Structure-from-Motion (SFM) photogrammetry to estimate biomass in cereal rye (Secale cereale L.) and winter wheat (Triticum aestivum) cover crops. Using a hand-held GoPro camera, we recorded videos over cover crop-planted fields in multiple locations (NC, IA, and MD) throughout the 2021-2022 growing season. We used SFM to digitally recreate the 3-D structure of the cover crop canopy with point clouds at multiple growth stages through maturity. Crop height, leaf area index (LAI), and photosynthetically active radiation (PAR) were also measured to characterize canopy structure throughout the season. For both species, biomass, height, and LAI were positively correlated. PAR penetrating the canopy was negatively correlated with biomass and crop height. We used point cloud pixel quantity and crop height to develop a model for predicting biomass in both species at different heights. Based on independent data validation, we were able to estimate cover crop biomass using 3-D point clouds with over 70% accuracy in both crops.

Cover Crops and Falls Residual Herbicides for Managing Multiple Herbicide-Resistant Italian Ryegrass.

Jose H. de Sanctis*, Charlie W. Cahoon, Wesley Everman, Travis Gannon, Zachary R. Taylor; North Carolina State University, Raleigh, NC (236)

Italian ryegrass (*Lolium perenne ssp. multiflorum*) is a troublesome weed that has evolved resistance to 7 herbicide modes of actions in the US. Furthermore, in North Carolina, populations resistant to herbicides from groups 1, 2, 9, and 22 have been reported. Italian ryegrass is a winter annual weed species that has been traditionally managed with burndown herbicide applications in the spring shortly before planting, which increases the selection pressure on postemergence herbicides. To mitigate the spread or selection of such biotypes and provide effective control of Italian ryegrass it is crucial to integrate alternative management approaches. The objectives of this study were to evaluate Italian ryegrass control with cover crops and fall applied residual herbicides and investigate the level of cover crop injury from residual herbicides. The study was conducted in Salisbury and Clayton, NC, during the fall/winter of 2021-2022. The study was designed as a 3x5 strip-plot with four replicates, where the strips were the three cover crop treatments (no-cover, cereal rye at 80 kg ha^{-1}, and crimson clover at 18 kg ha^{-1}) and the subplots consisted of five residual herbicide treatments (S-metolachlor at 1420 g ai ha^{-1}, flumioxazin at 60.6 g ai ha^{-1}, metribuzin 470 g ai ha^{-1}, pyroxasulfone at 119 g ai ha^{-1}, and no-pre). Cover crops were seeded on Oct/20/2021 and Oct/18/2021 in Salisbury and Clayton, respectively, and residual herbicides were applied immediately after planting. Data collection consisted of bi-weekly visual estimates of Italian ryegrass control and cover crop injury, cover crop stand, and Italian ryegrass density. Cover crop and Italian ryegrass biomass were collected in the spring of 2022, at 24 weeks after planting (WAP). S-metolachlor and pyroxasulfone resulted in the least amount of cover crop injury and at 24 WAP treatments resulted in <14% injury, regardless of the cover crop type and location. Metribuzin injured cereal rye and crimson clover 64% and 53% , respectively. At 8 WAP, cover crop treatment had little effect on Italian ryegrass control; cereal rye and crimson clover resulted in comparable control (83% and 81%, respectively). Furthermore, at the same timing, all herbicides provided satisfactory Italian ryegrass control, ranging from 73% to 88%. However, at 24 WAP Italian ryegrass control in plots with cereal rye was 82% compared with 47% from plots with crimson clover. In addition, at 24 WAP, metribuzin was the least effective treatment and it resulted in 34% Italian ryegrass control. Findings of this study suggest that although residual herbicides resulted in satisfactory control on Italian ryegrass initially, cover crop was more important to the final suppression. Which can be explained by the fact that cereal rye no herbicides have similar level of Italian ryegrass suppression to cereal rye plus herbicides, with the exception of metribuzin that had least amount of Italian ryegrass suppression at 24 WAP is because it injured the cover crop. This further evidence the importance of cover crop biomass for weed suppression.

Herbicides Affect White Clover Bloom Morphology and Pollinator Foraging in Lawn Turf. Navdeep Godara*, Shawn Askew; Virginia Tech, Blacksburg, VA (237)

The recent decline in pollinator abundance may impact global food production. Several common weeds of managed turfgrass systems attract pollinators, and turfgrass is often treated with insecticides harmful to pollinators. Field studies were conducted in 2021 and 2022 to evaluate the effect of various herbicides and herbicide-formulation constituents on pollinator foraging and white clover (*Trifolium repens* L.) flower morphology in managed tall fescue turf in an effort to develop best practices to prevent pollinators visiting turf that receives insecticide sprays. Three separate experiments were implemented in Blacksburg, VA, as randomized complete block designs with six replications and seven treatments. Treatments included a nontreated control; MCPP; 2,4-D; dicamba; Trimec Classic™ (2,4-D, MCPP, dicamba); Speedzone™ (carfentrazone, 2,4-D, MCPP, dicamba); and a formulation blank (inert ingredients of Speedzone™). Flower density in each plot was assessed each morning, starting one day before treatment and ending 6 days after treatment (DAT). All insect foragers observed for one minute were recorded for each plot three times daily (10:00am, 1:00pm, and 4:00pm). Each day, flower petal discoloration was calculated via digital image analysis for three flowers in each plot. The formulation blank did not alter white clover flower density or petal quality compared to nontreated plots; however, all herbicides reduced white clover flower density equivalently, with complete reduction occurring 4 to 5 DAT. Furthermore, honey bees and other insects vacate herbicide-treated areas in less than 2 DAT, even though the loss of floral density and petal discoloration were observed three days later. Overall, the results suggest that insecticides could be applied two days after herbicide application to white-clover-infested turf with a limited risk of pollinator exposure.

Effect of Metcamifen on Trifloxysulfuron Absorption, Translocation and Metabolism in Turfgrass Systems. Ronald R. Rogers*, Travis Gannon, Khalied Ahmed; North Carolina State University, Raleigh, NC (238)

Trifloxysulfuron is a post-emergent herbicide used to control broadleaf weeds, grassy weeds and sedges in select turfgrass systems. Many warm season turfgrass species are tolerant to trifloxysulfuron; however, St. Augustinegrass is sensitive. The safener metcamifen, is known to safen clodinafop-propargyl applications in rice and maize. The objective of this study was to evaluate the effect of the safener metcamifen on the absorption, translocation, and metabolism on ^{14}C-trifloxysulfuron in yellow nutsedge and St. Augustinegrass. In absorption and translocation studies plants treated with ^{14}C-trifloxysulfuron were harvested 0,4,12, 24, 48, 96 and 192 hours after treatment (HAT) and partitioned into six different plant parts. Translocation studies showed greater ^{14}C-trifloxysulfuron recovered in the leaf wash (41.2% > 36.3%) and treated leaf (39.0% > 33.7) for trifloxysulfuron + metcamifen compared to trifloxysulfuron alone in yellow nutsedge. St. Augustinegrass translocation data show greater ^{14}C-trifloxysulfuron recovered in the treated leaf (34.8% >28.4%) for trifloxysulfuron + metcamifen treatment compared to trifloxysulfuron alone. Plants used in the metabolism experiment were grown, maintained, and treated in a similar fashion as plants reserved for the absorption and translocation experiment. Metabolism studies showed a decrease in ^{14}C-trifloxysulfuron recovered for the trifloxysulfuron + metcamifen treatment 48 HAT (76.8 % < 84.4%), 96 HAT (67.2% < 74.5 %) and 192 HAT (54.8 % < 66.8%) compared to trifloxysulfuron alone in St. Augustinegrass while differences were not observed in yellow nutsedge.

Confirmation of Glufosinate Resistance and Inheritance in a North Carolina Palmer Amaranth (Amaranthus Palmeri) Population. Eric A. Jones*, Jeffrey Dunne, Diego J. Contreras, Charlie W. Cahoon, Katherine M. Jennings, Ramon G. Leon, Wesley Everman; North Carolina State University, Raleigh, NC (239)

A putative glufosinate-resistant *Amaranthus palmeri* population was reported in 2015 in Anson County, North Carolina. In the field, two different *A. palmeri* cohorts exhibited differential responses of death and survival suggesting the population is still segregating for resistance. Dose-response assays conducted in the greenhouse determined the Anson County population exhibited reduced susceptibility to glufosinate compared to three glufosinate-susceptible populations. The LD_{50} values (210-267 g ai ha^{-1}) for the Anson County populations were always significantly higher than the LD_{50} values (118-158 g ai ha^{-1}) for the tested glufosinate-susceptible population from the dose-response assays. Anson County plants that survived lethal glufosinate rates (267-450 g ai ha^{-1}) were reciprocally crossed with a glufosinate-susceptible plants to create F_1 genotypes and treated with a lethal rate of glufosinate (267 g ai ha^{-1}) to determine the distribution of injury and survival for each biparental cross compared to a cross including two glufosinate-susceptible parents. The distribution of injury was non-normal for the biparental crosses containing an Anson County plant compared to the cross with a susceptible male and female. Survival was 68-84% for biparental crosses containing an Anson County plant, while the survival was significantly reduced to 35% for the susceptible plant cross. Chi-square goodness of fit tests were used to test inheritance models to describe survival of the progeny of each biparental cross. The RM × SF and SM × RF_2 cross were best described with a heterozygous two loci with incomplete dominance model compared to the RM × RF_1 cross that was best described with a heterozygous single loci with incomplete dominance model. These results suggest that the Anson County *A. palmeri* population has evolved glufosinate resistance and the inherited mechanism is oligogenic with incomplete dominance from heterozygous parental genotypes with a possible maternal effect.

Factors That Influence Uniform Droplet Deposition from an Agricultural Spray Drone. Daewon Koo*[1], Navdeep Godara[1], John M. Peppers[1], Clebson G. Goncalves[2], Shawn Askew[1]; [1]Virginia Tech, Blacksburg, VA, [2]University of California, Agriculture and Natural Resources, Ukiah, CA (240)

Agricultural spray drones (ASD) have become increasingly accessible in recent years, but little is known regarding their use for pest control. Since pesticide labels do not currently specify use patterns for ASDs, commercial applicators are reliant on aerial application specifications that limit spray tip placement to no greater than 75% of the aircraft's rotor span. To improve efficiency and allow drones to span more than a few meters per pass, drone manufacturers equip drones with spray tips that generate fine droplets subject to dispersal by drone rotors. Our previous research showed that droplet deposition decreases up to 60% as drone height increases up to 10 m. Medical and pesticide sciences literature suggests these losses are due to droplet vaporization. Studies were conducted to further test this theory and examine potential drift reduction technologies to help conserve deposition. The trial was conducted at Virginia Tech in Blacksburg, VA, USA to examine the influence of TeeJet spray tips (XR11001, XR11002, AIXR11002) and drift reduction agent (DRA) (BAS-638, BAS-639, Intact™) on droplet vaporization. Twenty white craft papers (21.6 cm x 30.5 cm) were positioned parallel to the ground at vertical heights of 0.5-m increments. Blazon blue colorant (Milliken, USA) and water (1:1) were sprayed with a hand-held boom sprayer at 10-m above the lowest paper simulating the flow rate of the ASD and comparing spray tips supplied with the ASD (XR11001) with and without DRA to the other two spray tips without DRA. Papers were scanned and analyzed with customized python codes to count and measure droplets, and colorant was extracted and analyzed with a spectrophotometer to quantify colorant deposition. As expected, colorant spots less than 150 µm decreased from 19 to 1 spot cm^{-2} as the distance from a XR 11001 spray tip increased to 10 m. Larger spray tip orifices and DRA mixtures had no more than five < 150 µm colorant spots cm^{-2} regardless of height. These data support our hypothesis regarding small-droplet vaporization. Since the spray boom was operated at extreme heights and only two spray tips were utilized, large-droplets were dispersed due to their mass and resulting inertia along the angle of propulsion. This horizontal dispersion of large droplets varied by treatment and limited our ability to estimate deposition conservation.

The Confirmation and Characterization of Auxinic Herbicide Resistance in a Population of Green Pigweed (*Amaranthus powellii*) from Ontario, Canada. Isabelle Aicklen*[1], Peter Smith[1], Brendan Metzger[2], Todd A. Gaines[3], Mithila Jugulam[4], Darren E. Robinson[5], Peter H. Sikkema[5], Francois Tardif[1]; [1]University of Guelph, Guelph, ON, Canada, [2]BASF Canada, Morden, MB, Canada, [3]Colorado State University, Fort Collins, CO, [4]Kansas State University, Manhattan, KS, [5]University of Guelph, Ridgetown, ON, Canada (241)

Auxinic herbicides have been historically used for the selective control of dicot weed species in crop production. Repeated applications of these herbicides has been a strong selection pressure for herbicide resistance creating management challenges for farmers. A population of green pigweed with suspected resistance (R) to the auxinic herbicide MCPA has been reported in Ontario. The objective of this study is to confirm resistance of this population to MCPA and to determine the pattern of cross resistance to other auxinic herbicides. Dose response experiments were conducted to compare differences in GR_{50} values (dose causing growth reduction of 50%) between R and a known susceptible (S) green pigweed population using MCPA amine, mecoprop, 2,4-D ester, dichlorprop-p, halauxifen-methyl, dicamba, and aminocyclopyrachlor. Field trials were conducted using auxinic herbicides applied post emergence in corn to evaluate the control of green pigweed and to confirm herbicide resistance at the field level. Following completion of resistance confirmation studies, a second objective was addressed to confirm the mechanism of herbicide resistance in this weed species. Experiments using [^{14}C] MCPA were conducted to compare non-target site resistance mechanisms between R and S populations. [^{14}C] MCPA was applied to the fourth true leaf and sampled at 6, 24, 48, and 72 h after treatment. Following sampling of plant tissue, the % absorption, translocation, and metabolism was quantified for each time point and population and comparisons were made. The dose response results confirmed that R has 4.4-fold resistance to MCPA with cross resistance to aminocyclopyrachlor (3.0-fold), dichlorprop (2.5-fold), and mecoprop (2.4-fold). Field trials showed that dicamba provided superior control (>80%) at 56 days after application but control was 30% with MCPA alone, 46% with MCPA + fluroxypyr, and 36% with MCPA + fluroxypyr/halauxifen-methyl. This confirms that the level of resistance observed in the lab brings the effectiveness of MCPA or MCPA with other auxinic herbicides below commercially acceptable levels. The findings of the dose response and field study demonstrates the importance of determining the mechanism of herbicide resistance in this population. Preliminary results of the [^{14}C] study indicated no differences in the % absorption, translocation, and metabolism between populations R and S however, visual inspection of the populations indicates strong phenotypic differences. It is therefore suspected that there is a different mechanism of resistance and further research will be conducted to test this objective.

Speed, Angle and Spacing Effects on Finger Weeder Performance. Jordan W. Parks*, Margaret R. McCollough, Eric R. Gallandt; University of Maine, Orono, ME (242)

Organic farmers rely on physical weed control (PWC) to reduce weed density in most crops. Farmers adjust PWC tools based on trial and error, and prior experience. Depending on the type of tool, adjustments may include changes in tool depth, angle, distance to crop row, and speed. Research based information on optimizing tool settings would benefit these farmers, especially beginning farmers. In the present study, finger weeder angle, spacing, and speed were evaluated in both field and soil bin experiments. A test crop of red table beets were used in tool angle and spacing trials. Bush beans were used for tool speed trials. Surrogate weeds, condiment mustard (Brassica juncea) and garnet red amaranth (Amaranthus tricolor), were seeded to provide a uniform density of size. Soil bin experiments employed artificial crops (6 mm dia. by 152 mm-long wooden dowels) and weeds (70 mm-long wooden golf tees) for rapid testing of tool settings. The three tool angles (68°, 90°, and 108° relative to the soil surface) and three tool spacings (0.6 cm overlap, 0.0 cm fingers touching, and 2.5 cm gap) were selected to represent a relevant range of finger weeder settings. The three operational speeds evaluated (4, 7, and 9 km h^{-1}) were chosen to represent walking, tractor speeds, and camera guidance speeds, respectively. Tool angle adjustment had the most consistent effect among settings tested, in both the soil bin ($P < 0.001$) and the field ($P = 0.001$). The 68° setting, which caused soil hilling, achieved the greatest efficacy in the intra- and near-row zones, removing 60 and 65% of weeds, respectively. The 90° setting (in the middle of hilling and soil removal) and the 108° (causing soil removal) setting performed similarly, and at a lower efficacy than the hilling angle (achieving 42 and 33% efficacies in the intra-row and 45 and 41% in the near-row zones, respectively). The effect of tool spacing was significant in the soil bin ($P = 0.002$), as the space between fingers decreased from a 2.5 cm gap to a 0.6 cm overlap, efficacy increased from 27 to 43%. However, this effect was not detected in the field ($P = 0.12$). Red table beet yield was not affected by tool angle and spacing. In field experiments, tool speed did not affect efficacy ($P = 0.444$); however, a significant effect was observed in the soil bin ($P = 0.003$). In soil bin experiments, the 7 km h^{-1} tractor speed resulted in the greatest intra-row efficacy (76%), and the 4 km h^{-1} walking speed resulted in the lowest (43%). Overall, these results suggest farmers should prioritize adjusting tool angle as a way to increase efficacy in the intra- and near-row zones. However, some evidence from the trials suggests that using a medium speed and decreasing tool spacing can also increase efficacy in these zones. jordan.parks@maine.edu

Interaction of See and Spray^{TM} Model Sensitivity and Application Timing in Soybean. Diego J. Contreras*[1], Wesley Everman[1], Jason K. Norsworthy[2], Tristen H. Avent[2], Bryan G. Young[3], Marcelo Zimmer[4], Jasmine M. Mausbach[5], Lauren M. Lazaro[6], William L. Patzoldt[6]; [1]North Carolina State University, Raleigh, NC, [2]University of Arkansas, Fayetteville, AR, [3]Purdue University, Brookston, IN, [4]Purdue University, West Lafayette, IN, [5]University of Nebraska-Lincoln, Lincoln, NE, [6]Blue River Technology, Sunnyvale, CA (243)

New technologies provide benefits to stakeholders in agricultural productions, such as improved productive efficiency, increased food security from a productive standpoint, as well as reduced costs. Remote sensing has garnered special interest in the past decade; a very promising use for remote sensing is the use of the technology in combination with a sprayer to realize site-specific herbicide applications. See & Spray^{TM} technology, from Blue River and John Deere, can successfully detect, spray, and control weeds within a crop while specifically targeting weeds. See & Spray^{TM} users can select the spray sensitivity of the system, which adjusts the confidence threshold for when to trigger sprays. A lower sensibility requires a higher confidence to spray, a higher sensibility requires a lower confidence to spray. Theoretically, a lower sensitivity results in reduced pesticide output, however small weeds might be missed. Higher sensitivity may miss fewer weeds, if any, but at the cost of an increased pesticide output. An experiment using See & Spray^{TM} was set up at four sites (NC, AR, IN, MS) to test if overall weed control changes with the interaction of the sensitivity level (Low, Mid and High) used and application timing (Early [14 and 28 days after planting{DAP}], Mid [21 and 35 DAP] and Late [28 and 42]) in soybeans. There was no difference in weed control by species at all locations. Common species by locations had the same level of control regardless of sensitivity level and application timing. When grouped across locations, broadleaf weed species' control was the same across sensitivity levels and application timings. When grouped across locations, grass weed species control was the same 14 days after the first application, regardless of timing, however there was a slight grass control reduction 14 days after the second application (from 99% to 97% control, $p=0.0003$). Overall, See & Spray^{TM} sensitivity level and application timing did not affect weed control in soybean.

Residual Control of Palmer Amaranth Varies Across VLCFA-Inhibiting Herbicides. Jackson W. Alsdorf*[1], Amanda M. Foderaro[2], Eric A. Jones[1], Diego J. Contreras[1], Wesley Everman[1]; [1]North Carolina State University, Raleigh, NC, [2]Syngenta, Raleigh, NC (244)

Very long-chain fatty-acid-inhibiting herbicides provide residual preemergence control of grasses and small-seeded broadleaves in many crops. These herbicides are commonly applied to control herbicide-resistant weeds, such as Palmer amaranth (Amaranthus palmeri). Residual control of Palmer amaranth is important to reduce competition with the emerging crop and prolong the time until a needed postemergence herbicide application. Greenhouse studies were conducted to determine the length of control of acetochlor, dimethenamid-P, pyroxysulfone, and S-metolachlor. Each herbicide was applied with a respective discriminating LD90 dose. Two different populations of Palmer amaranth were utilized, a known susceptible and a difficult to control population. Seeds were planted into pots containing wakeland silt loam soil at 6, 4, 2, and 0 weeks after treatment. Results were combined between the two populations. Live counts were recorded 21 DAT and the length of control was calculated by determining the amount of time it took to reduce weed emergence by 50% (I50 value). S-metolachlor and pyroxysulfone had the longest residual with an I50 of approximately 6.5-7 weeks. Acetochlor and dimethenamid-P had the lower residual with an I50 of about 1-1.5 weeks. Pyroxasulfone and S-metolachlor demonstrated longer periods of residual control compared to acetochlor and dimethenamid-P at biologically effective rates, further evidence supporting the use of full labeled rates.

A Tour Through Corteva Headquarters: A Global Perspective of an Integrated Field Science Organization. Estefania Gomiero Polli*; North Carolina State University, Raleigh, NC (278)

Corteva Agriscience™ is a global seed and chemical company that was founded of Dow, DuPont, and Pioneer. It is the only major agriscience company completely dedicated to agriculture. The Travel Enrichment Experience (TEE) award provided me the opportunity to visit the Corteva Agriscience™ Headquarters in Indianapolis, IN in September 2022. I had the pleasure to spend three days under Dr. David Simpson's guidance. He carefully organized an agenda of activities considering my area of interest. Since day 1, I had the opportunity to talk to brilliant professionals who patiently and passionately shared their experiences and role at Corteva Agriscience™ with me. I also visited multiple greenhouses, laboratories, and buildings. As my doctoral dissertation is focused on herbicide off-target movement, I spent a day with the herbicide discovery, formulation, and application technology teams. The high-quality and innovative technology tools and processes used by them to develop new products are fascinating. During my visit, I was given the chance to learn from distinguished researchers and broaden my knowledge of the herbicide development and commercialization chain. They provided me with a life-changing experience that I will carry throughout my entire professional life. Many thanks to the Corteva Agriscience™ team, especially Dr. David Simpson, and the WSSA for this experience.

Travel Enrichment Experience at Beltsville Agricultural Research Center, Maryland. Sachin Dhanda*; Kansas State University, Hays, KS (279)

As a recipient of the 2022 WSSA Travel Enrichment Experience award, I got the opportunity to visit Dr. Steven Mirsky's Sustainable Agricultural Systems Lab at Beltsville Agricultural Research Center, Maryland.

It was one of my best experiences. I learned about different ongoing projects focusing on integrated weed management like cover crops, herbicides, and harvest weed seed control. I spent time with the Digital Weeds Team, where they were capturing pictures of different weed species to build high-resolution, open-access, annotated agricultural image repositories for species identification. I got to see BenchBot Technology which collects images of weeds rapidly and efficiently by running along a conveyor, capturing images of weeds. Also, I learned about ongoing cover crop experiments in interaction with weed, nutrient, and pest management. Dr. Mirsky and his team concluded the visit with a wonderful summary and wished me good luck for my future. I am grateful for the hospitality I received from Dr. Mirsky and his team during my visit. The WSSA Travel Enrichment Experience award is a great opportunity for weed science students to explore beyond their own research experiments. I encourage all graduate students to apply for this unique opportunity in the future.

Travel Enrichment Experience with the Weed Research Lab at Colorado State University. Joshua W. Miranda*; Oregon State University, Corvallis, OR (280)

During my visit with the weed research lab at Colorado State University I had the opportunity to work alongside lab members and the principal investigators, Drs. Todd Gaines and Franck Dayan. I started this experience by learning about DNA extraction from weeds, determining its concentration and purity prior to its use. With one of the lab members' projects, I had the opportunity to run the polymerase chain reaction (PCR) technique to synthesize new strands of DNA to later analyze by using the agarose gel electrophoresis method to separate DNA molecules by size. I also performed a real-time PCR method to measure gene expression. After learning the aforementioned methods and techniques, I had the opportunity to learn about primers design to amplify genes of interest, as well as the several parameters and considerations one has to be aware of to avoid failure and issues in the PCR, e.g., nonspecific amplification. Lastly, I observed how the Liquid Chromatography with tandem mass spectrometry (LC-MS-MS) machine works and learned the multiple ways this technique can be used in weed science to discover of mechanisms of action of herbicide resistance in weeds. I am very grateful for the Weed Science Society of America, Drs. Todd Gaines and Franck Dayan, and members of the weed research lab for giving me this amazing opportunity to learn and explore new fields to contribute to improving economic costs and environmental impacts of agriculture as I pursue my passion for weed science.E–mail: josh.miranda@oregonstate.edu.

From East to West: Biocontrol Musings from Montana. Emily Duenk*; University of Guelph, Ridgetown, ON, Canada (281)

Weed Science Research, Policy, and Partnerships with USDA-ARS: 2022 Travel Enrichment Experience.
Sarah E. Kezar*[1], Steve Young[2]; [1]Texas A&M University, College Station, TX, [2]USDA-ARS, Beltsville, MD
(282)

Science Policy Fellowship: A Unique Learning Opportunity for Aspiring Weed Scientists. Taylor M. Randell-Singleton*[1], Navdeep Godara[2], Lee Van Wychen[3]; [1]University of Georgia, Tifton, GA, [2]Virginia Tech, Blacksburg, VA, [3]Weed Science Society of America, Alexandria, VA (283)

Science Policy Fellowship: A Unique Learning Opportunity for Aspiring Weed Scientists Navdeep Godara, Taylor Randell Singleton, and Lee Van Wychen The Science Policy Fellowship is a unique opportunity for early career weed science professionals to assist Dr. Lee Van Wychen, Executive Director of Science Policy for the Weed Science Society of America (WSSA), while gaining experience on various weed science policy issues. The WSSA's Science Policy Committee strives to provide expertise on science-based information on weed science policies to the National and Regional Weed Science Societies. Taylor Randell Singleton, University of Georgia, and Navdeep Godara, Virginia Tech, were awarded Science Policy Fellowships for 2022-23. As a Science Policy Fellow (SPF), Taylor has drafted comments on the atrazine interim registration review decision, the Plant Protection Act (7 U.S.C. 104) Section 7721, the ESA workplan update, and a letter to the House and Senate Appropriations Subcommittees for Transportation to request funding for the new Invasive Plant Removal Program that was authorized in the 2021 Infrasture Law (P.L.117-58). Navdeep analyzed the national weed survey for the most common and troublesome weeds in the U.S. and Canada. He also analyzed and categorized the common and troublesome weeds for each of the Regional Weed Science Societies. Furthermore, the SPFs also participated in congressional visits and witnessed the importance of advocating for research priorities and regulatory policy at the federal level. They participated in meetings with congressional staff from Georgia, Iowa, Indiana, Mississippi, and Virginia, along with the presidents of the National and Regional Weed Science Societies. They discussed weed science funding priorities and advocated for changes in current Farm Bill policy to provide more research opportunities for weed scientists. The SPFs gained substantial leadership experience in public policy and advocacy, and are grateful to the WSSA and the Science Policy Committee for this opportunity. In conclusion, the SPFs highly recommend this opportunity to peers regardless of their interest in serving academia, industry, or federal agencies.

Diflufenican: a Tool for Managing *Amaranthus* Species in Corn and Soybean Cropping Systems. John Buol*, Carl Coburn, Richard Leitz, Eric Riley, Ananda Datta; Bayer, St. Louis, MO (307)

The continued development and spread of herbicide resistance constitutes a major threat to the efficiency and profitability of corn and soybean production. Weeds such as some *Amaranthus* species have developed resistance to multiple herbicide modes- and sites- of action and are among the most challenging broadleaf weeds in North America. Bayer CropScience is developing an herbicide platform that features the use of diflufenican, a new site of action for *Amaranthus* spp. control in corn and soybean production systems in North America, pending registration with the U.S. EPA and Canada PMRA. Diflufenican functions as a phytoene desaturase inhibitor classified by HRAC as a group 12 herbicide and has been used outside of the U.S. for control of broadleaf weeds in cereals, peas, lentils, lupins, clover pastures, and oilseed poppy. Given the increasing challenge of managing herbicide-resistant weeds, diflufenican is being evaluated in field trials in North America for residual activity on *Amaranthus spp.* and crop selectivity in soybean and corn. Pending registration with the U.S. EPA and Canada PMRA, diflufenican would enable a new weed management tool that should be used in combination with other weed management practices as part of an integrated weed management plan.

Depletion of Multiple-Herbicide-Resistant Waterhemp Seed in the Soil Seed Bank with Integrated Weed Management Strategies. Peter H. Sikkema*, Nader Soltani; University of Guelph, Ridgetown, ON, Canada (308)

The development of an Integrated Weed Management strategy, based on a more holistic approach, that includes crop rotation, cover crops, increased seeding rates, and efficacious herbicides for control of glyphosate-resistant waterhemp can provide field crop producers with a strategy to deplete waterhemp seeds in their farms. Field experiments were established on two commercial Ontario farms with multiple herbicide-resistant (MHR) waterhemp in 2017. The objective of the study was to document the depletion in the number of waterhemp seeds in the seed bank after years 3, 6, and 9 (the spring of 2020, 2023, and 2026) of this nine-year study. The number of waterhemp seeds in the seed bank at the Cottam and Walpole Island sites prior to establishing the experiments was 413 and 40 million seeds/hectare, respectively. At Cottam, after 3 years of this study, the waterhemp seed in the seed bank was not significantly affected by continuous soybean (37.5 or 75 cm spacing), soybean/wheat, or corn/soybean, but was reduced by 74% in corn/soybean/wheat, and 73% in corn/soybean/wheat plus cover crops rotations. In Walpole Island, ON, after 3 years of this study, the waterhemp seed in the seed bank was not significantly affected by the rotations evaluated. Results indicate that the Integrated Weed

Management (IWM) strategies implemented can provide acceptable control of GR waterhemp in corn, soybean, and wheat.

How is Soybean Biomass Partitioning to Vegetation, Floral Development and Pod Formation Affected by Exposure to Simulated Drift Rates of Auxin Herbicides? Koffi Badou-Jeremie Kouame*[1], Thomas R. Butts[1], Brad M. Davis[1], Leah M. Collie[2], Troy W. Dillon[1], Jeremy K. Sutterfield[1], Benjamin C. Thrash[1], Nick R. Bateman[3], John J. Adamcyzk[4], Gus M. Lorenz[2]; [1]University of Arkansas System Division of Agriculture, Lonoke, AR, [2]University of Arkansas, Lonoke, AR, [3]University of Arkansas System Division of Agriculture, Fayetteville, AR, [4]USDA, Poplarville, MS (309)

Auxin herbicide drift, a primary concern in modern agriculture, has been exacerbated following the introduction of dicamba and 2,4-D resistant soybean [*Glycine max* (L.) Merr.] and cotton cultivars. The impact of simulated drift rates of common auxin herbicides on soybean reproduction (biomass partitioning, total number of reproductive organs, pollen production, and yield) needs to be understood for soybean production and for understanding potential impacts on pollinators. Field experiments were conducted in 2022 at the University of Arkansas at Pine Bluff Small Farm Outreach Center near Lonoke, AR, to evaluate the impact of auxin herbicide simulated drift rates on soybean biomass partitioning to various organ groups, number of reproductive organs, pollen production, and grain yield. A randomized complete block design was installed with eight replications. The experiment consisted of four herbicides [florpyrauxifen-benzyl (Loyant), 2,4-D (Enlist One), dicamba (Engenia), and quinclorac (Facet)] used at two rates (1/100th and 1/1000th the labeled rate) except for quinclorac, applied only at 1/100th the labeled rate. A nontreated control was included in the experiment for a total of eight treatments. Four random soybean plants were collected from each treatment at R3, R4, R5, and R6 soybean growth stages. The reproductive organs were counted, plants were separated into leaves, stems, and reproductive organs, dried, and the biomass of each organ group was recorded. Soybean flowers were collected for pollen quantification on ten plants randomly selected per replicate (one flower per plant) 1 d before anthesis. Partitioning coefficients to stem, leaf, and reproductive organs, the total number of reproductive organs, pollen production, and grain yield data were subjected to ANOVA using the GLIMMIX procedure in SAS version 9.4 (SAS Institute Inc, Cary, NC, United States). Results showed that soybean did not partition biomass differently to stem, leaf, or reproductive organs between sampling dates (R3 to R4, R4 to R5, and R5 to R6) across herbicides. However, total soybean reproductive organs, pollen production, and grain yield were all decreased by multiple auxin herbicide by simulated drift rate treatments compared to the nontreated control ($P < 0.05$). At the R3 reproductive stage, dicamba and florpyrauxifen-benzyl applied at 1/100th of the labeled rate induced a 35 and 39% reduction in the number of reproductive organs compared to the nontreated control. Similarly, the application of dicamba and florpyrauxifen-benzyl at these two rates induced a 28 and 18% decrease in pollen grains produced per anther, respectively. Likewise, dicamba and florpyrauxifen-benzyl applied at 1/100th of the labeled rate provoked a 38 and 24% reduction in grain yield compared to the nontreated control. These results illustrate that in addition to grain yield reduction, simulated auxin herbicide drift rates can negatively impact pollinator foraging sources by decreasing pollen production and the number of reproductive organs.

Effectiveness of Tirexor® and Voraxor® on Glufosinate-Resistant Palmer Amaranth. Pamela Carvalho-Moore*[1], Jason K. Norsworthy[1], Ty Smith[1], Casey H. Arnold[1], Matthew C. Woolard[1], Maria Leticia Zaccaro-Gruener[1], Tom Barber[2]; [1]University of Arkansas, Fayetteville, AR, [2]University of Arkansas System Division of Agriculture, Lonoke, AR (310)

Glufosinate resistance was confirmed in Palmer amaranth [*Amaranthus palmeri* (S.) Wats.] populations from Arkansas. This confirmation brought an additional apprehension level to farmers and researchers across the US since this weed has evolved resistance to herbicides from nine different sites of action, and glufosinate was among the few over-the-top options still effective. Alternative control options are highly sought. Therefore, this study aimed to evaluate the efficacy of the novel protoporphyrinogen IX oxidase (PPO) inhibitor herbicides Tirexor® (trifludimoxazin) and Voraxor® (trifludimoxazin plus saflufenacil) on the control of glufosinate-resistant Palmer amaranth. The experiment was conducted in Fayetteville, Arkansas, and organized as a two-factor factorial with four replicates. Factor A consisted of herbicide type and rates. These were: 1) nontreated, 2) Tirexor at 25 g ai ha^{-1}, 3) Tirexor at 50 g ai ha^{-1}, 4) Tirexor at 75 g ai ha^{-1}, 5) Voraxor at 37.5 g ai ha^{-1}, and 6) Voraxor at 75 g ai ha^{-1}. Factor B was the presence or absence of glufosinate. Adjuvants were added to applications with Tirexor or Voraxor. This experiment was conducted in an area geographically isolated from agricultural lands, and seeds of the highly glufosinate-resistant Palmer amaranth accession were spread and incorporated on the field prior to the preemergence application. Palmer amaranth control (%) was evaluation at 21 days after treatment. The interaction between herbicide types and rates and presence of glufosinate was significant. Alone applications of the lowest rate of Tirexor or glufosinate obtained the lowest control levels, 48 and 50%, respectively. The other treatments were similar statistically with control ranging from 76 to 98%. New herbicide chemistries are strong allies in controlling the advancement of glufosinate resistance across Palmer amaranth populations. However, it is necessary to combine these chemistries with a varied range of pre- and postemergence herbicides to avoid the development of additional herbicide resistance.

Agriculturally Significant Weeds and Associated Soil Nematodes of South-Central New Jersey: the Good, the Bad, the Ugly. Melissa A. Bravo*; Rutgers, The State University of New Jersey, Woodstown, NJ (311)

Combines are a significant source of long-distance transport of difficult to control noxious, invasive, and poisonous to livestock weed seed. Many weed species are also alternate host to deleterious plant pest nematodes that do well in higher sand content soils. One problematic animal parasitic nematode in the county, especially on sandy soils is barberpole worm (*Haemonchus contortus*). The L3 larvae of barberpole worm is ubiquitously associated with all genera if the weeds are present in the grazing system. A field-crop weed observance was conducted in Fall 2020 and Spring, Summer, and Fall 2021 in Salem County during the Covid-19 Pandemic in person program restrictions. There are approximately 568 road kilometers (km) in 83 designated routes in the county, and over 718 km in 200 municipal routes. The county contains more than three dozen combines and a thousand individual fields making up some 39,659 hectares (ha) of farmland. Annually, approximately 8,093 county ha are planted to field corn (*Zea mays* L.), 8,093 ha to soybeans (*Glycine max* L.), and 2,832 ha to small grains (primarily Triticum L.). Vegetables (4,856 ha), all hay (4,047 ha), pasture and orchard acreage make up the balance. Soils are coastal plain highly permeable sandy clay loam, sandy loam, loam, and silt loams with little to no slope situated less than 90 feet above sea level. Soils with greater than 55% sand are common. Pastured livestock is estimated (2017 Census Data) at 6,300 head bovine, 1,400 head equine, 1,300 head ovine, and increasing numbers of commercial flocks of pasture grazed Aves. Over 1,609 km were driven to conduct the drive-by survey on thirty-two occasions.Forty-eight problematic weeds (65% dicots) were documented in the afore mentioned grain crop, hay, and pasture fields in the county. Seventy-nine percent (63% dicots) of observed problematic weed species have plant parts injurious to livestock or can contain toxins detrimental or deadly to poultry, cattle, horses, sheep, or goats. Sixty-nine percent of weeds (60% dicots) are detrimental to hay handlers or are toxic to humans by ingestion. Twenty-two plant parasitic nematode genera, host to or are associated with, the observed problematic weeds in the reviewed literature and databases, are as follows: Forty-five weeds (64% dicots), root–knot (*Meloidogyne* spp.). Forty-one weeds (61% dicots), cyst (*Heterodera* spp.). Thirty-eight weeds (61% dicots), lesion (*Pratylenchus* spp.). Thirty-one weeds (74% dicots), reniform (*Rotylenchulus* spp.). Thirty-one weeds (71% dicots), stem (*Ditylenchus* spp.). Nineteen weeds (58% monocots), pin (*Paratylenchus* spp.). Eighteen weeds (50% monocots), stubby root–knot (Trichodorids) family. Sixteen weeds (56% monocots), spiral (*Helicotylenchus* spp.). Fifteen weeds (57% monocots), ring (*Mesocriconema* spp.). Fourteen weeds, (57% dicots), dagger (*Xiphinema* spp.). Fourteen weeds (57% monocots), stunt (*Tylenchorhynchus* spp.). Fourteen weeds (64% monocots), needle (*Longidorus* spp.). Eleven weeds (78% monocots), lance (*Hoplolaimus* spp.). Eleven weeds (55% dicots), cactus (*Cactodera* spp.). Nine weeds (67% monocots), sting (*Belonolaimus* spp.). Seven weeds (72% monocots), grass cyst (Punctodera spp.). Four weeds (100% monocots), sheath (*Hemicycliophora* spp.) Three weeds (100% dicots), false root–knot (*Nacobbus* spp.). Three weeds (66% monocots), ring (*Criconemoides* spp.). Three weeds (66% dicots), (*Scutellonema* spp.). One weed (100% monocot), (*Quinisulcius* spp.). One weed (100% monocot), awl (*Dolichodorus* spp.).

A Survey of Twelve Common Rowcrop Weeds and Associated Soil Nematodes of Salem County, NJ.

Melissa A. Bravo*; Rutgers, The State University of New Jersey, Woodstown, NJ (312)

Populations of uncontrolled velvetleaf (*Abutilon theophrasti* L.), giant ragweed (*Ambrosia trifida* L.), wild garlic (*Allium ursinum* L.), green pigweed (*Amaranthus hybridus* L.), palmer amaranth (*Amaranthus palmeri* S. Watson), spiny pigweed (*Amaranthus spinosus* L.), jimsonweed (*Datura stramonium* L.), ivyleaf morningglory (*Ipomoea hederacea* L.), a perennial morningglory (*Ipomoea* spp.), curly dock (*Rumex crispus* L.), shattercane (*Sorghum bicolor* L. Moench), and cocklebur (*Xanthium strumarium* L.) were identified in crop fields on nine different farms in late summer 2022 from predominantly sandy loam soils. The weed root zone soil (approximately a quart) from three to five plants (populations) for each weed (if it occurred in any of the nine different fields) were combined and a sub-sample (approximately a quart) analyzed for fifteen plant parasitic nematodes. Cyst (*Heterodera* spp.), dagger (*Xiphinema* spp.), lance (*Hoplolaimus* spp.), lesion (*Pratylenchus* spp.), ring (*Criconemoides* spp.), root–knot (*Meloidogyne* spp.), spiral (*Helicotylenchus* spp.), sheath (*Hemicycliophora* spp.), stubby–root knot (*Paratrichodorus* spp.), and stunt nematodes (*Tylenchorhynchus* spp.) were detected (individuals/100 cc soil) from thirty-seven soil samples. Five genuses were recovered from ivyleaf morningglory field populations; four each from velvetleaf field populations and cocklebur field populations; three each from jimsonweed field populations and palmer amaranth field populations; and one or two genuses from each of the remaining weeds field populations. Spiral nematodes (n=732) were detected in all fields and by population: (100%) of curly dock; (80%) of cocklebur, (66%) each of palmer amaranth, giant ragweed, and wild garlic; (60%) of ivy leaf and jimsonweed; (40%) of velvetleaf; one of two shattercanes; in the green pigweed population which was comingled in a palmer infestation; and in the perennial morningglory population. Spiral was not detected in the spiny pigweed root zone soil. Palmer amaranth and giant ragweed had the highest spiral counts (266, 106) followed by jimsonweed, shattercane, cocklebur, ivyleaf morningglory and green pigweed (78, 66, 50, 46, 44). Spiral nematodes have a wide host range but are seldom considered a major pest when counts are below six hundred. But as palmer amaranth infestations continue to expand in the county, spiral may become more prevalent. The infective juvenile stage (J2) of cyst nematodes (n=580) were detected in 66% of fields and by population: (66%) of wild garlic; (60%) of jimsonweed; (60%) of ivyleaf morningglory; (40%) each of cocklebur and velvetleaf; and (33%) each of palmer amaranth and giant ragweed root zone soil; and in one of two shattercane root zone soil. Giant ragweed, jimsonweed, and cocklebur had the highest cyst counts (360,110,82). J2 infective stage is a good predictor of yield loss. Less than two hundred J2's is considered low-level infestation. Because even one cyst detection is damaging for soybean or vegetable production, producers may want to consider these weedy escapes as an alternate host for soybean cyst nematode. Root–knot nematodes (n=2) were detected in two populations of cocklebur; and from the perennial morning glory (n=20)–all from different fields. Sampling the root zone soil of these weeds may be an early detection method of further crop root assay if levels reach into the hundreds (corn), fifties (soybeans) or just one for vegetable producers. Stubby–root knot nematodes (n=4) were detected in one population of ivyleaf morningglory. More than forty stubby–root knot detections is considered damaging for corn yields. Just one detection of stubby–root knot is of concern for vegetable growers. Stunt nematodes (n=4,2) were detected from velvetleaf populations from two different fields. No literature records of stunt associated with velvetleaf were noted. Stunt are of concern for corn if over two hundred, and greater than thirty for vegetable producers. Dagger nematodes (n=4) were collected from one cocklebur population. Asteraceae are rarely associated with dagger and no cocklebur literature reference in association with dagger was found. Dagger threshold is quite low, counts less than ten, for all crop species compared to other genuses due to its ability to spread viruses. Sheath nematodes (n=4). were collected from one velvetleaf population. Sheath are not a significant pest of agronomic crops. Lesion nematodes (n=2) were collected from one ivyleaf morningglory population. Lesion counts over ten are of concern to vegetable producers and of concern for corn over two hundred. Young trees, especially apples, infected with counts of more than twenty exhibit poor growth. Lance nematodes (n=2) were collected from one palmer amaranth population. Counts less than sixty are considered a low threshold for lance in corn fields. But even one detection is significant for soybeans and vegetable crops. Ring nematodes (n=4) were collected from the only perennial morningglory population. Ring nematodes (.5 per gram of soil) are of concern to vineyard production. Other nematodes considered widely prevalent in New Jersey that were not observed include: Needle (*Longidorus* spp.) associated with wild garlic; Grass cyst (*Punctodera* spp.) associated with curly dock and shattercane; (*Paratylenchus* spp.) and (*Tylenchorhynchus* spp.) both associated with wild garlic and shattercane.

The Potential of Inter-row Non-selective Herbicide Application to Manage Wild Oat (*Avena fatua* L.) in Wheat. Eric N. Johnson*, Christian Willenborg, Steve Shirtliffe; University of Saskatchewan, Saskatoon, SK, Canada (313)

Herbicide resistant wild oat (*Avena fatua* L.) is prevalent in the Canadian Prairies, with surveys reporting multiple resistance to ACCase and ALS inhibiting herbicides in 25 to 40% of sampled fields. The use of inter-row (IR) weed management such as cultivation or non-selective herbicide application for managing wild oat in wheat crops has not been widely researched. Previous research by our lab found that IR cultivation integrated with other agronomic practices successfully controlled broadleaf weeds in pulse crops. To assess non-selective IR herbicide application, a shrouded sprayer was constructed and attached to a steerable cultivator. Through much trial and error, the sprayer was able to target IR weeds and minimize, but not eliminate, crop injury. A study investigating the potential of IR weed management to control wild oat in a spring wheat management system was conducted at three sites near Saskatoon in 2021-22. The study consisted of three factors: 1) fall triallate application (none, 1400 g ai ha^{-1}); 2) wheat planting density (150, 300, 450 seeds m^{-2}); and 3) IR weed management (none, IR glufosinate at 600 g ai ha^{-1}, IR cultivation). Fall triallate application, increased planting density, IR cultivation, and IR herbicide application reduced wild oat biomass production by 79, 63, 30, and 43%, respectively. The effect of planting density and IR weed management on wild oat biomass was much greater in treatments that did not receive triallate. The best combination of the 3 factors resulted in >90% reduction in wild oat biomass. Triallate application increased wheat yield by as much as 35%, while planting density and IR weed management resulted in small, inconsistent yield responses. IR weed management has the potential to control residual wild oats that escape PRE herbicide application or other cultural practices. Further refinement of non-selective IR herbicide application is required, and alternative experimental designs may be required to properly assess this technology.

Discovery of Tetflupyrolimet: A New Mode-of-Action Herbicide Interfering with De Novo Pyrimidine Biosynthesis for Effective Management of Herbicide-Resistant Grass Weeds Globally. Atul Puri*, Thomas Selby, Steve Gutteridge, Mark Holliday, Adam Prestegord; FMC Corporation, Newark, DE (314)

Control of Multiple-Herbicide-Resistant Waterhemp (*Amaranthus tuberculatus*) with Acetochlor-based Tank Mixtures in Soybean. Hannah E. Symington*[1], Nader Soltani[2], Allan Kaastra[3], David C. Hooker[2], Darren E. Robinson[2], Peter H. Sikkema[2]; [1]University of Guelph, Rr#3 Watford, ON, Canada, [2]University of Guelph, Ridgetown, ON, Canada, [3]Bayer Crop Science, Guelph, ON, Canada (315)

Waterhemp has evolved resistance to Group 2, 5, 9, 14, and 27 herbicides in Ontario, Canada, making control of this challenging weed even more difficult. Acetochlor is a Group 15, chloroacetanilide herbicide that has activity on many small-seeded annual grasses and some small-seeded annual broadleaf weeds including waterhemp. The objective of this study was to ascertain if acetochlor tank mixtures with broadleaf herbicides (dicamba, metribuzin, BCS 0478, sulfentrazone, or flumioxazin), applied preemergence (PRE), increases multiple-herbicide-resistant (MHR) waterhemp control in soybean. Five trials were conducted over two years (2021-2022). The acetochlor tank mixtures caused =2% soybean injury, except acetochlor plus flumioxazin, which caused 19% soybean injury. Acetochlor applied PRE controlled MHR waterhemp 82% at 12 weeks after application (WAA). Dicamba, metribuzin, BCS 0478, sulfentrazone, or flumioxazin controlled MHR waterhemp 37, 53, 38, 55, and 81%, respectively, at 12 WAA. Acetochlor applied in a tank mixture with dicamba, metribuzin, BCS 0478, sulfentrazone, or flumioxazin provided good to excellent control of MHR waterhemp; control ranged from 89 to 97%, but similar to acetochlor applied alone. Acetochlor alone reduced MHR waterhemp density and biomass 98 and 93%; acetochlor + flumioxazin reduced waterhemp density and biomass by an additional 2 and 7%, respectively. This research concludes that acetochlor applied in a tank mixture with flumioxazin was the most efficacious tank mixture evaluated for MHR waterhemp control.

Spatiotemporal Changes in Glyphosate Efficacy on Key Weed Species in the US and Canada.

Christopher A. Landau*[1], Kevin W. Bradley[2], Erin E. Burns[3], Ryan P. DeWerff[4], Michael L. Flessner[5], Karla L. Gage[6], Aaron G. Hager[7], Joseph T. Ikley[8], Prashant Jha[9], William G. Johnson[10], Paul O. Johnson[11], Sarah Lancaster[12], Dwight Lingenfelter[13], Mark M. Loux[14], Eric J. Miller[15], Scott A. Nolte[16], Debalin Sarangi[17], Peter H. Sikkema[18], Christy L. Sprague[3], Mark VanGessel[19], Rodrigo Werle[20], Martin Williams[21], Bryan G. Young[22]; [1]University of Illinois, Crop Sciences, Urbana, IL, [2]University of Missouri, Columbia, MO, [3]Michigan State University, East Lansing, MI, [4]University of Wisconsin, Madison, WI, [5]Virginia Tech, Blacksburg, VA, [6]Southern Illinois University Carbondale, Carbondale, IL, [7]University of Illinois, Urbana, IL, [8]North Dakota State University, Fargo, ND, [9]Iowa State University, Ames, IA, [10]Purdue University, West Lafayette, IN, [11]South Dakota State University, Brookings, ND, [12]Kansas State University, Manhattan, KS, [13]Penn State University, University Park, PA, [14]The Ohio State University, Columbus, OH, [15]Southern Illinois University, Carbondale, IL, [16]Texas A&M AgriLife Extension, College Station, TX, [17]University of Minnesota, St. Paul, MN, [18]University of Guelph, Ridgetown, ON, Canada, [19]University of Delaware, Georgetown, DE, [20]University of Wisconsin-Madison, Madison, WI, [21]USDA-ARS, Urbana, IL, [22]Purdue University, Brookston, IN (317)

The adoption of glyphosate-tolerant crops beginning in 1996 led to an unprecedented reliance on glyphosate that facilitated the evolution of glyphosate-resistant weed populations. Currently 16 weeds species are reported to have glyphosate-resistant weed populations in the US and Canada. In areas where glyphosate-tolerant crops dominate, how rapidly does glyphosate efficacy change over time? The objectives of this research were to: 1) quantify spatiotemporal changes in glyphosate efficacy on troublesome weed species, and 2) compare efficacy of glyphosate used alone to that of glyphosate preceded by an effective, soil-residual herbicide applied close to crop planting. A database of herbicide evaluation trials was constructed from 24 individual datasets, spanning as many as 25 years, from locations across the US and Canada and harmonized across various file formats. Two subsets were created; one with in-crop foliar-applied glyphosate as the only treatment and, the other with in-crop glyphosate following a soil-residual herbicide application. Within each subset, mean and variance of weed control ratings for seven problematic weed species were regressed over time for 14 locations. With few exceptions, mean weed control with glyphosate alone decreased over time, whereas variation in weed control increased over time. When glyphosate application was preceded by a soil-residual herbicide effective against the target weed species, there was little to no change in mean weed control, or variation in weed control, over time. Additionally, for several cases, there was a noticable decrease in mean weed control prior to the species' first report of glyphosate resistance for the state/province from which the data were obtained. While it comes as no surprise that utilizing a soil-residual herbicide in tandem with glyphosate would provide better, more stable weed control compared to glyphosate alone, our experimental approach provides a means to quantify such dynamics within field populations from various sites throughout North America across a broad temporal range.

Chaff Lining for Management of Weeds in Winter Wheat. Matthew P. Spoth[1], Michael L. Flessner*[1], Kevin Bamber[1], Wykle C. Greene[1], Cynthia Sias[1], Eli C. Russell[1], Vipin Kumar[2], Vijay Singh[3]; [1]Virginia Tech, Blacksburg, VA, [2]Virginia Polytechnic Institute and State University, Painter, VA, [3]Virginia Tech, Painter, VA (318)

Chaff lining is a form of harvest weed seed control that concentrates the chaff fraction of harvest residues and weed seed therein into a narrow line using a chute installed on the back of the combine. We examined chaff lining for the management of wild mustard (*Sinapis arvensis* L.) and Italian ryegrass (*Lolium perenne* L. ssp. *multiflorum* (Lam.) Husnot) in wheat due to their problematic status and seed retention at harvest. The objective was to determine how varying wheat yield and the associated chaff amount will affect chaff lining weed control and the subsequent crop (double crop soybean followed by wheat) compared to a conventional harvest. Both weed species were evaluated in separate experiments across six site years with no prior history of the weeds. Since chaff amount increases with crop yield, chaff lines were created to mimic a range of wheat yields, with equal weed seed additions based on existing fecundity and seed retention data to each chaff line. A conventional harvest (control) and an outside-the-chaff-line treatment were included, where total fecundity or weed seed rain occurring prior to harvest based on weed species were broadcast, respectively. Despite reductions in stand and yield in the chaff line, crop yield across treatments at the field scale (which accounts for both chaff lines and outside-the-chaff-line), was not affected in double-crop soybean following wheat chaff lining. Similarly, field scale wheat yield compared to conventional harvest was not different at 8 site-years, increased in 3 site-years, and decreased in 1 site-year. Wheat chaff lining reduced total weed emergence over the combined double-crop soybean and winter wheat growing seasons by 43-54% at the field scale. Greater amounts of chaff caused greater reductions in weed emergence. This research indicates strong potential for chaff lining in wheat and justifies further work at commercial scale.

Changes in Weed Population Density After Harvest with a Seed Impact Mill. Eli C. Russell*, Michael L. Flessner, Matthew P. Spoth, Wykle C. Greene, Kevin Bamber; Virginia Tech, Blacksburg, VA (319)

Seed impact mills are modifications that are mounted directly to the back of a combine. They kill weed seeds that exit the combine in the chaff fraction, and they are one way to implement harvest weed seed control (HWSC). By killing the weed seeds, seed impact mills prevent seeds from being added to the soil seed bank. Over time the soil seed bank should decrease resulting in less weeds emerging in subsequent years. The purpose of this experiment was to track emergence patterns of common ragweed (*Ambrosia artemisiifolia* L.) and Italian ryegrass (*Lolium perenne* L. ssp. *multiflorum* (Lam.) Husnot) in soybean and wheat fields, respectively, after harvesting with a seed impact mill. We tracked common ragweed density in several soybean fields in commercial production, which are categorized as follows: (1) After one harvest with the seed impact mill evaluate density after planting the following year, (2) After one harvest with the seed impact mill evaluate density before harvest the following year (3) Evaluate density after one harvest with the seed impact mill followed by one year of soybeans without harvest with a seed impact mill, and (4) Evaluate density after two harvests with the seed impact mill. We tracked two wheat fields in commercial production and tracked Italian ryegrass density the following season after one harvest with the seed impact mill. For testing, each field was divided into two sections, the ON section was harvested with the seed impact mill and the OFF section was harvested without it. Density reductions for common ragweed after planting ranged from -56% to 51%, with an average reduction of 11% across all 6 sites after one harvest using the seed impact mill. Density reductions before harvest ranged from -11% to 81% with an average reduction of 50% across all 5 sites after one harvest using the seed impact mill. After one year of using the seed impact mill followed by one year of soybeans harvested conventionally, density reductions were -36% and 47% with an average of 5.7%. After two years of using a seed impact mill, common ragweed density reductions were -19% and 29%, with an average reduction of 5.2%. Density for Italian ryegrass after one harvest ranged from 29% to 34% with an average reduction of 32%. Density reductions are highly variable because of the soil seed bank. Although using the seed impact mill reduced seed inputs into the soil for one or two years, it will not prevent the seeds that are already in the soil from emerging the next growing season. Previous research indicates that it could take 5 to 7 years of implementing HWSC before the soil seed bank is minimized. More research is needed, but our data indicate that there is promise for seed impact mills to be an effective tool for reducing the soil seed bank.

Reduction in Palmer Amaranth (Amaranthus Palmeri) Emergence and Growth Rate in Response to Cover Crop Termination Timing: Applications for Optimizing Herbicide Programs. Cynthia Sias*, Michael L. Flessner, Kevin Bamber; Virginia Tech, Blacksburg, VA (320)

The evolution of herbicide resistance, led by Palmer amaranth (Amaranthus palmeri S. Wats.), has highlighted the need for non-chemical weed control options. Cover crops are known to suppress Palmer amaranth, but questions remain regarding termination practices to maximize weed control benefits and if herbicide inputs can be reduced in the presence of a cover crop. An experiment was conducted in 2021 and 2022 to determine the effect of cover crop termination timing on Palmer amaranth emergence patterns and growth rate. Treatments included planting brown (cereal rye terminated two weeks before soybean planting), planting green (cereal rye terminated at soybean planting), and a no cover (winter fallow) check. Palmer amaranth emergence counts and growth rate were evaluated from new emergence cohorts every two weeks up to ten weeks after soybean planting. To explore herbicide program optimization, simulated herbicide programs were compared consisting of all relevant combinations of residual and postemergence (POST) herbicides from 1-, 2-, and 3- pass programs that resulted in Palmer amaranth control until soybean canopy. Residual herbicides were assumed to suppress Palmer amaranth for 28 days, after which the number of days to reach 10 cm in height was used to determine when and if a POST1, POST1 + residual, or POST2 herbicide application(s) were necessary. Simulated herbicide program analysis indicated that the simplest and most cost-effective cover crop and herbicide regime was winter fallow with a single POST herbicide without residual. Among cover crop termination options (planting green versus brown), planting green with a single POST herbicide optimized inputs. This finding contradicts previous research and survey data, likely due to the abnormally dry May experienced in this study, which delayed and reduced Palmer amaranth emergence compared to winter fallow. Although more site-years are necessary, this information will provide better recommendations for herbicide programs when integrating cover crops.

The Cover Crop Evolution/Revolution- Past, Present, Future. William S. Curran*; Penn State University, Bozeman, MT (321)

Cover crops have been used for millennia. Ancient civilizations used certain plant species to enhance the growth of their food crops. During American colonial times, farmers grew crops to eat and sell and other crops to replenish the soil. Prior to WWII, green manures were commonly used to provide N to rotational grains and sometimes used as forage for livestock. Grass covers such as cereal rye (*Secale cereale L.*) were also employed to reduce soil erosion. With the introduction of synthetic N fertilizer as well as herbicides and other pesticides after WWII and a decrease in the number of integrated crop-livestock farms, larger farms, and more specialization, cover crop use declined. By the 1980's, only smaller more specialized farms such as organic operations employed cover crops where they were an important component to raising crops without synthetic inputs. With the increased interest in conservation tillage that began some years earlier, by the 1990's farmers and academic researchers were testing different no-till practices that often-included cover crops mostly to help prevent soil erosion over the winter. The introduction of paraquat and then glyphosate were break-through technologies that allowed farmers to burndown weeds and cover crops using no-till. The first National No-tillage Conference was held in 1993 attracting more than 800 attendees. That conference has grown to well over a thousand attendees in its 31st year in 2023. By the 21st century, adoption of no-till with the aid of herbicide resistant crops was becoming common place in regions of the corn and cotton belts. Although interest in cover crops increased, adoption by most US farmers was not wide-spread. The soil health and sustainability movement took off during the first decade of the 21st century and by 2010 and still today, every ag conference was talking about soil health and cover crops. Several new organizations evolved around the country that included passionate farmers that strongly believed in the merits of no-till, but also were looking to advance their production systems by integrating cover crops and other more sustainable practices. By 2012, the US Ag Census reported that only about 2% of the US row crop production acres included cover crops. However, over the last ten years, cover crop adoption by farmers in some regions of the US has at least doubled (USDA-ERS report). According to a 2017 USDA-NASS survey, states like Maryland help lead the charge using cover crops on almost one-half million of their 2-million acres (25%), while Pennsylvania reports about 600,000 acres with about 2 million acres of corn and soybean. The Iowa Nutrient Research & Education Council (INREC) reported that Iowa planted 2.2 million acres of cover crops on its 23 million acres of corn and soybean in 2019 (9.5%) and Indiana set a conservation record in 2021 by planting an estimated 1.5 million acres on about 11 million acres (13.7%) of corn and soybean (Indiana Conservation Partnership survey). Finally, across the Midwest, cover crop adoption averaged 7.3% in 2021 up from just 1.8% a decade prior according to a University of Illinois study that used satellite-based remote sensing to detect cover crops across 140 million acres (*Geophysical Research Letters* (2022). DOI: 10.1029/2022GL100249). The interest in improving soil health has been a primary driver along with increased financial incentives offered by both state and federal partnerships, but several other benefits have also played a role. One of those perceived benefits has been improved weed control and the spread of herbicide-resistant weeds which is helping to drive adoption. This presentation will further explore some of these benefits, areas of concern, and opportunities to advance cover crop adoption in the future.

Cross Resistance to Acetyl-CoA Carboxylase-Inhibiting Herbicides in a Johnsongrass [(*Sorghum halepense* (L.) Pers.] Accession Collected in Arkansas. Fidel Gonzalez Torralva*, Jason K. Norsworthy; UNIVERSITY OF ARKANSAS, Fayetteville, AR (322)

Johnsongrass is a troublesome weed species in cropping systems including corn (*Zea mays* L.), cotton (*Gossypium hirsutum* L.), and soybean [*Glycine max* (L.) Merr.]. A putative-resistant johnsongrass accession to ACCase-inhibiting herbicides was collected in eastern Arkansas. The objective of this research was to assess this accession for resistance to ACCase-inhibiting herbicides. Screening of this accession to 1× field rate of fluazifop (1× = 210 g ai ha^{-1}) and pinoxaden (1× = 70 g ai ha^{-1}) showed a 75 and 83% plant survival, respectively at 21 d after herbicide treatment. A susceptible johnsongrass accession treated under the same conditions displayed 100% mortality using both herbicides. Initial dose-response experiments with field seeds revealed a resistance index >16 using dry weight. These results demonstrated the resistance of johnsongrass to fluazifop and pinoxaden herbicides. Experiments to describe the resistance mechanisms in this johnsongrass accession are underway.

Georgia's Endangered Species Pilot Program: A Team Approach to Help Protect Species and the Farm. Taylor M. Randell-Singleton*[1], Jenna C. Vance[1], William J. Chism[2], A Stanley Culpepper[1]; [1]University of Georgia, Tifton, GA, [2]US EPA, Point Of Rocks, MD (323)

Weed management in Georgia and throughout the U.S. is challenged by herbicide label restrictions that prohibit the use of certain products in either sections of agricultural fields, entire fields, or whole counties. These restrictions were established out of an abundance of caution to protect threatened and endangered species under the federal Endangered Species Act (ESA), and are in place where listed species may be present. While efforts to protect listed species are critically important, preliminary information suggest that these restrictions may be more prohibitive than necessary, limiting the ability to effectively manage pests, reducing land suitable for agricultural production, and therefore impacting food, feed, and fiber supplies. The Georgia Pilot Program was developed to address ESA restrictions placed on practical herbicide use in the state, and is a cooperative effort between the University of Georgia Cooperative Extension Service (UGA), the Georgia Department of Agriculture (GDA), the U.S. Fish and Wildlife Service (FWS), the U.S. Environmental Protection Agency (EPA), and Georgia farmers. Initially, the Pilot Program has focused on addressing the EPA's restrictions placed on Enlist Duo in Georgia, where applications are prohibited in 11 counties due to historical record of two endangered salamander species. Through collaboration with Southern Regional FWS, information was collected to correlate the potential interaction of these salamanders with farm fields to create refined range maps, as well as to identify previously unknown habitats where additional populations may exist. If a farm field is identified as having the potential to damage these species or their habitats when treated with Enlist Duo, mitigation plans will be implemented to protect the species, thereby eliminating concern. Upon completion of this effort, a request for the returned use of Enlist Duo to the appropriate Georgia counties through cooperation with the GDA and EPA will be made. Once efforts are completed with Enlist Duo, a similar process will be initiated for glyphosate and atrazine. Through cooperation with multiple agencies, farmers, and other partners, the Georgia Pilot Program hopes to contribute to the ongoing efforts of finding timely solutions to preserve listed species while protecting the practical use of pesticides and farm sustainability.

Survey of Herbicide Resistance in Pigweed *(Amaranthus* Spp.) Escapes in North Dakota Row Crops.
Joseph T. Ikley*[1], Stephanie DeSimini[1], Michael J. Christoffers[1], Megan O'Neil[2], Zack Bateson[2]; [1]North Dakota State University, Fargo, ND, [2]National Agricultural Genotyping Center, Fargo, ND (324)

Pigweeds (*Amaranthus* spp.) are some of the most problematic weeds found across the Midwest in agronomic crops, and are among the many weeds that are developing herbicide resistance. To verify the extent of existence of herbicide resistant (HR) pigweed populations in North Dakota, The National Agricultural Genotyping Center (NAGC) and North Dakota State University (NDSU) initiated a statewide project to screen pigweed samples for resistance to three herbicides: glyphosate, imazamox, and fomesafen. Pigweed samples were collected from 16 counties in North Dakota from September through November 2021 and sent to NDSU for a greenhouse trial. Pigweed species in this study included Palmer amaranth (*Amaranthus palmeri*), waterhemp (*Amaranthus tuberculatus* (Moq.) Sauer), redroot pigweed (*Amaranthus retroflexus L.*), Powell amaranth (*Amaranthus powellii*), and tumble pigweed (*Amaranthus albus L.*) The greenhouse trial was a RCBD with four replicates. Each population was sprayed with two rates of each herbicide (1X and 3X) as follows: glyphosate (1260 g ae ha^{-1} and 3780 g ae ha^{-1}), imazamox (35 g ae ha^{-1} and 105 g ae ha^{-1}), and fomesafen (198 g ae ha^{-1} and 594 g ae ha^{-1}). Twenty-one days after application (DAA) plants were visibly rated on a scale of 0 to 100% (0=no symptoms, 100=plant death). Prior to herbicide application, one leaf per plant was excised and sent to the NAGC for DNA extraction and genotyping. Greenhouse and genotyping results revealed waterhemp resistance to glyphosate, imazamox, and fomesafen in seven counties surveyed. The NAGC used high-throughput genotyping of EPSPS copy number to confirm glyphosate resistant waterhemp in eight counties, and Palmer amaranth in four counties. The PPO-210 deletion was genotyped in waterhemp, confirming resistance in seven counties. Resistance to imazamox was determined in populations by NAGC confirmation of two target-sites of interest, amino acid sites ALS-574 and ALS-653. The ALS-574 substitution was found in a resistant Palmer amaranth and the ALS-653 has been confirmed in Powell amaranth, waterhemp, and tumble pigweed samples. These results confirm that pigweed populations in North Dakota have resistance to glyphosate, imazamox, and fomesafen.

Does Interseeded Cover Crop Seed Size Influence Tolerance to Residual Herbicide Programs? Tosh R. Mazzone*[1], John M. Wallace[1], Matthew R. Ryan[2], Chris J. Pelzer[2]; [1]Penn State University, University Park, PA, [2]Cornell University, Ithaca, NY (325)

Cover cropping success can vary widely based on environmental and management conditions in the Northeast US. Use of long-lived residual herbicides can reduce cover crop establishment, particularly when interseeded into standing corn at early growth stages, and there is interest in developing herbicide recommendations for use of diverse cover crop mixtures. Understanding relative sensitivity of various cover crop species to common corn residual herbicides may improve grower decision making when incorporating cover crops into a conventional grain cropping system. A field experiment was conducted in 2021 in corn grain to assess interseeded cover crop mixture response within functional groups such as plant family to residual herbicide programs. The experiment was designed as a RCBD with four replications and three site years, including Russel E. Larson Agricultural Research Center (RS; Pennsylvania Furnace, PA), Southeast Agricultural Research and Extension Center (LV; Manheim, PA) and Musgrave Research Farm (CN; Aurora, NY). Main plots consisted of cover crop treatments seeded at V5/6 corn growth stage using a high clearance no till drill Interseeder, and included a (1) small-seeded mix composed of annual ryegrass + medium red clover + rapeseed seeded at 22, 4, 1 kg ha^{-1}, respectively; (2) a large-seeded cover crop mixture included cereal rye, cowpea, and tillage radish seeded at 67, 6, 1 kg ha^{-1}, respectively; and (3) a cereal rye monoculture seeded at 101 kg ha^{-1}. Split plots included four PRE herbicide treatments applied at corn planting using the 1X label rate, including (1) saflufenacil + dimethenamid-P (Verdict, 0.63 kg ai ha^{-1}), (2) acetochlor + atrazine (Harness, 1.96 kg ai ha^{-1} + AAtrex, 1.12 kg ai ha^{-1}), (3) S-metolachlor + atrazine + mesotrione + atrazine (Lumax EZ, 2.78 kg ai ha^{-1} + AAtrex, 0.42 kg ai ha^{-1}), and a control consisting of a preplant burndown followed by a glyphosate only POST (Roundup Powermax, 1.54 kg ai ha-1) treatment at interseeding. Cover crop and weed abundance were estimated by harvesting aboveground biomass at ground level within an 0.5-m^2 quadrat at 30- and 120-days post interseeding (DPI). Biomass was sorted to species and oven dried until constant weight. Plant biomass variable was log-transformed to fit assumptions of normality. Relative response indices

(RRI) were calculated to standardize plant responses to the POST glyphosate control within cover cropping treatments. Models were fit using linear mixed model approach using package *nlme* and post-hoc means comparisons were estimated using Tukey's LSD from package *emmeans* in R. Data were analyzed first in the POST glyphosate only control at 30 DPI to quantify interseeded cover crop biomass potential. Total cover crop biomass ranged from 66-94 kg ha^{-1} at LV and RS, which was 32-46% lower than at site CN (207 kg ha^{-1}; $p < 0.001$). These results suggest that corn competition is an important factor in cover crop growth early season. At 30 DAI, cereal rye and large-seeded cover crop mix produced higher dry biomass at 140-193 kg ha^{-1} compared to the small-seeded mix ($p < 0.001$). This suggests that large-seeded mixtures can accumulate biomass more quickly across a wide latitudinal range. Based on relative response ratio (RRI) analysis, cover crop mixture was not significant. At RS and LV, Lumax + Atz, Harness + Atz, and Verdict RRI of cover crop biomass was not significantly different than the POST glyphosate control treatment, suggesting that these cover crop mixtures have some herbicide tolerance. RRI of Verdict at RS site exhibited the greatest cover crop response relative to the POST glyphosate control (0.37), 20% to 31% higher than the response to Harness + Atz (0.17) or Lumax + Atz (0.06). These results suggest that Verdict may improve cover crop response over a POST glyphosate control. At the CN site where Verdict (RRI -0.91, $p = 0.001$) provided the best weed control, 40% better than Lumax + Atz. This relationship may result from cover crop injury using Lumax + Atz, whereas cover crop was stimulated in Verdict treatment, resulting in an additive weed suppressive effect of herbicide + cover crop competition. At RS, Verdict weed management was 61-75% less effective (0.07, $p = 0.0021$) than other treatments and did not differ from the POST glyphosate control. At LV, both Verdict (0.72) and Harness + Atz (0.59) weed biomass RRI was higher than that of Lumax + Atz, indicating that both treatments broke at LV while Lumax + Atz remained effective through 30 DPI. This is likely why Verdict and Harness + Atz cover crop RRI was no different than the POST glyphosate control treatment at LV. Additional field trials evaluating cover crop response to alternative MOA will be instructive to develop management recommendations based on cover crop functional groups in future research.

A23980B: Corn Weed Control from Four Complimentary Active Ingredients. Craig Austin*[1], Seth A. Strom[2], Mark Kitt[1], Thomas Beckett[1]; [1]Syngenta, Greensboro, NC, [2]Syngenta Crop Protection, Monticello, IL (326)

A23980B is a new selective herbicide coming soon for weed control in field corn, seed corn, popcorn, and sweet corn containing s-metolachlor, pyroxasulfone, mesotrione, bicyclopyrone, and the safener benoxacor. The combination of four active ingredients in A23980B was designed to deliver residual control of difficult to manage weeds. Field trials were conducted to determine the benefit of two Group 15 herbicides and two Group 27 herbicides in a premix for long-lasting residual weed control. Results demonstrated that the active ingredients in A23980B work better together to deliver consistency and efficacy resulting in control of key weeds such as *Amaranthus* spp. and grasses. Overall, A23980B provides the foundation needed for growers facing the most problematic broadleaf and grass weeds in corn agronomic cropping systems.

Introducing Golden Pennycress (*Thlaspi arvense*) into the Corn-Soybean Rotation. Mark L. Bernards*, Brent S. Heaton; Western Illinois University, Macomb, IL (327)

Field pennycress (*Thlaspi arvense*) is being domesticated as a winter annual oil-seed crop. Pennycress will provide environmental services such as reducing erosion and nutrient losses. Oil extracted from the seed is expected to be used as a biofuel for the aviation industry, and meal will be incorporated into livestock feed rations. It is anticipated that pennycress will provide growers an additional revenue-generating crop that fits within a corn-soybean rotation. One model envisions pennycress being planted in the fall after corn, but pennycress establishment in corn residue has been inconsistent. Pennycress planted into soybean may establish better, but delaying corn planting until after pennycress harvest in late May might cause an unacceptable yield penalty. A portion of our research 1) simulated herbicide carryover and evaluated pennycress response in greenhouse studies, and 2) evaluated pennycress stand and yield following field applications of common corn herbicides in the preceding corn crop. In greenhouse studies, herbicide doses were based on half-life values and expected herbicide residual concentrations in the soil up to 90 days after herbicide application, and herbicides were applied as single active ingredient products. At 6 weeks after treatment in greenhouse studies, clopyralid, pendimethalin and trifluralin caused no stand loss and minimal injury at labeled use rates. Based on published half-life data and the corresponding simulated doses, pennycress did not show biomass reduction from doses expected: 30 days after application from 2,4-D, dicamba, rimsulfuron, isoxaflutole, tolpyralate and atrazine (in a soil with a history of atrazine use); 60 days after application of topramezone and dimethenamid-P; and 90 days after application of flumioxazin, thiencarbazone-methyl, mesotrione, and tembotrione. Four herbicides, saflufenacil, S-metolachlor, acetochlor, and pyroxasulfone, caused mild biomass reduction at doses representing concentrations expected 90 days after application. In field studies, commercial corn premixes were applied at 1X and 2X rates for the soil type in May and June. Corn was harvested as silage, and pennycress was drilled into the plots shortly after harvest. In the 2021-2022 field trial, only one herbicide combination (a 2X rate of isoxaflutole+thiencarbazone-methyl followed by a 2X rate of tembotrione+thiencarbazone-methyl) caused pennycress yield loss relative to the untreated control. In average weather years, degradation process in field soils appear to be adequate to minimize pennycress injury from most corn herbicides applied in May and June, even at 2X rates.

Investigating the Genetic Control of Interspecific Hybridization Between *Sorghum bicolor* and *S. halepense* Using Recombinant Inbred Lines (RILs). Usha R. Pedireddi*, Nithya K. Subramanian, George Hodnett, Patricia Klein, William Rooney, Muthukumar V. Bagavathiannan; Texas A&M University, College Station, TX (328)

Classification of Herbicide Symptomology in Cotton Using Digital Images and Deep Learning. Ubaldo Torres*[1], Joe Johnson[1], Shefali Sarang[1], Bishwa B. Sapkota[1], Scott A. Nolte[2], Muthukumar V. Bagavathiannan[1]; [1]Texas A&M University, College Station, TX, [2]Texas A&M AgriLife Extension, College Station, TX (329)

Herbicide-induced crop stress from off-target movement is a growing concern in cotton production, the severity of which is influenced by several factors, including crop sensitivity, herbicide chemistry, and dose. The specific herbicide-induced response (i.e. symptomology) and the level of damage are typically assessed through visual evaluations, which require expert knowledge and can at times be subjective and inconsistent. With the advances in computer vision, there is a potential to automatically classify herbicide-induced symptomology in cotton using digital images and deep learning. Field experiments were conducted in 2021 and 2022 at the Texas A&M Research Farm near College Station, TX, to classify specific herbicide symptomology and assess cotton injury using digital images and deep learning. Cotton injury from eight herbicides (nicosulfuron, 2,4-D, dicamba, atrazine, isoxaflutole, glyphosate, glufosinate, and paraquat) at three simulated drift rates (high, moderate, and low, all below 1X) were evaluated at three (15-, 30-, and 60-cm-tall) growth stages, along with a non-treated control. Herbicide treatments were applied with a CO_2 pressurized backpack sprayer to conventional cotton using a TTI11002 (for auxins) or a TT11002 (all other herbicides) nozzle, at a spray volume of 140 L ha^{-1}. Following the herbicide application, digital images were collected weekly using a Cannon® DSLR camera. Five pre-trained classification models (Densenet121, InceptionV3, VGG19, ResNet50, and ViT) were implemented. The dataset used in the study consisted of 21171 training images and 2514 testing images, belonging to 9 classes. Overall, the models effectively classified herbicides pertaining to different modes of action. The Densenet121 and InceptionV3 models provided similar results, with the non-treated control having the highest classification accuracy at 80%, followed by nicosulfuron and isoxaflutole. The standalone ViT model lacked overall accuracy, however once its backbone network is replaced by a convolution neural network, the ViT model produced the best results with classification accuracy between 74 and 96%. Findings show the great prospects of using image-based classification models for diagnosis of herbicide injury on crops. Research is ongoing to further improve the classification accuracy by fine-tuning the models, detecting injury on individual leaves, and modeling with other convolutional neural networks.

Evaluation of Crop Herbicide Tolerance Using UAV Obtained Multispectral Imagery. Ryan D. Langemeier*, Livia Pereira, Justin T. McCaghren, Steve Li; Auburn University, Auburn, AL (330)

The use of unmanned arial vehicles in agriculture has increased rapidly in recent years. While use is often focused on field scale usage, opportunity also exists for data collection in research trials. One area of potential focus is crop tolerance which does not require the use of artificial intelligence to identify weeds. In the 2021 and 2022 growing seasons four trials were conducted in central and southern Alabama on both corn (*Zea maize*) and upland cotton (*Gossypium hirsutum*). Both crops received herbicide treatments both pre-emerge 0 weeks after planting (WAP), and post-emergence 4 WAP. Each crop had herbicides from five different modes of actions (MOA) applied; ALS, HPPD, PSII, PPO, and synthetic auxin. Rates were chosen to cause moderate crop injury. Data collection involved visual injury and drone imagery collection 2, 4, 6, and 8 WAP. Heights were recorded at 4 and 8 WAP. Imagery was then stitched using Pix4D mapper, and analysis was conducted using QGIS 3.20. Green pixel count after thresholding was used to estimate stunting using a multispectral VDVI (visible difference vegetative index). UAV derived stunting values correlate well with data derived from imagery analysis, with most R^2 values in the 0.75 to 0.95 range. In addition, a similar process was done with a modified VDVI using RGB images, and results were similar as those derived from multispectral images. These results suggest that the use of UAVs could potentially add an objective data measurement to crop tolerance evaluations while reducing labor and time demands relative to current practices.

Comparison of Weed Control Outcomes When Using Precision Application Versus Traditional Approaches in Corn (*Zea mays*). Isaac H. Barnhart*[1], Greg Kruger[2], Kalvin Andrew Miller[3], Chris Proctor[4], Rodrigo Werle[5], Anita Dille[1]; [1]Kansas State University, Manhattan, KS, [2]University of Nebraska - Lincoln, North Platte, NE, [3]BASF, Seymour, IL, [4]University of Nebraska-Lincoln, Lincoln, NE, [5]University of Wisconsin-Madison, Madison, WI (331)

With the rise of artificial intelligence and weed recognition technology, precision herbicide applications are becoming a feasible option for producers. Precision sprayers allow for rapid on-the-go detection and treatment of weeds where they are in the field. The objective of this study was to evaluate weed control when applying current chemical weed control programs with a precision sprayer in corn [*Zea mays* (L.)]. A field study was conducted near Manhattan, Kansas in 2022. A dual-boom precision research sprayer was used to apply both preemergence (PRE) and postemergence (POST) herbicide treatments. Boom 1 was used to spot-spray (SS) foliar-applied herbicides and boom 2 was used to broadcast (BDCST) soil-applied (residual) herbicides. The experiment was arranged as a split-plot design, with the main plot factor being herbicide treatment program and the split-plot factor being weed detection sensitivity. Four herbicide treatment programs were evaluated: 1) SS and BDCST PRE followed by (fb) SS POST, 2) SS PRE fb SS POST, 3) SS and BDCST PRE only, and 4) SS and BDCST PRE fb SS and BDCST POST. Boom 1 applied a tank mixture of topramezone (12.3 g ai ha^{-1}) and glyphosate (840 g ae ha^{-1}) and boom 2 applied a dimethenamid-P (843 g ai ha^{-1}) and atrazine (2,244 g ai ha^{-1}) when applicable. The weed detection levels were high sensitivity (1), balanced sensitivity (2), a low sensitivity (3), and no sensitivity equal to a broadcast application (BC). PRE and POST applications were made on May 19 and June 17, 2022, respectively. Visual measurements (percentage of the plot that was weed-free) were taken 2, 4, and 6 weeks after the POST treatment (WAPT) on a scale between 0 and 100, with 0 indicating a complete weed infestation and 100 indicating no weeds within the plot. Generalized linear mixed models were fitted to the data using a beta distribution to analyze differences in the percentage of the plots that were free of weeds. The resulting models were analyzed using ANOVA, and means were separated with a Tukey honest significant difference test (a = 0.05). Results showed that at 6 WAPT, treatments with residual herbicides were less weedy than the treatments with no residual herbicides applied. At the same time frame, sensitivities 2 and 3 differed significantly from the BC application, but sensitivity 1 was not statistically different than BC. These results suggest that applying current chemical weed control programs in corn with precision application sprayers have potential in providing the same level of weed control as a traditional broadcast application.

Weed Control in Soybeans (Glycine Max) Using Precision Sprayers. Isaac H. Barnhart*[1], Proctor Chris[2], Thiago H. Vitti[2], Kalvin Andrew Miller[3], Greg Kruger[4], Anita Dille[1]; [1]Kansas State University, Manhattan, KS, [2]University of Nebraska-Lincoln, Lincoln, NE, [3]BASF, Seymour, IL, [4]University of Nebraska - Lincoln, North Platte, NE (332)

Effects of PRE Herbicides on Precision Post Emergence Herbicide Application in Plasticulture Production. Ana Claudia Buzanini*[1], Arnold W. Schumann[2], Nathan Boyd[3]; [1]University of Florida, Wimauma, FL, [2]University of Florida, Lake Alfred, FL, [3]University of Florida, Balm, FL (333)

To reduce off-target applications by banded or broadcast applications of postemergence (POST) herbicides, the University of Florida developed smart spray technology for weed control plasticulture systems. In the fall of 2021 and spring of 2022, one experiment was conducted on the plasticulture banana pepper field at the Gulf Coast Research and the Education Center in Balm, FL. The objective of this study was to evaluate the efficacy of a precision sprayer for POST-herbicide application in plasticulture pepper row middles in the presence or absence of pre-emergent (PRE) herbicide. The application of flumioxazin resulted in lower density for broadleaf and overall total in both seasons, as well as lower density of grasses in spring. For nutsedge, the two precision applications reduced the nutsedge density in spring. No significant pepper damage was observed by the spray methods tested. Using the smart spray technology, the herbicide volume used in treatments after the PRE-herbicide application was reduced by 84% and 54% for fall and spring respectively, and reduced by 30% and 45% for fall and spring respectively when no PRE-herbicide was applied. No reduction in weed control or pepper yield was observed using the precision application. Overall, the precision sprayer developed by the University of Florida reduced the sprayer volume with no reductions in weed control, no significant injuries on pepper, or negative effects on yield.

Herbicide Options in Industrial Hemp. Wyatt J. Stutzman*, Kevin Bamber, Cynthia Sias, Eli C. Russell, Michael L. Flessner; Virginia Tech, Blacksburg, VA (334)

With the increasing value in hemp production for cannabidiol, there is a need for efficacious weed management options. Current options are limited to manual and non-chemical options, as no herbicides are currently labeled for use in hemp production. To alleviate this issue, an experiment was established in 2020 and 2021 to evaluate potential herbicide candidates for pre-emergent (PRE), delayed pre-emergent (DPRE), and post-emergent (POST) use in transplanted, raised bed CBD hemp. A total of 8 PRE, 7 DPRE, and 6 POST herbicides were applied according to the label for other similar specialty crops. Data collection included visible phytotoxicity over the growing season as well as harvest biomass.. The trials were in an RCBD design with four replications in Blackstone and Blacksburg, Virginia. A total of 4, 3 and 3 site-years were completed for each PRE, DPRE and Post, respectively. Data was subjected to ANOVA in the SAS software JMP ($p<0.05$) and mean separation using Fisher's Protected LSD. Treatment by site-year interactions occurred in all data. Visible phytotoxicity >10 % was observed from dimethenamid (Outlook), acetochlor (Warrant) and pyroxasulfone (Zidua SC) in 1 site-year each at 28 days after application (DAA). At 56 DAA, however, Blackstone in 2020 was the only data in which phytotoxicity differences were observed. DPRE herbicide application at 14 DAA of clomazone (Command 3ME) showed 26% phytotoxicity damage in 1 site year. At 28 and 56 DAA, no differences in phytotoxicity were observed in DPRE applications. Bromoxynil (Moxy 2EC) caused 5% - 8% phytotoxicity 7 DAA in POST applications in Blacksburg for both 2020 and 2021. Differences were not observed in any other locations or ratings. Harvest biomass in relation to the nontreated checks were not influenced by treatments. This research suggests potential for herbicide labels for hemp production under certain circumstances. Herbicides such as S-metolachlor, ethalfluralin, pendimethalin, linuron and flumioxazin PRE, S-metolachlor, acetochlor, dimethenamid, ethalfluralin, pendimethalin and pyroxasulfone DPRE, and all POST treatments were safe on hemp, with less than 10% observed phytotoxicity and no reduction of biomass. Future research is needed to evaluate residue of herbicide treatments and retention in harvested buds.

Assessing Deposition and Plant Responses Following Dicamba Applications in Cotton and Soybean Using Hooded and Open Boom Sprayers. Devin Hammer*, Naresh Pai, Allen Olmstead; Bayer Crop Science, St. Louis, MO (335)

Primary movement or physical drift is among the greatest contributors to off-target movement of pesticides. Adoption of practices and technologies that mitigate physical drift are critical not only for continued protection of sensitive species and habitats, but also for proper control of troublesome pests. This study was conducted to further demonstrate the ability of hooded sprayers to decrease the amount of spray that is subject to physical drift away from the intended application area. Applications were made using both hooded and open boom setups spraying tank-mixes with a dicamba herbicide formulated for use in dicamba-tolerant cropping systems. Eleven different sites with multiple university academics were utilized that showcase areas of high agricultural activity. These comprised of 8 soybean and 3 cotton sites.Spray applications were made in a direction perpendicular to the prevailing wind direction (3 swaths at 6 m/swath). Three transects were established perpendicular to the sprayer travel in the downwind direction and filter papers placed from 3-50 m. Upwind filter papers were also placed in line with these transects at 30 m. Filter papers were analyzed for dicamba residue according to the current version of the ME-1871 method. The crop surrounding the spray block included dicamba-sensitive varieties of either soybean or cotton depending on location which were rated for visual symptomology at 14 and 28 d following application at distances that matched the filter paper placement along each transect as well as in the upwind direction.In 2021, average dicamba spray drift deposition across 10 of the 11 locations was below the no observable adverse effect levels at 5 m from the edge of field with a hooded application. At 7 of the 8 soybean locations, downwind visual symptomology averaged <10% at the nearest evaluated distance of 3 m from the hooded spray block at 28 d after application. Average injury remained below 10% for all subsequent distances rated. Email for Primary Author: Devin.Hammer@bayer.com

A Novel Tool for Improving Selection of Nozzles and Adjuvants, CPDA's New *Application Enhancement Certification* Program. James Reiss*; Precision Laboratories, LLC, Lake Geneva, WI (336)

A Novel Tool for Improving Selection of Nozzles & Adjuvants, CPDA's New Application Enhancement Certification Program. James D. Reiss*[1], J. Susan Sun[2], Bradley K. Fritz[3], Greg R. Kruger[4],[1]Precision Laboratories, LLC, Waukegan, IL, [2]Croda, Inc., New Castle, DE, [3]USDA-ARS, College Station, TX, [4]BASF, Durham, NC For over twenty years applicators have been searching for guidance in making more efficacious and on-target applications. To meet this industry need, and at the urging of academia, the

CPDA (Council of Producers and Distributors of Agrotechnology) has developed the Application Enhancement Certification Program. This program recognizes that drift control without biological efficacy is not a successful application and is intended to provide applicators a more complete picture of the likely outcome of an application based on the interactions between pesticide formulation type, nozzle design and adjuvant selection. To accomplish this objective, the test protocol was designed to compare the pesticide formulation type alone, to the inclusion of a single adjuvant and use-rate, across a combination of four different nozzle type designs and four pesticide formulations known to influence spray droplet spectra. The four nozzle design types chosen are: a single orifice flat fan, an air-induction flat fan, a turbulence chamber and an air-inducted turbulence chamber nozzle. The four pesticide formulations used in the test protocol are: an emulsifiable concentrate (EC), a soluble liquid (SL) formulated with anionic surfactant, a soluble liquid (SL) formulated with cationic surfactant, and a soluble liquid (SL) with minimal inerts/adjuvancy). All nozzles were 11004 in size and solutions were sprayed at 276 kPa and analyzed with Sympatec Helos Vario KR laser diffraction particle size analyzer in a low speed (24.14 km/h air flow) wind tunnel. With the R7 lens installed, it can detect particle sizes ranging from 18 to 3500 microns. For ease of use the data distribution set is segregated into three classification categories based on the actual reference nozzle data. The three categories are: percentage of spray volume less than ~160 m (too small); percentage of spray volume between ~160 m to ~840 m (effective); and the percentage of spray volume greater than ~840 m (too large). The test results, from the two different adjuvants presented here, demonstrate how the interactions between pesticide formulation type, nozzle and adjuvant selection resulted three different types of outcomes. These outcomes included, a negative impact of a 13% reduction in the spray volume determined to be in the effective droplet size range; a neutral or do no harm effect; and a notable 32% increase in the spray volume determined to be in the effective droplet size range. Of course, this information and program are not intended to supersede any pesticide manufacturer's label requirements but rather, when possible, to help applicators make better decisions relative to the variables involved in a specific application.

Dicamba Applications Using Drop Nozzles to Reduce Drift in Texas Cotton. Hayden R. Taylor*[1], Scott A. Nolte[2], Peter A. Dotray[3], Joshua A. McGinty[4]; [1]Texas A&M University, College Station, TX, [2]Texas A&M AgriLife Extension, College Station, TX, [3]Texas Tech University, Texas A&MTexas Tech University, Texas A&M AgriLife Research, Texas A&M AgriLife Extension, Lubbock, TX, [4]Texas A&M AgriLife Extension, Corpus Christi, TX (337)

Dicamba provides Texas cotton growers with a useful tool when battling herbicide resistant weeds. With the onset of glyphosate resistance, the mode of action of dicamba has proven to be a practical rotation method. However, due to the potential of off-target movement in dicamba, growers need to be well equipped to manage this drift-prone herbicide. In this study, we will observe alternative spray applications to reduce the potential for off-target movement deposition, from both physical drift and volatility. The study was conducted at three Texas A&M locations in Lubbock, Corpus Christi, and College Station. Two sites of dicamba tolerant cotton were planted and applied with dicamba and a drift reducing agent to simulate a practical early-season application. At Lubbock only, a third testing site was established for a third treatment. The sites were compared using different boom application methods of a standard open nozzle boom, a modified post-direct drop nozzle boom, and a hooded boom (Lubbock only). The sites were sprayed simultaneously under wind displaying off-label wind speeds. To detect physical drift, potted non-dicamba tolerant indicator species were placed in three transects downwind of the spray in the College Station and Corpus Christi locations. In Lubbock, non-dicamba tolerant cotton was field-planted surrounding the treated area. The indicator cotton was evaluated at distances of 30m upwind, 3m, 5m, 10m, 20m, 30m, 40m and 50m downwind where a filter paper was placed for deposition analysis. Weather data at all locations was implemented at 5-minute intervals. In conclusion of the application, the filter papers were collected and the indicated species were moved to a location off site. An injury percentage rating was taken of the indicator species at 14 and 28 days after treatment. Results showed a significant reduction in dicamba deposition when comparing a modified post-direct drop nozzle to all other boom types at most locations. Providing Texas growers with these data will be useful with increasing restrictions on dicamba use. With further research on application methods, cotton growers can confidently apply dicamba despite drift issues associated with the variety of Texas climates.

Developer of Novel Adjuvant Systems for Herbicides. Jim T. Daniel*[1], Chase T. Boman[2]; [1]Daniel Ag Consulting, Keenesburg, CO, [2]AgraSyst, Spokane, WA (338)

Evaluation of Soybean and Cotton Injury to Low Volume Applications as Influenced by Application Type and GPA. Livia Pereira*, Steve Li, Ryan D. Langemeier, Justin T. McCaghren; Auburn University, Auburn, AL (339)

Spray drones are still a new method of application to deliver pesticides and are being adopted worldwide. One challenge of spray drones is that due to logistical concerns, low volumes must be used for large fields. Additionally, the effect of rotor wash driving spray into the canopy may also affect crop herbicide response. Spray tank mix and volume interactions for drone applications still need to be studied in order to understand their effect on crop response. The objective of this study was to evaluate soybean injury to low-volume application as influenced by spray drones or ground sprayer and spray volumes. The study was conducted at two sites in Macon and Limestone counties, Alabama. The experiment was a randomized complete block design with three replications. Plot sizes at both sites were 5.5 m by 45.7 m for drone applications and 3.6 m by 45.7 m for ground sprayer applications. Treatments consisted of a combination of glyphosate, pyroxasulfone, COC, and AMS either with or without fomesafen applied with a machine-mounted CO_2 pressurized sprayer at different spray volumes. Application occurred when soybean was at the 5-6 trifoliate stage. Data collection included visual injury at 3, 7, 14, and 21 Days After Treatment (DAT). Soybean heights were taken at 7, 14, and 21 DAT. Additionally, NDVI was measured using a UAV sensor at 0, 7, 14, and 21 DAT. All data were analyzed in SAS 9.4. The results showed a site-by-treatment interaction at 3 DAT. At Macon County, treatments with fomesafen applied with ground sprayer at 93.5 L ha^{-1} presented the highest percentage injury, and it was significantly higher compared to drone applications at 18.7 L ha^{-1} for both treatments with and without fomesafen. At Limestone Co., treatments with fomesafen applied with ground sprayer at 46.7, 93.5, and 140 L ha^{-1} and without fomesafen at 140 L ha^{-1} also presented more injury than 17.7 L ha^{-1} both with and without fomesafen. For both sites, treatments applied with ground sprayer at 93.5 L ha^{-1} with fomesafen had a higher injury when compared to both drone applications at 18.7 L ha^{-1} at 7, 14, and 21 DAT. Overall, treatments without fomesafen tended to show less crop injury. Soybean heights were not statistically different from the NTC across all ratings. Spray application with ultra-low volume does not always mean more crop injury. The study showed that ultra-low volumes will not necessarily increase crop injury, however, questions on efficacy remain.

What's New at the Winfield United Innovation Center? Gregory K. Dahl*[1], Eric P. Spandl[2], Ryan J. Edwards[2], Joshua J. Skelton[3], Steven A. Fredericks[2], Aszhia K. Albrecht[4], Elizabeth R. Alonzi[2], Elizabeth Buescher[2], Cody Hoerning[2], Derick Jiwan[2]; [1]Winfield United, Eagan, MN, [2]Winfield United, River Falls, WI, [3]WinField United, Saint Paul, MN, [4]Land O' Lakes - Winfield United, River Falls, WI (340)

The Winfield United Innovation Center in River Falls, Wisconsin opened in September 2017. It was designed and constructed to address research, size, capabilities, and support the business needs of Winfield United. We regularly conducted tours and meetings for people interested in herbicides, adjuvants, application methods, fertilizers, crops, and other topics. Many features and capabilities were incorporated into the Winfield United Innovation Center. Chemistry lab areas, equipment and chemists were increased to analyze, develop, optimize, and scale up manufacturing of products. Greenhouses and growth chambers were added to conduct year-round research and studies that are difficult to conduct in the field. Wind tunnel spray analysis systems were created or reengineered to improve capabilities, analyses, and imaging of spray droplets formation, characteristics, movement, and fate. Research and imaging that were recently conducted at the Winfield United Innovation Center included spray drift and deposition comparisons of dicamba and glyphosate containing mixtures in a large, closed, wind-tunnel system using susceptible soybean plants and spray collectors. Another study used plant phenotyping methods to evaluate control of corn plants treated with clethodim or clethodim plus adjuvants. High speed imaging equipment was used to observe droplet formation with Pulse Width Modulation (PWM), elevated wind speed, simulated aerial application conditions, and other demonstrations.

Single Tank vs Dual Tank See and Spray™ Programs in Soybean. Wesley Everman*[1], Diego J. Contreras[1], Jason K. Norsworthy[2], Tristen H. Avent[2], Bryan G. Young[3], Marcelo Zimmer[4], Jasmine M. Mausbach[5], Michael M. Houston[2], Lauren M. Lazaro[6], William L. Patzoldt[6]; [1]North Carolina State University, Raleigh, NC, [2]University of Arkansas, Fayetteville, AR, [3]Purdue University, Brookston, IN, [4]Purdue University, West Lafayette, IN, [5]University of Nebraska-Lincoln, Lincoln, NE, [6]Blue River Technology, Sunnyvale, CA (341)

During the last few years, the implementation of machine learning for weed control has made significant progress. The first step were site-specific herbicide applications using green-on-brown technology to target weeds in fallow fields. The latest development has been green-on-green technology, where specialized sprayers can detect weeds within crops, as seen with Blue River and John Deere's See & Spray™ technology. See & Spray™ technology allows users to successfully detect, spray and control weeds within their crop productions. The See & Spray™ technology also includes a dual-tank system, which consists of two separate tank and spray systems, allowing a simultaneous application of different herbicides. The dual tank system provides benefits such as reduced herbicide output and joint application of herbicides that are incompatible as single tank mixes. An experiment was set up in soybeans at four locations (AR, IN, IL, MS) using an Agronomy Test Machine (ATM) with See & Spray™ technology to test how does a dual tank system compare to a single tank system regarding weed control. Herbicide treatments included a preemergence application and two postemergence applications with either a systemic or a contact product. Across all locations, dual tank system applications provided similar control as single tank system applications in soybeans.

Integrated Strategies of Weed Control in Semiarid Central Great Plains. Vipan Kumar*[1], Rui Liu[2], Sachin Dhanda[1], Prashant Jha[3], Jason K. Norsworthy[4]; [1]Kansas State University, Hays, KS, [2]Washington State University, Prosser, WA, [3]Iowa State University, Ames, IA, [4]University of Arkansas, Fayetteville, AR (342)

The widespread evolution of herbicide-resistant (HR) weeds poses a serious production challenge for producers in no-tillage (NT) semiarid Central Great Plains (CGP). The repetitive and sole reliance on herbicides with the same site of action (SOA), and lack of diversity in weed control practices resulted in evolved resistance to herbicides in major cropland weed species, including horseweed (*Erigeron canadensis* L.), kochia (*Bassia scoparia* L.), Palmer amaranth (*Amaranthus palmeri* L.), common waterhemp (*Amaranthus tuberculatus* L.), and Russian thistle (*Salsola tragus* L.). For instance, glyphosate resistance has widely spread among kochia and Palmer amaranth populations in the CGP region. In addition, several biotypes of these weed species have also evolved with multiple resistance to 4- to 5 different herbicide SOAs. Managing these multiple herbicide-resistant (MHR) weed populations is complex and varies both within and between regions. Nonetheless, increasing herbicide costs to manage these MHR weeds in combination with low commodity prices necessitates the development of ecological-based, alternative weed control strategies in the region. Several nonchemical strategies, including cover crops, occasional/strategic tillage, alteration in agronomic practices, and harvest weed seed control (weed seed destruction and chaff lining) have shown promising results in managing HR weeds in recent years. This paper aims to highlight the role of these nonchemical strategies as important component of integrated weed management (IWM) strategies for MHR weed control in the CGP region, with emphasis on ecological, economic, and agronomic benefits. We will also illustrate current research gaps, propose new research, and outreach needs for IWM adoption in the region.

Development of a Novel Electrocution-Based Robotic Platform for Weed Control in Vegetables and Field Crops. Ran Nisim Lati*[1], Avital Bechar[2]; [1]Department of Plant Pathology and Weed Research, Newe Ya'ar Research Center, Agricultural Research Organization (ARO) - Volcani Center, Ramat Yishay, Israel, [2]Institute of Agricultural Engineering, Agricultural Research Organization, The Volcani Center, Bet-Dagan, Israel, Bet Dagan, Israel (343)

Herbicides are the central element of most weed management programs in industrial countries. However, the growing concerns over environmental and human health aspects and the rising demand for pesticide-free foods have put great pressure on farmers to reduce the amount of herbicides being used and to adopt non-chemical weed control tactics. Electrical energy (weed electrocution) is an alternative weed control method that considered a sustainable mean as it neither causes soil erosion nor leaves chemical residue or contamination in the soil or the plant. The objectives of this study are to: 1) develop and assess a new, low-energy, electrocution-based weed control devices, 2) evaluate the impact of various operational and biological factors on the control performance and 3) develop autonomous prototype for weed electrocution for field crops and for orchards and side roads. To this end, two prototypes were developed for high-voltage weeding. The use of the electrostatic systems succeeds to damage and control weeds and can be considered as a breakthrough in the ability to use electrostatics in safe and efficient way for weeding. The damage to the weeds increased with increasing the exposure time for the electrostatic treatment and the electrostatic potential. The tolerance of the weeds to the electrostatic treatment increased with the weed growth stage. However, for at early stages, most species were damage until complete control level following 2-5 s of exposure time of the electrostatic treatment. As in other weed control means, the sensitivity to the treatment will vary for different weed species.

Investigating Novel Integrated Weed Management Systems on the Canadian Prairies. Breanne D. Tidemann*[1], K. Neil Harker[1], Steve Shirtliffe[2], Christian Willenborg[2], Eric N. Johnson[2], Robert Gulden[3], Newton Lupwayi[4], T. Kelly Turkington[1], Emma Stephens[4], Charles M. Geddes[4]; [1]Agriculture and Agri-Food Canada, Lacombe, AB, Canada, [2]University of Saskatchewan, Saskatoon, SK, Canada, [3]University of Manitoba, Winnipeg, MB, Canada, [4]Agriculture and Agri-Food Canada, Lethbridge, AB, Canada (344)

Investigating Novel Integrated Weed Management Systems on the Canadian Prairies. Tidemann, B.D.*[1], Harker, K.N.[2], Shirtliffe, S.J.[3], Willenborg, C.J.[3], Johnson, E.N.[3], Gulden R.H.[4], Lupwayi, N.Z.[5], Turkington, T.K.[1], Stephens, E.[5], Geddes, C.M.[5] [1]Agriculture and Agri-Food Canada, Lacombe, AB. [2]Retired. [3]University of Saskatchewan, Saskatoon, SK. [4]University of Manitoba, Winnipeg, MB. [5]Agriculture and Agri-Food Canada, Lethbridge, AB. Using integrated weed management systems can aid in management and delayed evolution of herbicide resistant weeds. This five-year rotational study was conducted at six locations in the Canadian Prairies: Lacombe, Beaverlodge, and Lethbridge, AB; Scott and Saskatoon, SK, and Carman, MB. Treatments included combinations of rotational diversity, increased seeding rates, reduced herbicide applications, chaff collection, and crop silaging. Wild oat (*Avena fatua*), and wild buckwheat (*Fallopia convolvulus*) populations were supplemented at each site to increase consistency in density. Other common problem weeds were also supplemented, specific to each location. In the final year of the study all treatments were seeded into a 2x seeding rate of wheat and no herbicides were applied to allow for even comparison of the impact of the preceding 4 years of treatments. Weed counts were conducted by species, weed biomass collected by broadleaves and grasses, and crop yield as well as other agronomic data measured and collected. IWM treatments were compared to a continuous canola-wheat rotation, with baseline seeding rates and full herbicide applications. Wild oat, across locations, only showed significant increases in density compared to the control in some of the treatments without herbicide applications. Wild buckwheat showed no significant differences in density in any treatment compared to the control treatment. Wheat yield in the final year was significantly increased in treatments preceded by 3 years of alfalfa, or when chaff collection was added to base seeding rates and herbicide use. In some cases broadleaf weed biomass was reduced, not due to control imposed by the treatment, but due to lack of control of grass weeds in those treatments which then outcompeted the broadleaf weeds. Overall this study shows that some combinations of integrated weed management strategies can be as effective in managing weed populations as a traditional herbicide only based management system, however the efficacy of those strategies will also be species specific.breanne.tidemann@agr.gc.ca

Soil Residual Herbicide Degradation as Influenced by Soil Microbial Activity in Cover Cropping Systems.
Lucas Oliveira Ribeiro Maia*[1], William G. Johnson[1], Bryan G. Young[2], Eileen J. Kladivko[1], Shalamar Armstrong[1]; [1]Purdue University, West Lafayette, IN, [2]Purdue University, Brookston, IN (345)

The use of residual herbicides at cover crop termination has been proposed as one strategy to improve the weed control during the critical weed-free period. Previous research suggests that the use of cover crops can increase soil microbial activity which, in turn, is known as the main pathway for herbicide degradation in the soil. Limited research has been published on the interaction of soil microbial activity and degradation of soil residual herbicides in cover cropping systems. Long-term research trials were established in 2019 and conducted in a corn (*Zea mays* L.)-soybean (*Glycine* max) rotation to investigate the influence of cereal rye (*Secale cereale* L.) and crimson clover (*Trifolium incarnatum* L.) cover crops on soil microbial activity and degradation of soil residual herbicides. Trials were established in a split-plot design with cover crops and fallow as the main plot and three herbicide programs as the subplot. The three herbicide programs were: no residual (glyphosate + glufosinate), medium residual (glyphosate + atrazine + *S*-metolachlor), and heavy residual (glyphosate + atrazine + *S*-metolachlor + mesotrione). Herbicides were applied at cover crop termination, two weeks before corn planting. Cover crop biomass was determined the day before termination. ß-glucosidase (BG) activity, dehydrogenase (DHA) activity, and herbicide concentrations were measured from soil samples taken at 0, 10, 14, 28, 56, 84, and 112 days after termination (DAT) at 0 to 5 cm depth. Weed biomass was determined in four weeks after corn planting. The use of cereal rye for three years increased ß-glucosidase and dehydrogenase activities by an average of 23 and 76%, respectively, compared to the fallow control. However, the degradation of atrazine and mesotrione was not affected by the increased soil microbial activity measured in cover cropping systems. The initial concentrations of atrazine and mesotrione in the soil were reduced by up to 41 and 36%, respectively, in the presence of cereal rye compared to the fallow control. The inclusion of three residual herbicides at cereal rye termination provided 95% reduction in weed biomass compared to the termination without residual herbicides. E-mail for contact: oliveirl@purdue.edu

Incorporating Winter Camelina or Canola into Rotational Cropping Systems of Colder Climates to Reduce Weed Pressure. James V. Anderson*[1], Wun S. Chao[1], Xuehui Li[2], Russ W. Gesch[3], Marisol T. Berti[2], David P. Horvath[1]; [1]USDA-ARS, Fargo, ND, [2]North Dakota State University, Fargo, ND, [3]USDA-ARS, Morris, MN (346)

Winter oilseed cash cover crops are gaining popularity in integrated weed management programs for suppressing weeds. A RCBD with three replications of winter canola (*Brassica napus* L.), winter camelina [*Camelina sativa* (L.) Crantz], and fallow was used at two field sites (Fargo, ND and Morris, MN) to determine the freezing tolerance and weed-suppressing traits of these oilseed crops under field conditions in the upper Midwestern USA. The top 10 freezing tolerant accessions from a phenotyped population of winter canola (under controlled growth chamber conditions) were bulked and planted at a rate of 5.0 kg ha^{-1} with 30.5-cm row spacing and winter camelina (cv. Joelle) was planted at 6.7 kg ha^{-1} with 15-cm row spacing. To phenotype our entire winter canola population (621 accessions) for freezing tolerance under field conditions, seeds were also bulked and planted at 3.7 kg ha^{-1} with 30.5-cm row spacing. All canola and camelina were no-till seeded at Fargo directly and no-till seeded into wheat stubble at Morris at two planting dates, late August (PD1) and mid-September (PD2) 2019. Sampling date (SD) for winter survival of oilseed crops (plants m^{-2}) and their corresponding weed suppression (plants m^{-2} and dry matter m^{-2}) were collected between April and June 2020. Crop and SD were significant (P <0.05) for crop plant counts at both locations, and PD in Fargo and crop x PD interaction in Morris were significant for weed dry matter. At Morris and Fargo, PD1 produced better winter canola survival (28% and 5%, respectively) and PD2 produced better camelina survival (79% and 72%, respectively). Based on coefficient of determination (r^2), ~50% of weed plant count was explained by camelina density, whereas <20% was explained by canola density at both locations. Camelina from PD2 suppressed weed dry matter by >90% of fallow at both locations, whereas weed dry matter in canola was not significantly different from fallow at either PD. Genotyping of overwintering canola under field conditions identified nine accessions that survived at both locations, which also had excellent freezing tolerance under controlled conditions. These accessions are good candidates for improving freezing tolerance in commercial canola cultivars.james.v.anderson@usda.gov

Cereal Rye Cover Crop Significantly Delays Weed Emergence in Cotton. Gustavo Camargo Silva*, Muthukumar V. Bagavathiannan; Texas A&M University, College Station, TX (348)

The most popular cover crop species in recent times in the United States is cereal rye, which apart from offering soil and water conservation benefits, has great potential as an integrated weed management tool. There is evidence that cereal rye can delay weed emergence timing in summer cash crops, potentially by decelerating soil warming in the spring and by physically suppressing weeds with a dense biomass. This project seeks to determine the effect of cereal rye biomass levels on soil temperature, water, and in turn weed emergence patterns in a cotton crop. Cereal rye was planted at four seeding rates (0, 20, 40, and 80 kg ha^{-1}) and terminated at three timings (6, 4, and 2 weeks before planting the cash crop), totaling 12 treatments to create a wide range of biomass. Soil water and temperature were continuously monitored using automatic sensors. Cereal rye biomass production was determined at termination, which was carried out using glyphosate (870 g ae ha^{-1}). Cotton was planted into the cereal rye residues using a no-till drill. Weed density and emergence were assessed during early summer. Preliminary results show that high cereal rye biomass levels reduce the soil thermal amplitude by 5°C at the cash crop planting time. The cereal rye residue also reduced the evaporation of soil water by 0.2 m^3 m^{-3}. Weed emergence was both reduced and delayed in cereal rye plots compared to the fallow ones, the extent of which had a positive relationship with cereal rye biomass levels. A reduction in the amplitude of soil temperature is intrinsically tied to a reduction in soil water evaporation, and this mechanism plays a significant role in weed suppression by cereal rye cover crops.

Effects of Cover Crop Management on Weed Emergence Across the Eastern United States. Eugene P. Law*[1], Mark VanGessel[2], Steven Brian Mirsky[3], Victoria Ackroyd[4], Muthukumar V. Bagavathiannan[5], Kevin W. Bradley[6], William S. Curran[7], Adam S. Davis[8], Jeffrey Evans[9], Wesley Everman[10], Michael L. Flessner[11], Nicholas R. Jordan[12], Lauren M. Lazaro[13], Ramon G. Leon[10], John Lindquist[14], Jason K. Norsworthy[15], Lovreet S. Shergill[16]; [1]University of Delaware, Beltsville, MD, [2]University of Delaware, Georgetown, DE, [3]USDA ARS, Beltsville, MD, [4]University of Maryland, College Park, MD, [5]Texas A&M University, College Station, TX, [6]University of Missouri, Columbia, MO, [7]Penn State University, Bozeman, MT, [8]University of Illinois, Urbana, IL, [9]Five Rivers Conservation Trust, Concord, NH, [10]North Carolina State University, Raleigh, NC, [11]Virginia Tech, Blacksburg, VA, [12]University of Minnesota, St. Paul, MN, [13]Blue River Technology, Sunnyvale, CA, [14]University of Nebraska, Lincoln, NE, [15]University of Arkansas, Fayetteville, AR, [16]Montana State University, Huntley, MT (350)

Cover crops, i.e., crops that are planted to cover the soil and provide ecosystem services rather than to be harvested, have been shown to provide weed suppression in a wide variety of cropping systems, alongside other services such as improved soil health and nutrient retention. Cover crops compete with weeds for resources while living, and physically and chemically suppress weeds after they are terminated. While cover crops alone are not often able to provide satisfactory weed management, they can be a useful tool alongside other tactics in integrated weed management systems. In particular, there is evidence that cover crops can delay weed seedling emergence after cash crop planting, thereby reducing weed competition during critical periods of crop establishment and extending the window of opportunity for effective POST herbicide applications. Cover crop management has been shown to impact the cover crop's weed suppressive potential due to management effects on cover crop biomass and quality (i.e., C:N ratio) relative to environmental factors such as soil type, temperature, and moisture. To investigate the effects of delaying cover crop termination on weed seedling emergence across a wide range of environmental conditions, a field experiment was carried out at 12 locations in the Eastern United States between 2017 and 2020. At each location, cereal rye cover crops were grown and terminated six (Early termination), four (Intermediate), and two (Late) weeks prior to planting soybean, alongside a fallow (No cover) control. All plots received a burndown herbicide application at the time of soybean planting. Cover crop biomass was measured at the time of termination, and weed seedling emergence for a minimum of three weed species was counted in two 0.5 m2 quadrats at 2, 4, 6, 8, and 10 weeks after soybean planting (WAP). After each seedling count an effective non-selective herbicide was applied to isolate sequential cohorts of emerging weeds. Cover crop biomass increased linearly with later termination. Across all site years, later cover crop termination was associated with lower cumulative seedling emergence across the growing season. Fewer seedlings of both broadleaf and grass species emerged in plots with Late terminated cover crops than in plots with Intermediate or Early termination, and all termination timings had less total weed emergence than plots without cover crops. Seedling emergence decreased as WAP increased, but an interaction between WAP and Late and Intermediate termination indicated that emergence timing was delayed by those treatments. Future analysis of this robust dataset will explore the impacts of cover crop management on emergence of individual troublesome weed species, particularly those that frequently exhibit herbicide resistance, and will also examine how environmental conditions might mediate the observed effects of cover crops on weed emergence timing.

Using "Stacked" Cultivation Implements to Improve Weed Control Selectivity. Bryan Brown*[1], Daniel C. Brainard[2], Percy Marshall[3]; [1]New York State Integrated Pest Management Program, Cornell University, Geneva, NY, [2]Michigan State University, East Lansing, MI, [3]University of Arkansas, Fayetteville, AR (351)

Many different types of cultivation implements may be used to control weeds in the intra-row zone. These include flat sweeps, finger weeders, torsion weeders, and harrows. Previous work has suggested that these implements can be effectively used in combination, or stacked, and they have greatest efficacy when they undercut, then uproot, and then finally bury weeds. But damage to crops was unacceptably high in previous trials. Therefore, we tested various combinations of one, two, or three implements in beets and snap beans to determine if adjustments could be made to the stacked setups of previous work to reduce crop damage while maintaining high efficacy. Most of the implements were set farther from the crop than in previous work, but we found that the implements could still disturb the soil at the base of the crops. Likewise, we added a small set of disks to some setups and found they threw a steady amount of soil on the base of the crop to bury small weeds. GPS guidance was also used in all the trials, which ensured straight rows and optimal positioning of the implements. We used a randomized complete block design with four replications per treatment, and most crop-treatment combinations were repeated several times over our eight site-years. We assessed weed control after each cultivation by comparing weed density in the 10-cm intra-row zone to untreated plots. Crop mortality was determined using pre- and post-cultivation counts of each plot. Overall, our changes resulted in improved weed control selectivity compared to previous work. In our New York snap bean trials, mean efficacy in the intra-row zone was 52%, 71% and 87%, while crop mortality was 1.0%, 1.8%, and 3.8%, for one-, two-, and three-implement combinations, respectively. Although the ratio of efficacy to crop mortality was least in the three-implement combinations, our farmer advisors indicated that since any remaining weeds would remain uncontrolled through harvest, the three-implement combinations were most favorable. While intra-row cultivation caused unacceptable damage in our trials with two-leaf beets, efficacy in four-leaf beets was 13%, 38%, and 61% with crop mortality at 9%, 15%, and 12% for one-, two-, and three-implement combinations, respectively.

Integrating Electrocution as a Supplemental Weed Control Tool in No-till Organic Soybean. Madhav Dhakal*, Carolyn Garrity, Reza Afshar; Rodale Institute, Kutztown, PA (352)

Weed control is one of the major and enduring challenges in cover crop-based no-till organic farming. Zapping weeds with high-voltage electricity could eliminate late-season intra-row weeds as part of the integrated weed control systems to supplement the available mechanical tools in organic systems, however, information on electrical weed control is lacking. A plot-scale agronomic experiment was initiated in 2022 at Kutztown, PA, to demonstrate the influence of high-residue cultivation (HRC) and electrocution (once and twice during season) on weed community composition, weed biomass, and crop yield under cereal rye (*Secale cereale*) based no-till organic soybean (*Glycine max*). The cover crop biomass yield before termination was 7.43 ± 3.78 Mg ha^{-1}. Soybean (370,500 seeds ha^{-1}) was no-till planted into the rolled-crimped rye residue in a single operation at 0.76 m spacing. A tractor-mounted equipment called The Weed Zapper™ was used to electrocute weeds taller than crop canopy in July and August 2022. Up to 24 weed species were identified and categorized as broadleaf, grasses, and sedges between June and September 2022. The species ground cover was determined using the Grid method and percentage green cover was measured using ImageJ analysis. Electrocution changed the functional composition of weed communities as it increased the species richness by 21 and 7% over the control and HRC only, respectively. Electrocution significantly reduced the abundance of *Seteria faberi* and *Ambrosia artemisifolia* and increased the Shannon Hill diversity and evenness over the control and HRC only. The total weed biomass did not differ between the treatments although the HRC + two electrocutions had 35% more green cover than other treatments. None of the weed control tactics had grain yield more than the control (1.49 Mg ha^{-1}) and even lower for HRC + electrocutions (1.21 Mg ha^{-1}). This indicates that although electrocution may kill tall-growing key weeds, it may not be conducive to soybean productivity.

A New Enemy in Town: Resistant Ryegrass Heading North in Europe. Alberto Collavo*, Bianca Assis Barbosa Martins, Bodo Peters; Bayer AG, Frankfurt Am Main, Germany (353)

A new enemy in town: herbicide-resistant ryegrass heading north in EU During the last decade thousands of weed samples have been sent to Bayer's Weed Resistance Competence Center in the frame of complaint monitoring. Herbicide resistance testing is part of Bayer's stewardship measures and an important step for understanding the spread, pattern, and level of herbicide resistance. Early detection of resistant weeds provides the farmer time to limit the spread and mitigate severe resistance problems. Since 2005 weed samples of more than 20000 fields collected in the European geographical area have

been tested. Based on the field history, infestation levels and the weed spectrum of each characterized location a continuously increasing number of ryegrass samples (*Lolium* spp.) has been observed. Such weed samples originate from complaints and are not randomly collected. Previous field occurrences of ryegrass species in northern locations cannot be excluded; however, complaints indicate that ryegrass is an emerging problem in those agricultural environments. Herbicide resistance has been detected mainly to post-emergence herbicides belonging to the ACCase- and ALS-inhibitor groups. Pre-emergence herbicides are generally still an option; however, few ryegrass samples showed shifts in sensitivity. A closer look at the collection areas confirms an increase in herbicide-resistant ryegrass samples originating from Belgium, Denmark, Germany, and United Kingdom, areas historically infested by other grasses, such as blackgrass (*Alopecurus myosuroides*) or silky bent grass (*Apera spica-venti*). The increase in infestation levels of ryegrass species in the northern latitudes is likely due to their overall adaptation to the increasing temperatures and rainfall reduction in those areas over the last decades.
alberto.collavo@bayer.com

Effect of Winter *Brassica carinata* Production on Weed Control in Subsequent Peanut. Pratap Devkota*, Sithin Mathew, Navjot Singh; University of Florida, Jay, FL (354)

Ethiopian mustard (*Brassica carinata*) is recently introduced as a winter biofuel crop in the southeastern United States. As a new crop in the region, it is important to understand the effect of carinata production on weed control in the summer crop. Field research was conducted from November 2021 to October 2022 at the University of Florida, West Florida Research and Education Center near Jay, FL to evaluate the influence of growing winter carinata for weed control on subsequent summer peanut production. The main plot factor was winter cropping history: a) carinata, and b) weedy fallow. The sub-plot factor was preemergence (PRE) applied herbicide programs: a) fluridone, b) acetochlor, c) fluridone+acetochlor, d) no PRE herbicide. PRE herbicides were followed by postemergence (POST) application of *S*-metolachlor+2,4-DB+imazapic at 4 weeks after PRE (WA-PRE). At 2 WA-PRE, the peanut stand with winter carinata treatment was >7 plants m^{-1} compared to weedy fallow. Results at 4 WA-PRE illustrated that peanut height was greater with carinata compared to weedy fallow. Likewise, grass weed species (barnyardgrass, Texas panicum, large crabgrass) control was 20% or greater and density was about 3 times lower with carinata production compared to weedy fallow. However, there was no effect of winter cropping history and PRE herbicide programs for control and density reduction of broadleaf weed species, such as sicklepod, morningglory spp., and yellow nutsedge control. At 4 WA-POST herbicide applications, winter cropping history resulted in 5 and 8 cm greater peanut height and canopy width, respectively, with winter carinata compared to weedy fallow. There was no effect of winter cropping history and herbicide programs on the grass or broadleaf weed control and density. The peanut yield was not different with winter cropping history and herbicide program treatments. Overall, the study suggests that winter carinata production has the potential to complement early-season grass weed control for peanut production in the subsequent summer.

Automated Intra-row Weeding Informed by Sugar Beet Response to Hoeing and Flaming Across Early Growth Stages. Margaret R. McCollough*[1], Frank Poulsen[2], Bo Melander[3]; [1]University of Maine, Orono, ME, [2]F. Poulsen Engineering Aps., Hvalsø, Denmark, [3]Aarhus University, Department of Agroecology, Slagelse, Denmark (355)

Automated intelligent intra-row weeding machines offer effective intra-row weed control in slow-to-establish direct-seeded crops like sugar beet. These machines identify individual crop plants and selectively hoe or flame weed in the between-crop zone, leaving the intra-row close-to-crop zone untreated. To characterize crop tolerance to intra-row weeding and describe optimal tool working distances, five pot experiments were performed with the test crop sugar beet. Variables tested included hoeing distance (0, 10, 20, 30, 40, and 50 mm), flaming distance (0, 5, 10, 15, 20, 25, 30, 35, and 40 mm), and flaming intensity (0.19, 0.37, 0.74, 1.49, 2.97, and 5.95 kg propane fuel consumption km^{-1} of crop row); crop growth stages ranged from cotyledon to the six-leaf stage. At cotyledon to the four-leaf stage, hoeing could be performed as close as 1 cm to sugar beet plant centers before adverse crop effects were observed. Based on crop injury measured, direct flaming of sugar beet at the cotyledon stage is not advised at any intensity, although flaming tolerance was observed to increase with crop growth stage. Direct flaming did not affect sugar beets when the propane dose was = 0.74 kg km^{-1} at the two-leaf stage, = 1.49 kg km^{-1} at the four-leaf stage, and = 5.95 kg km^{-1} at the six-leaf stage. Sugar beets at the two-leaf growth stage were not damaged by flaming at distances = 1.5 cm at a propane dose of 3.72 kg km^{-1}. Field experiments will be necessary to confirm these operational guidelines for next-generation automated intelligent weeders.

HRAC Global Validation Criteria for Unique Herbicide Resistance Cases. Harry J. Strek*[1], Mark A. Peterson[2], Marcel S C de Melo[3], Caio V S Rossi[4], Ian M. Heap[5]; [1]HRAC Global, Frankfurt Am Main, Germany, [2]Dow AgroSciences LLC (Ret.), Indianapolis, IN, [3]Bayer Brasil, Paulínia, Brazil, [4]Corteva, Mogi Mirim, Brazil, [5]The International Herbicide-Resistant Weed Database, Corvallis, OR (356)

Pairing Nanoparticles with FANA Technology for Spray-on Gene Silencing in Weeds. Katie Martin*[1], Todd A. Gaines[1], Veenu Aishwarya[2]; [1]Colorado State University, Fort Collins, CO, [2]AUM Lifetech, Philadelphia, PA (357)

Agricultural production relies on effective control of pests to produce top yields. Herbicide resistant weeds are a major threat to crop yield loss and growers can spend up to 65% of their annual pest control budget on herbicides and/or weed control measures. With resistant weed populations on the rise since the late 1970s there is an evident need for new herbicidal technologies. Fluoroarabinucleic acid (FANA) is a modified sugar that can be incorporated into the DNA backbone. This enhancement causes the DNA to become more stable in aqueous environments, making it an ideal candidate for spray-on gene silencing. Gene silencing targets the mRNA of organisms and can be used to target specific sites within biological pathways. Recent research highlights the ability to use RNA-targeting technologies in a spray solution for uptake by plants and insects. Existing spray-on gene silencing methods in plants involve the use of plasmids or RNAi, both of which are unstable for extended time periods at ambient temperature. A major challenge spray-on gene silencing methods face in plant matrices is crossing the cell wall. Researchers have found that nanoparticles can assist in cell uptake of the gene silencing trigger, such as single-stranded RNA, and significantly increase down-regulation of the target gene. There are numerous types of nanoparticles that can be used to deliver genetic material into plants. Most of these particles utilize weak surface interactions between the particle (or functionalized particle) and the nucleic acid. This research aims to pair FANA technologies with nanoparticles for efficient and effective uptake of FANA antisense oligonucleotides to silence genes in various weed species resulting in a new management method for herbicide resistant weeds.

Species-Level Weed Biomass Estimation from Video Imagery Using 3D Point Clouds. Daniel J. Ginn*[1], Paula J. Ramos-Giraldo[2], Michael A. Alcorn[3], Maria L. Cangiano[3], April M. Dobbs[2], Søren K. Skovsen[4], Matthew Kutugata[1], Ramon G. Leon[2], Chris Reberg-Horton[2], Muthukumar V. Bagavathiannan[1], Steven Brian Mirsky[5]; [1]Texas A&M University, College Station, TX, [2]North Carolina State University, Raleigh, NC, [3]USDA, Beltsville, MD, [4]Aarhus University, Aarhus, Denmark, [5]USDA ARS, Beltsville, MD (358)

Hard Water Effect on Glyphosate Grass Control in the Enlist Weed System. Justin T. McCaghren*, Steve Li, Ryan D. Langemeier, Livia Pereira; Auburn University, Auburn, AL (359)

The Enlist weed control system allows growers to tank mix glyphosate, 2,4-D, and/or Liberty (glufosinate) to control a broad spectrum of weeds. It has been documented that certain salts of 2,4-D, including the choline salt (Enlist One), can antagonize glyphosate control of grassy weeds. Additionally, hard water ions such as calcium and magnesium are also known to reduce glyphosate efficacy. Previous literature states that with the addition of water conditioners, such as diammonium sulfate (AMS) antagonism by both 2,4-D and hard water ions can be overcome, however little information is available about when both antagonists are present. A study was created with the objectives of evaluating the effects of annual grass control based on 1) tank mixtures of glyphosate + Enlist One and glyphosate + Enlist One + Liberty compared to glyphosate alone, 2) comparing tank mixtures of each treatment in both hard or distilled water, 3) analyzing the effects AMS in overcoming antagonism. The study was conducted in summer of 2021 in Elmore and Henry counties in Alabama. This study was repeated in summer of 2022 in Henry County, Alabama. This study included a randomized complete block design with four replications at each site. Visual ratings were taken at 7, 14, 21, and 28 days after treatment (DAT). Biomass was recorded at 28 DAT. Glufosinate had little overall effect on grass control. The results indicate that both hard water and Enlist One can antagonize glyphosate annual grass control. However, AMS is able to overcome all antagonism even when all products are combined in hard water.

Multistate Assessment of Cowpea Cover Crop Germplasm, Seeding Rate, and Termination Timing Effects on Weed Suppression. Carlene A. Chase*[1], Daniel A. Boakye[1], Parmeshwor Aryal[1], Nicholas T. Basinger[2], Lisa Fultz[3], Audrey V. Gamble[4], Erin Haramoto[5], Nithya Rajan[6], Vitalis Temu[7], Elide Valencia[8]; [1]University of Florida, Gainesville, FL, [2]The University of Georgia, Athens, GA, [3]Louisiana State University, Baton Rouge, LA, [4]Auburn University, Auburn, AL, [5]University of Kentucky, Lexington, KY, [6]Texas A&M University, College Station, TX, [7]Virginia State University, Petersburg, VA, [8]University of Puerto Rico, Mayaguez, Mayaguez, PR (360)

Cowpea [*Vigna unguiculata* (L) Walp.] is a legume that is used for food and forage and to provide agroecosystem services as a warm season cover crop and green manure. The indeterminate germplasm lines, US-1136, US-1137 and US-1138, were selected for use as cover crops due to their soft seed trait that reduces volunteer incidence compared to the commonly used 'Iron Clay' cowpea. The objective of this experiment was to compare the efficacy of weed suppression by the three cowpea germplasm lines and determine the optimum seeding rates of the germplasm lines for use as cover crops in the southern region. In summer 2021, the three cowpea lines were planted at 40, 60, 80, 100 and 120 lb/acre along with a no cover crop, weedy control in seven states (Alabama, Florida, Georgia, Louisiana, Kentucky, Texas, Virginia) and Puerto Rico. The experimental design was a randomized complete block with four replications. Data were collected on cowpea shoot biomass and weed biomass at 6, 8, and 10 weeks after planting (WAP) to simulate three cover crop termination dates. Delaying termination until 10 WAP resulted in the maximum amount of biomass at 85% of the locations. All three cowpea germplasm lines had considerable shoot biomass accumulation, but their performance differed with location. At most locations, cowpea lines significantly suppressed weed biomass compared to the weedy control with no difference among the germplasm lines. There was a significant decrease in weed biomass as seeding rates increased at all locations except for Georgia. All three lines show adaptability for weed suppression across the region with significant weed suppression occurring even at the lowest rate of 40 lb/acre.

Integrated Weed Management Practices to Control ALS and PPO-Inhibitor Resistant Palmer Amaranth.
Justin T. McCaghren*, Steve Li, Ryan D. Langemeier, Livia Pereira; Auburn University, Auburn, AL (361)

Palmer amaranth (*Amaranthus palmeri*) is an economically damaging weed found throughout the southeastern United States. Over reliance of herbicides in management has caused development of herbicide resistance in Palmer amaranth including ALS-inhibitors and PPO inhibitor herbicides potentially limiting peanut producers' options to control Palmer. Evaluation of alternative pre-emergent herbicides and cultural practices (high residue cover crops) to control Palmer amaranth is imperative for peanut production. This project evaluates integration of high residue cover crops (CC) versus conventional tillage (CT) and pre-emergent herbicide programs in peanut to control resistant Palmer amaranth without using PPO-inhibitor herbicides. Cereal rye was the only CC species using in CC system. Acetochlor, fluoridone, norflurazon, pendimethalin were used as PRE-treatments in different tank mixes. Paraquat was used in CT system to clean up Palmer with 2,4-DB. Imazapic, 2,4-DB and S-metolachlor were applied in CC system to provide morning glory and residual control of Palmer amaranth. This study was conducted in summer of 2021 in Henry, Elmore, and Baldwin counties in Alabama. This study was repeated in summer of 2022 in Baldwin County, Alabama. Approximately 7000-8000 kg ha^{-1} CC residue was present at peanut planting and 2000-3000 kg ha^{-1} residue still remained on soil surface at 56 days after peanut planting (DAP) in all locations. All herbicide treatments provided effective control of Palmer amaranth with more than 90% control compared to the non-treated check (NTC) in all planting systems through 70 DAP. The CC NTC had fewer Palmer amaranth compared to the NTC of CT at 70 DAP. Data indicated that cover crop residue allowed for quicker canopy closure and as well as effective Palmer amaranth suppression throughout the growing season. These results suggest that alternative approach of residual herbicides plus CC residue is an effective method to control ALS and PPO resistant Palmer amaranth while maintaining sufficient peanut crop growth.

Cultural Tactics Prevail in the War on Multiple Herbicide-Resistant Kochia (*Bassia scoparia*). Charles M. Geddes*[1], Louis J. Molnar[1], Cindy A. Gampe[2], Prabhath Lokuruge[2]; [1]Agriculture and Agri-Food Canada, Lethbridge, AB, Canada, [2]Agriculture and Agri-Food Canada, Scott, SK, Canada (362)

Kochia [*Bassia scoparia* (L.) A.J. Scott] management is a growing issue for many farmers in western North America. High genetic diversity combined with efficient cross-pollination of this tumbleweed result in rapid evolution and spread of herbicide resistance. Kochia populations exhibiting resistance to acetolactate synthase inhibitors, glyphosate, and synthetic auxins are increasing in occurrence in the Canadian prairies, leaving limited effective herbicide options. Two different fully-phased four-year crop rotation studies were conducted to assess whether (a) crop row spacing and seeding rate, and (b) crop diversity and crop life cycle diversity, augment management of multiple herbicide-resistant kochia while also implementing herbicide layering. The former study was conducted in Lethbridge, AB from 2018–2021, while the latter was conducted in Lethbridge, AB and Scott, SK from 2019–2022. Doubling crop seeding rates in a spring wheat–canola–spring wheat–lentil rotation decreased kochia biomass in the 3rd and 4th year by 64% overall. Reducing row spacing from 46–23 cm decreased kochia biomass by 56%. When combined, narrow rows and double seeding rates decreased kochia biomass by 80% compared with wide rows and recommended seeding rates; equivalent to the threshold considered control by herbicide regulators. This suggests that optimizing crop spatial arrangement can result in kochia management equivalent to a new effective herbicide site of action. The Lethbridge site of the crop diversity study had sufficient kochia densities to derive treatment differences. Crop rotations that replaced spring wheat with winter wheat had 74% and 64% lower kochia density and biomass, respectively, by the third year. Kochia density and biomass were reduced by 89% and 99%, respectively, by perennial alfalfa/meadow brome compared with the rotations containing spring wheat. Together, these results suggest that narrow row spacing, increased seeding rates, and crop life cycle diversity contribute strongly to integrated management of multiple herbicide-resistant kochia.

Cover Crop Monocultures and Mixtures Affect Weed Seedbank Diversity in the Southeast United States. Aniruddha Maity*[1], Audrey V. Gamble[1], Annu Kumari[1], Anna Murry Johnson[1], Andrew J. Price[2]; [1]Auburn University, Auburn, AL, [2]USDA-ARS, Auburn, AL (363)

Community Management of Herbicide Resistance: Insights from the Pacific Northwest. Nick Bergmann*[1], Ian Burke[1], Chloe Wardropper[2]; [1]Washington State University, Pullman, WA, [2]University of Illinois, Champaign, IL (364)

A community management approach to addressing herbicide resistance – grounded in common pool resource theory – has emerged as a popular conceptual framing within the social dimensions of weed management literature. This presentation critically analyzes common pool resource theory and its limitations for herbicide resistance management. Specifically, we argue for a more expansive approach to community management that moves beyond a singular focus on common pool resource theory. In particular, we argue that a co-production of knowledge framing offers an alternative approach for community management that may yield greater benefits for weed management. To support our argument, we present early results from a co-production of knowledge approach to herbicide resistance in the Pacific Northwest.

IR-4 Weed Science Update - Food Crops. Roger B. Batts*, Jerry Baron, Venkat Pedibhotla; IR-4 Project HQ, NC State University, Raleigh, NC (365)

IR-4 Weed Science Update – Food Crops. Roger B. Batts, Jerry Baron, Venkat Pedibhotla. IR-4 Project, NC State University, Raleigh, NC Residue projects IR-4 data submitted to EPA led to over 600 new uses in 2022. Of these, nearly 280 uses were for herbicides (tribenuron and glufosinate) in many specialty crops, crop groups or subgroups. Glufosinate approvals include avocado, bushberry subgroup, cottonseed subgroup, fig, vining small fruit subgroup, hops, melon subgroup, pepper/eggplant subgroup, rapeseed subgroup, squash/cucumber subgroup, tomato subgroup, tropical and subtropical small fruit-edible peel subgroup, tuberous and corm vegetable subgroup. Tribenuron approvals involve dried shelled bean subgroup, dried shelled pea subgroup, rapeseed subgroup, cottonseed subgroup, proposed wheat subgroup, proposed barley subgroup, proposed field corn subgroup, proposed grain sorghum and millet subgroup, and proposed rice subgroup. IR-4 submitted three herbicide data petitions to EPA in 2022 (acifluorfen, triclopyr and saflufencail). These submissions could potentially lead to more than a dozen new uses. Ten new herbicide magnitude-of-residue studies began in 2022, which could result in more than 50 new uses. Twenty-one new herbicide and PGR residue studies will begin in 2023. Product Performance projects Generating Product Performance (efficacy and crop safety) data to support registration of pest management tools in specialty crops continues to be an important and expanding part of the IR-4 annual research plan. This data is often required by registrants and/or states to complete the registration process. The number of on-going herbicide Product Performance studies in 2022 was twenty (60 individual trials), with twelve of them beginning in 2022. The 2023 field research plan for herbicides and plant growth regulators includes twenty-nine (>90 individual trials) continuing or new Product Performance studies. Integrated Solutions projects IR-4's Integrated Solutions (IS) Program is structured to assist specialty crop growers outside of the traditional single product/single crop residue and product performance research. IS research efforts focus on crop-pest combinations to address solutions in these four areas, 1) pest problems without solutions, 2) resistance management, 3) products for organic production and 4) pesticide residue mitigation. In 2022, there were nine active IS projects with herbicides and plant growth regulators (22 individual trials), three of which will continue in 2023. Four new weed control IS studies will begin in 2023 (10-12 individual trials), including quinoa, stevia, date palm, and dry bulb onion.

Weed Control and Crop Tolerance as Influenced by Pendimethalin Rate and Timing of Application in Cranberry Production. Thierry E. Besancon*[1], Marcelo L. Moretti[2], Katherine Ghantous[3], Jed Colquhoun[4], Cyrus A. Smith[5]; [1]Rutgers University, Chatsworth, NJ, [2]Oregon State University, Corvallis, OR, [3]UMass Amherst, UMass Cranberry Station, East Wareham, MA, [4]University of Wisconsin-Madison, Madison, WI, [5]Ocean Spray Cranberries, Inc., Bandon, OR (366)

Pendimethalin is a WSSA group 3 herbicide that provides excellent preemergence control of various annual grasses as well as small seeded annual broadleaf weeds. An encapsulated pendimethalin formulation (Satellite HydroCap) has recently added cranberry to its label. In 2017, the use of a similar pendimethalin formulation was associated with widespread highbush blueberry phytotoxicity in New Jersey. High use rate, blueberry shallow root system, sandy soil with low organic matter content, and applications timed around or after budbreak increased the risk of pendimethalin injury. Given the genetic proximity of blueberry and cranberry, and the lack of reference regarding, multistate field studies were conducted in 2022 to assess cranberry tolerance to chemigated (400 GPA) or boom applied (30 GPA) pendimethalin at different growth stages and rates. In NJ, MA, and WI, pendimethalin was applied at 800 (½X) or 1,600 (1X) g ha^{-1} on cranberry vines at the spring dormant (SD) or cabbage head (CH) phenological stages. Applications were either chemigated (MA) or boom-applied (NJ, WI). Additional NJ treatments included 1/2X and 1X applications at the rough neck (RN) stage and split-application at ½X at CH and RN stage. In OR, pendimethalin was chemigated or boom-applied at 1,600 (1X) or 3,200 (2X) g ha^{-1} on cranberry vines at SD or CH stage. In MA and NJ, pendimethalin did not affect development of cranberry uprights but decreased yield by 25% when applied at the 1X rate at the CH stage, as compared to untreated control or applications at the SD stage. Lower cranberry tolerance at later growth stage was shown in NJ with 56% and 70% yield reduction following pendimethalin applied at the RN stage at the ½X and 1X rate. Cranberry vine stunting was 12% 28 DAT with RN application at the 1X rate. In OR, all applications at the CH stage reduced fruit yield 66% to 99%, regardless of pendimethalin rate or spray volume. The 1X rate chemigated at the SD stage was the only treatment not affected by yield reduction. Greater injury 70 DAT was noted with applications made at the CH (34%) than at the SD (6%) stage, and at the 2X (27% than at the 1X (13%) rate. To a lesser extent, similar effect of rate and timing of application was noted again 100 DAT when boom-applied pendimethalin showed greater injury (16%) than chemigation (8%).

Reducing Herbicide Inputs in Plasticulture Peppers Using Spring-seeded Grass Cover Crops. Dwayne D. Joseph[1], Thierry E. Besancon*[2], Cerruti Rr Hooks[3], Alan W. Leslie[3], Kurt M. Vollmer[4]; [1]University of Maryland, Chestertown, MD, [2]Rutgers University, Chatsworth, NJ, [3]University of Maryland, College Park, MD, [4]University of Maryland, Queenstown, MD (367)

The objective of this study was to evaluate the use of spring-seeded grass cover crops as an integrated weed management approach in a bell pepper plasticulture system. The study was conducted in 2021 and 2022 at the Central Maryland Research and Education Center in Upper Marlboro, MD and was arranged in a split-plot factorial design with four replications. Factor A (cover crop termination) consisted of clethodim, paraquat or no herbicide/roller-crimped. Factor B (cover crop) consisted of: 1) cereal rye (*Secale cereale*), 2) spring oats (*Avena sativa*), 3) cereal rye + spring oats, or 4) no cover crop. Factor C (residual herbicide) consisted of a residual herbicide (s-metolachlor + fomesafen) or no residual application. Cover crops were seeded in the bare-ground area between plastic rows approximately 7 weeks before the peppers were transplanted and terminated 4 weeks after transplanting. Plots were evaluated for weed control and density, and crop yield. Cover crop presence, termination method and residual herbicide application influenced weed control and density. All cover crop plantings significantly increased overall weed control compared to no cover crop. In 2021, the oats treatment was the best with 94, 74, 60 and 55% weed control 1, 3, 5 and 8 weeks after cover crop termination (WATerm), respectively. Whereas, rye was the most effective treatment in 2002 with 95, 83 and 55% weed control 1, 3 and 5 WATerm, respectively. Plots terminated with paraquat showed significantly better weed control compared to clethodim and roller-crimped plots. Similarly, plots that received a residual herbicide application had significantly less weeds than plots without a residual. Cover crop treatments had significantly greater pepper yield than the no cover crop treatment. In 2021, the highest yielding treatment was rye + oats with 93% more yield than the no cover treatment. In 2022, the rye treatment was the highest yielding with 79% more pepper yield than the no cover crop treatment.

Can Vegetable Crop Seed Treatments Enhance Competitiveness with Weeds? Jed Colquhoun*, Jacklyn Slivicke, Emma Schmidt; University of Wisconsin-Madison, Madison, WI (368)

Integrating Spring-seeded Grass Cover Crops for Weed Management in Cucurbit Production. Kurt M. Vollmer*[1], Thierry E. Besancon[2], Alan W. Leslie[3], Dwayne D. Joseph[4], Cerruti Rr Hooks[3]; [1]University of Maryland, Queenstown, MD, [2]Rutgers University, Chatsworth, NJ, [3]University of Maryland, College Park, MD, [4]University of Maryland, Chestertown, MD (369)

Row-middle weed control in plasticulture vegetable production is becoming increasingly difficult due to a lack of effective herbicides, a need for multiple cultivations, or hand labor. Alternative solutions that integrate multiple weed control tactics are needed to address these issues. One solution is to use cover crops to aid in weed suppression along with an effective herbicide program. Studies were conducted in 2021 and 2022 to assess the effects of integrating spring-seeded grass cover crops with herbicide treatments for weed control in cucumber and watermelon. The study consisted of a three-factor factorial arranged in a split-split plot design with four replications. The whole plot consisted of cover crop management method (clethodim, paraquat, no-herbicide/rolled). Subplots consisted of cover crop species (cereal rye, spring oats, cereal rye + spring oats, no cover crop), and sub-subplots consisted of residual herbicide treatment (fomesafen + *S*-metolachlor) or no residual herbicide treatment. Prior to termination, cover crops alone reduced weed density 41% in the cucumber trials and 90% in the watermelon trials. At least 4 weeks into the growing season, terminated cover crops continued to reduce weed density by 56% and weed biomass by 58% in the cucumber studies, and weed density by 46% and weed biomass by 86% in the watermelon trials. Treatments that contained a cover crop terminated with paraquat showed the highest overall weed control in both the cucumber (95%) and watermelon studies (87%). Treatments that included a cover crop had 48% higher cucumber yields compared to no cover treatment regardless of cover crop management. Treatments with cereal rye or spring oats had 36% higher watermelon yields compared to cereal rye + spring oats and no cover crop, and 19% higher yields when cover crops were managed with paraquat. These results show that spring seeded grass cover crops can successfully be integrated with effective herbicide programs for improved weed control between plastic beds.

Flumioxazin Application Rate and Timing for Improved Crop Safety While Maintaining Efficacy in Potato. Harlene M. Hatterman-Valenti*, Collin Auwarter, Avery Shikanai; North Dakota State University, Fargo, ND (370)

Most nightshade species are alternate hosts for insects and diseases that attack potatoes such as Colorado potato beetle and late blight. Eastern black nightshade (*Solanum ptycanthum* Dun.) has been recognized by growers as one of their worst weed problems. Flumioxazin provides excellent control of nightshades but has been shown to also cause injury to potato. The label restricts its use in potato to only 25 states that includes North Dakota to alleviate most of the potential potato injury. The label further restricts and limits the herbicide application timing and states: Many weather-related factors, including high wind, splashing or heavy rains or cool conditions at or near potato emergence, may result in potato injury in fields treated with Chateau EZ Herbicide. On occasion this has resulted in a delay in maturity. Understand and accept these risks before using Chateau EZ Herbicide. Furthermore, the label states: In areas with historically higher amounts of rainfall during the time of preemergence herbicide applications, including the Red River Valley, Minnesota and North Dakota, the requirement for 2 inches of settled soil is critical to avoid crop injury. The objective was to determine the effect of flumioxazin application rate and timing on weed control and potato safety under irrigation. Treatments included the labeled use rate and half that rate at three application timings: two days after planting (DAP) and no hilling, after regular hilling (9 DAP), hilling 2 DAP and then applying, and then various application timings of flumioxazin and the combination product of metribuzin + metolachlor along with a standard of metribuzin + metolachlor after regular hilling. Results indicated that none of the treatments caused >20% visible potato injury and all treatments provided >85% control of grass and broadleaf annual weeds even though plots treated with 0.5X or 1X flumioxazin and no hilling provided greater potato injury and less weed control. Total, marketable and grade yield differences indicated that plots treated with 0.5X or 1X flumioxazin without hilling had the lowest yields while plots treated with 1X flumioxazin no hilling followed by 0.5X metribuzin + metolachlor after regular hilling, plots treated with 1X flumioxazin after an early hilling followed by 0.5X metribuzin + metolachlor after regular hilling, or plots treated with 0.5X or 1X flumioxazin after early hilling had the greatest yields. Yield and grade differences were attributed to the number of tubers produced and tuber size. Overall, the research suggested several application rate and timing combinations that resulted in a 3-6% increase in marketable yields when compared to the standard practice. Research also reinforced the need to have at least 2 of settled soil as the no hilling treatments resulted in a 13-21% decrease in marketable yields when compared to the standard practice.

An Update on NY Research in Snap Beans, Table Beets, and Carrots. Lynn M. Sosnoskie*, Elizabeth C. Maloney; Cornell University, Geneva, NY (371)

New York (NY) state is ranked 9th in the US in the production of fresh market and processing vegetables including carrots (Daucus carota), table beets (Beta vulgaris), and snap beans (Phaseolus vulgaris). To maintain yield quantity and quality, growers must manage weeds that are direct competitors with crops for water, nutrients, and light. Weed control is also imperative as unwanted vegetation can (1) serve as alternate hosts for insect pests and pathogens, (2) alter microclimates around the crop, which may facilitate disease development, (3) interfere with the deposition of crop protective chemicals, and (4) reduce harvest efficiency. Because of slow and variable emergence and growth, carrots do not compete well with many weed species. In addition to reducing yields, weeds can impact crop quality as roots can become misshapen in response to interference. A reliance on linuron has resulted in the subsequent development of resistant weeds. The long-term utility of linuron is further threatened by regulatory action, which necessitates the identification of additional weed control options. 2022 trials focused on evaluating acifluorfen (as Ultrablazer at 94 ml/ha) in May-planted, muck carrots ('Siroco' at the 2-4 leaf and 6-8 leaf crop growth stages) for summer broadleaf (primarily linuron-resistant Amaranthus powellii) weed control. Acifluorfen recently received emergency registration in Ontario for weed control in carrots. Averaged across developmental stages, acifluorfen injury was minor to carrots (maximum 11%

leaf bronzing); significant pigweed control, relative to an untreated check, was observed both in the field and in the greenhouse. 2023 trials will further evaluate the potential for use of acifluorfen in US carrot production. Oxyfluorfen (as GoalTender at 1.17 l/ha) and pyridate (as Tough EC at 0.5 l/ha) were also evaluated for pigweed control, although both were injurious to the crop under trial conditions. Table beets are not competitive with weeds, especially early in the growing season. As such, the crop must be intensively managed to prevent against losses to yield quantity and quality. Few herbicides are available for use in beets, and many are rated fair to poor for the control of many species including pigweeds, lambsquarters (Chenopodium album), nightshades (Solanum spp.), velvetleaf (Abutilon theophrasti), and ragweed (Ambrosia artemisiifolia). Consequently, novel herbicide tools are desperately needed. Acetolchlor, which is the active ingredient in the herbicide Warrant, is commonly applied for pre-emergence weed control in corn. Warrant can also be used in sugar beets. According to label recommendations, Warrant can be applied to sugar beets at 2.9 to 4.7 l/ha (depending on soil type) between the 2 leaf and 8 leaf (4 leaf is optimum) growth stages to extend residual weed control. 2022 trials evaluated 'Ruby Queen' beet safety to Warrant applied at 0, 2.34, 4.68, 9.36, and 18.7 l/ha to beets at the 2-4 and 6-8 leaf growth stages. Stunting was observed for all treatments with the greatest responses occurring at the 9.36 and 18.7 l/ha rates. Stunting responses were higher for plants treated at the 6-8 leaf stage (up to 20%) as compared to the 2-4 leaf stage (up to 12.5%). This injury was transient, and no stunting was observed at the time of harvest. Leaf distortion was also observed for all treatments, with the greatest amount of injury occurring at the 9.36 and 18.7 l/ha treatments, particularly for the 6-8 leaf growth stage (up to 33.8%). Despite the observed injury, marketable (USDA Grade 1) beet numbers and weights, and weights per beet did not differ significantly from the untreated check. Because of the crop's short stature and the comparatively short window between planting and harvest, season-long weed control is necessary to maximize snap bean production. Weeds that compete with the crop can also reduce harvest efficiency, harbor pests and pathogens, and interfere with the deposition of other pesticides. Herbicide options are limited in snap beans with respect to 1) spectrums of control and 2) use patterns/rotation restrictions. The evolution of bentazon resistance in snap beans has reduced the utility of an important herbicide tool for weed management in the crop. In 2022, a trial was conducted at Cornell, Penn State, and the University of Delaware to evaluate the safety of pyridate (as Tough EC) in snap beans; pyridate is registered for use in chickpeas/garbanzo beans and has been/is being evaluated in lentils, peanuts, and peas. Preliminary results indicate that pyridate applied at the first trifoliate may not be an effective tool for use in snap beans; observed injury was substantial (up to 41% chlorosis/necrosis and 48% stunting) and marketable yields (USDA Grade 1) were reduced 15% to 35% relative to the untreated check. Pyridate is rated as an effective tool for the control of common lambsquarters although the safety of the chemical does not appear to be sufficient for possible use in snap beans.

Evaluating In-season Management Strategies for Branched Broomrape Control in California Processing Tomatoes. Matthew Fatino*, Bradley D. Hanson; University of California, Davis, Davis, CA (372)

Recent detections of branched broomrape (*Phelipanche ramosa*) in California tomato (*Solanum lycopersicum*) fields have led to increased interest in management strategies to control this regulated noxious weed. Broomrapes (*Phelipanche spp.* and *Orobanche spp.)* are parasitic weeds that pose a significant risk to the processing tomato industry for several reasons: California's Mediterranean climate is similar to that of branched broomrape's native range, California agronomic practices (wide variety of host species cultivated, successive tomato crops, shared equipment) make the proliferation and spread of broomrape within and among fields highly likely, and broomrape's phenological development makes it difficult to monitor and inaccessible to conventional weed control practices. In addition, California's regulatory environment make soil fumigation difficult, while branched broomrape's regulatory status as quarantine pest does not incentivize accurate reporting.A study was conducted in 2022 to evaluate broomrape control with several herbicide programs based on the PICKIT decision support system developed in Israel for Egyptian broomrape control. These programs included chemigated imazamox and rimsulfuron alone and paired with preplant incorporated sulfosulfuron, as well as foliar rimsulfuron alone. Additionally, a planting date experiment was nested within this trial. This trial was conducted in a commercial tomato field heavily infested with branched broomrape in Yolo County, CA.There was significant broomrape emergence throughout the trial and there were differences among treatments. In general, chemigated treatments had lower broomrape emergence than non-chemigated or control treatments. Chemigated imazamox at all rates and timings resulted in severe injury on tomatoes. Chemigated rimsulfuron alone and paired with sulfosulfuron were safe on tomatoes and tended to have lower broomrape emergence than control treatments. The late planted treatment (May 20) tended to have lower broomrape emergence than the control treatment (May 3), and although more work will need to be done to quantify the effects of planting date on broomrape parasitism, it is a promising first step. Given the injury from imazamox in this experiment, the best treatment was chemigated rimsulfuron paired with preplant incorporated sulfosulfuron. This is very promising considering the recent CDPR approval of a 24c label for chemigated rimsulfuron which means that growers with suspected or at-risk fields can use this treatment protocol during the 2023 season. Further research will need to be conducted to quantify and understand imazamox injury before it can be recommended for registration for broomrape management.

Weed Control and Hazelnut (*Corylus avellana* L.) Tolerance with Tiafenacil. Joshua W. Miranda*, Marcelo L. Moretti; Oregon State University, Corvallis, OR (373)

Herbicide–resistant Italian ryegrass (*Lolium perenne* L. ssp. *multiflorum*) is a significant problem for hazelnut production because of its increased management costs and fewer herbicide options. Tiafenacil is a new protoporphyrinogen IX oxidase (PPO)–inhibiting herbicide on the market with activity against grass and broadleaf weeds. Given that no resistance to PPO–inhibiting herbicides has been reported in any weed in the western US, tiafenacil could be a valuable tool for Oregon's hazelnut industry. Field studies were conducted in commercial hazelnut orchards during 2021 and 2022 to evaluate tiafenacil efficacy against Italian ryegrass and prostrate knotweed (*Polygonum aviculare* L.), as well as hazelnut tolerance to tiafenacil. In the 2021 study, tiafenacil at 50, 75, or 150 g ai ha^{-1} reduced weed biomass by 62 to 79 % and was the most efficacious of all tested treatments compared to a nontreated control. Weed biomass reduction with tiafenacil (50 to 150 g ai ha^{-1}) was comparable to glufosinate (1,150 g ai ha^{-1}), and the combinations of tiafenacil (50 g ai ha^{-1}) with clethodim (135 g ai ha^{-1}), 2,4–D (1,060 g ai ha^{-1}), glufosinate, and tolpyralate (40 g ai ha^{-1}). Tiafenacil combined with clethodim or glufosinate provided the greatest Italian ryegrass control (91–94%). In the 2022 study, tiafenacil (50 g ai ha^{-1}) exerted the lowest Italian ryegrass control (22%) and biomass reduction (18%), while glufosinate (1,150 g ai ha^{-1}) and the combination of tiafenacil with glufosinate exerted the greatest Italian ryegrass control (55–68%) and biomass reduction (24–55%). Tiafenacil plus glufosinate reduced Italian ryegrass inflorescence weight by 87%, and was significantly better than glufosinate, which reduced inflorescence weight by 70% or tiafenacil alone (54% reduction). In a separate 2022 study, tiafenacil at 12.5, 25, or 50 g ai ha^{-1} controlled 80–100% of prostrate knotweed at 14 DAT compared to only 45% control with carfentrazone. Prostrate knotweed regrowth in most treatments by 35 DAT indicated a decline in control, except that glufosinate and glufosinate plus tiafenacil (50 g ai ha^{-1}) exerted >90% control over prostrate knotweed at 35 DAT. The first year of crop tolerance studies indicates that four basal–directed applications of tiafenacil 50 to 200 g ai ha^{-1} each do not injure young hazelnut trees, affect tree growth nor yield. Tiafenacil also effectively controlled hazelnut suckers. These results suggest that tiafenacil is an effective herbicide to manage Italian ryegrass in hazelnut orchards, aiding resistance management and it is safe to hazelnuts.Email: josh.miranda@oregonstate.edu.

Speculations on Weed Management and Herbicide Use Under Agrivoltaic Panels in Cranberry Production. Hilary A. Sandler*[1], Giverson Mupambi[2], Dwayne Breger[3], Samantha Glaze-Corcoran[4]; [1]UMass Cranberry Station, East Wareham, MA, [2]University of Massachusetts Amherst / Cranberry Station, East Wareham, MA, [3]UMass Clean Energy Extension, Amherst, MA, [4]University of Massachusetts, Amherst, MA (374)

The Massachusetts Department of Energy Resources (DOER) launched the Solar Massachusetts Renewable Target (SMART) program in 2018. It was created to financially incentivize the co-location of solar photovoltaics (PV) and farming; these ag-solar systems are termed agrivoltaic or dual-use. Research to investigate the potential impacts on yield and pest issues for specialty crops, such as cranberry, growing under solar panels is underway in MA, largely supported by recent grant funding from the U.S. Department of Energy Solar Energy Technologies Office[1]. Study of dual-use solar on grazing and hay is also part of the study. The construction of an agri-photovoltaic (APV) system involves the insertion of poles (typically steel) in an array over the cranberry production area. APV arrays can have panels that are fixed-tilt or in an articulating tracking system. As per MA DOER regulations, the solar panels are situated on the poles at a height of approximately 3 m (10 ft). Typical spacing between rows is approximately 6.5 m (21 ft). The presence of the solar panels will increase the duration and intensity of shading that the crop and weeds experience on a daily and seasonal basis. Although previous research has generated data on short-term shading of cranberry vines during the season, this project will be the first to produce multi-year, season-long environmental, yield, and pest management data. Since cranberries are a perennial crop, the weeds commonly found in the production system are also perennial plants, such as poison ivy (Toxicodendron radicans), poverty grass (Andropogon virginicum and Schizachyrium scoparia), moss (Polytrichum spp.), dewberry (Rubus spp.), and yellow loosestrife (Lysmachia terrestris). These plants, including cranberry, generally have low-light saturation points and should be able to tolerate periodic shading. As such, it is presumed that the suite of weed populations that occur in conventional cranberry production systems will continue to be problematic under dual-use solar. Application technology (e.g., chemigation, boom sprayers, and wick applicators) is a mainstay of weed management in cranberry production. The development of new use-pattern recommendations for weed management under APV will be a high priority. The presence of poles will impact the distribution patterns and efficacy of chemigated (e.g., delivery by low-statured, solid-set sprinkler heads) herbicides. Managing weeds at the base of the poles may require extra labor input. We anticipate that, in addition to shading impacts on general weed physicology, the overall weed suite may shift to favor shade-tolerate weeds. Growers will need to utilize/retrofit boom sprayers to fit between the rows of poles and increase the deployment of targeted weeding (especially around the poles). We speculate that the use of unmanned aerial systems (UAS) for PRE and POST herbicides will develop into an efficient mode of delivery for crops growing under APV. Unique herbicide rigs have been built by cranberry growers for use within this specialized cropping system; continued reliance on innovation from within the industry will be important to secure long-term weed management success. hsandler@umass.edu [1] This material is based upon work supported by the U.S. Department of Energy's Office of Energy Efficiency and Renewable Energy (EERE) under the Solar Energy Technologies Office, Award Number DE-EE0009374.

Development of Sustainable Weed Management Strategies for Michigan Cherry Growers by Identify Promising New Herbicides. Sushila Chaudhari*, Christopher G. Galbraith, Monique Mose; Michigan State University, East Lansing, MI (375)

Weed Encroachment into a Fine Fescue Mixture Affected by Different Seeding Timings. Ross C. Braun[1], Aaron J. Patton[1], Eric Watkins[2], Andrew B. Hollman[2], James A. Murphy[3], Bradley S. Park*[3], Alexander R. Kowalewski[4], Emily T. Braithwaite[4]; [1]Purdue University, West Lafayette, IN, [2]University of Minnesota, St. Paul, MN, [3]Rutgers University, New Brunswick, NJ, [4]Oregon State University, Corvallis, OR (376)

Turfgrass establishment vigor and initial growth rate can influence turfgrass competitiveness with germinating weeds. Research is lacking on best establishment practices of fine fescues (*Festuca* L. spp.), including optimal seeding timings. Our research objective was to determine optimal seeding timings and document weed encroachment into a fine fescue mixture seeded across multiple cool-season climate zones in the United States. A field experiment was replicated at four locations in the northern United States (Minnesota [MN], Indiana [IN], New Jersey [NJ], and Oregon [OR]). A fine fescue mixture was seeded at 200 kg pure live seed ha^{-1} during March through November (nine seeding timings) during 2019 into plots where glyphosate was used to maintain bare soil until seeding. Grid counts were taken to assess fine fescue cover at the termination of the study in August 2020. Weed cover was visually estimated for summer annuals (broadleaves and grasses) and perennial broadleaves in August 2020. Fine fescue cover was greatest in plots seeded in August and September at all four locations. The poorest establishment was observed in March-seeded plots (OR) and March- and April-seeded plots (MN, IN, and NJ). Weed encroachment varied depending on the specific weed, weed life cycle, weed seedbank at the site, and seeding time of year. Summer annual weed cover was greatest in plots seeded in March and April (NJ and IN) and October (NJ and MN). Summer annual weed encroachment was not observed at the OR site. Perennial broadleaf cover was greatest (> 30% plot cover) in March-seeded plots (MN, NJ, and OR). Perennial broadleaf cover at the IN location did not exceed 2% for any seeding timing in August 2020. Seeding a fine fescue mixture in August or September will increase probability of establishment success with less weed encroachment in cool-season climate zones in the United States with a wider seeding window in Mediterranean climates.

Influence of Mowing Heights and Post-treatment Irrigation on Annual Bluegrass Control with Methiozolin. John M. Peppers*, Shawn Askew; Virginia Tech, Blacksburg, VA (377)

Methiozolin is a newly registered herbicide for annual bluegrass control that offers high levels of safety as well as both PRE and POST weed control. The methiozolin label indicates that a higher rate is required to effectively control annual bluegrass at mowing heights greater than that typical of golf course putting greens. Additionally, the herbicide label recommends post-application irrigation for effective annual bluegrass control that is difficult to timely administer due to demands of golf play. Previous research has reported excellent annual bluegrass control by methiozolin on golf greens without post-treatment irrigation, suggesting that an evaluation of irrigation impacts on product performance is needed. Mowing height restrictions are based on observations between separate studies that suggest annual bluegrass control by a given rate of methiozolin is inversely related to mowing height, but this phenomenon has not been tested in replicated research. Since methiozolin translocates acropetally, root absorption maximizes weed control, and may be hindered by the vegetative density of higher-height turf. Thus, the need for and quantity of irrigation may likewise be dependent on turf height. The objective of this research was to determine the effect of mowing height and post-application irrigation on annual bluegrass control with methiozolin. Two field studies were initiated at the Glade Road Research Facility in Blacksburg, VA on creeping bentgrass and Kentucky bluegrass research fairways. Trials were arranged in a split-plot design, with three mowing heights (3.8, 7.6, and 15.2 mm) as main plots. Subplots were arranged as a four by two factorial with four levels of methiozolin rate (500, 1000, 2000, and 4000 g ai ha^{-1}) and two levels of post-application irrigation (0 and 12mm). Herbicides were applied twice biweekly in late fall and annual bluegrass control was evaluated in the following spring. Annual bluegrass control was assessed via grid count at the conclusion of the trial ~May 1. Rate-dependent effects of methiozolin on annual bluegrass control were subjected to linear regression and regression parameters were used to calculate the methiozolin rate required to control annual bluegrass 90% (I_{90}). Resulting I_{90} values were subjected to ANOVA with sums of squares partitioned to account for the split plot design and trial effects. Annual bluegrass control was highly dependent on mowing height but not dependent on irrigation. I_{90} values of annual bluegrass maintained at 3.8, 7.6, and 15.2mm were 864, 1396 and 1779 g ai ha^{-1}, respectively. Although there were numerical differences that indicated that post application irrigation may increase annual bluegrass control in higher-mown turf, these trends could not be separated with our study design. Results from this study support product label recommendations regarding increased use rates with higher heights of cut, but can't confirm the need for post-application irrigation to achieve optimal annual bluegrass control.

Seedbank Persistence of Annual Bluegrass (*Poa annua* L.) Across Multiple Climatic Zones in the United States. Andrew W. Osburn*, Muthukumar V. Bagavathiannan; Texas A&M University, College Station, TX (378)

Annual bluegrass is one of the most common and troublesome weeds found in turfgrass systems, greatly increasing management costs and decreasing aesthetic value. This species not only occurs during the cool-season, but can also exhibit perennial characteristics, extending growth well into the summer in several environments. Annual bluegrass is a prolific seed producer and establishes a soil seedbank from which it recruits annually. Understanding the seedbank longevity of this species is vital for developing effective management strategies. To this end, a study was conducted between the Fall seasons of 2020 and 2022 in seven locations across the U.S. along a climatic gradient encompassing five USDA plant hardiness zones. Annual bluegrass seed originating from 10 unique populations were placed in polyethylene mesh bags and buried in each of these seven locations at two depths: 0 (surface) and 5 cm. At six-month intervals (for up to 18 months), bags containing the seed of each population were retrieved from each depth and shipped to College Station, TX for conducting germination and viability tests on the extracted seed. Germination and viability assays were conducted in tandem for the seed kept in dry storage for comparison. There were differences in the longevity of seed viability among the seed origins during the 18-month retrieval, where annual bluegrass populations originating from the northern locations appeared to retain greater viability than those from the southern locations. At the 18-month retrieval, the greatest viability (24.8%) was observed with the seed originating from New Jersey (zone 7a), whereas the lowest viability (1.0%) was documented for the seed collected from Alabama (zone 8a). A high degree of seed viability was maintained in the dry storage treatment, indicating the strong influence of field environmental conditions on annual bluegrass seed viability loss. Overall, our findings indicate a relatively short (~one year) seedbank persistence of annual bluegrass, which presents opportunities for targeting soil seedbank through appropriate management practices.

Weed Control in Christmas Trees with Fluridone. Marcelo L. Moretti*; Oregon State University, Corvallis, OR (379)

Oregon leads the United States in producing Christmas trees, with almost 8.5 million trees sold in 2015. Herbicides are the primary weed management strategy for Christmas tree growers, who need an increased diversity of herbicide modes of action to manage weeds and mitigate herbicide resistance selection. The present study evaluated the efficacy and crop tolerance of fluridone alone or in combination with herbicides labeled for Christmas trees. Four field studies were initiated in the winter of 2022 in commercial Christmas tree farms in Western Oregon, two sites producing Douglas fir (*Pseudotsuga menziesii*), and two producing Nordman fir (*Abies nordmanniana)* in their second year after transplanting. All treatments were applied over the top of the trees. No crop injury was observed with fluridone at 450 or 900 g ai ha^{-1} up to 180 days after treatment (DAT). Treatments did not affect the growth of leader shoots, but growth was affected by species and experimental sites averaging between 17 to 33 cm in Douglas fir and 7 to 17 cm in Nordman fir. Fluridone (450 g ai ha^{-1}) controlled 58% of rattail fescue (*Vulpia muyros*), and it did not control false dandelion (*Hypochaeris radicata)* (28%) at 120 DAT. Control exerted by fluridone did not differ from the simazine plus oxyfluorfen, the standard treatment. Fluridone mixed with simazine plus oxyfluorfen improved control of false dandelion (73%) and rattail fescue (93%). Fluridone did not affect the efficacy of indaziflam, flumioxazin, or flazasulfuron compared to these herbicides used alone. Results indicate that fluridone is safe for both Douglas fir and Nordman fir applied over-the-top in the dormant season and that fluridone can improve the control of important broadleaf and grassy weeds when mixed with simazine and oxyfluorfen. This ongoing research will evaluate Christmas tree response to repeated fluridone treatments.

Factors Associated with the Exposure of Sub-canopy Stolons and Shoots to Herbicides in Dormant Zoysiagrass. Shawn Askew*, Jordan M. Craft, Clebson G. Goncalves; Virginia Tech, Blacksburg, VA (380)

Zoysiagrass can be treated with nonselective herbicides during winter dormancy, but such treatments often cause turf injury that varies from year to year. Several studies were conducted to show that ambient temperatures, at and following treatment, impact zoysiagrass response to glyphosate and glufosinate by increasing speed of herbicidal activity and zoysiagrass green shoot density. The question still remains why injury is sometimes observed following treatments to dormant turf that doesn't have any green leaves visible among the canopy. It was noted that zoysiagrass will have some subcanopy green leaf portions and stolons regardless of the level of dormancy or severity of winter temperatures. We hypothesized that green leaves or stolons that subtend dormant, brown shoots may absorb glyphosate or glufosinate, yet herbicide absorption into zoysiagrass stolons has not been previously reported. We also hypothesized that spray droplets will penetrate zoysiagrass canopies more when droplet size, droplet velocity, and application volume is increased. At 3 d after treatment, Zoysiagrass absorbed radiolabeled glyphosate and glufosinate 25 and 56%, respectively, through treated leaves and 80 and 78%, respectively, through treated stolons. Increased glufosinate absorption through zoysiagrass leaves may partially explain increased zoysiagrass injury caused by glufosinate compared to glyphosate. High absorption rates of both herbicides by zoysiagrass stolons suggest that stolon exposure within the canopy may contribute to herbicide-induced plant response. Spray droplet velocities ranged from <5 to >20 m sec^{-1} and were nonlinearly related to droplet diameter and dependent on pressure, boom height, and spray tip. Droplet velocities from air induction spray tips were generally less dependent on pressure and slower at a given droplet diameter than that of flat fan spray tips, which were highly dependent on pressure. At a boom height of 25 cm, 17 to 18% of recovered colorant reached the bottom third of dormant zoysiagrass canopies with no difference between spray tips or pressure. When the boom height was increased to 61 cm above the turf, 19% of colorant from air induction spray tips penetrated to the bottom canopy level and significantly more than that of flat fan spray tips (11%). Although significant, this difference in penetration could likely be improved further by manipulating application volume. Additional work determined that flat fan spray tips applying several volumes between 47 and 470 L ha^{-1} delivered 2% more colorant and twice as many droplets to the lower half of dormant zoysiagrass canopies for each 94 L ha^{-1} increase in spray volume. Thus, applying herbicide to dormant zoysiagrass at 47 L ha^{-1} (5 GPA) will deliver 10% less product and, more importantly, ten times less droplets compared to a more conventional 470 L ha^{-1} (50 GPA).

Common Ragweed Control in Southern Appalachian Fraser Fir Christmas Tree Production. Joe C. Neal*[1], Christopher D. Harlow[1], Jeff Owen[2], Brad Edwards[3]; [1]North Carolina State University, Raleigh, NC, [2]North Carolina State Univeristy, Raleigh, NC, [3]North Carolina Cooperative Extension, Jefferson, NC (381)

Herbicide-tolerant common ragweed (*Ambrosia artemisiifolia*) populations have increased in Southern Appalachian Fraser fir Christmas tree production fields. In this production region, growers are encouraged to maintain 100% vegetated ground cover to prevent erosion and insulate Fraser fir tree roots from high soil temperatures in the summer. The preferred ground cover is dominated by white cover (*Trifolium repens*). However, control options for common ragweed that do not damage white clover are lacking. On-farm experiments were conducted to assess preemergence and postemergence herbicide efficacy on common ragweed and safety to Fraser fir and white clover ground cover. Herbicide treatments included products reported to have efficacy on ragweed or related species in other cropping systems, as well as products reported to be safe on white clover but with unknown efficacy on ragweed. PRE and POST experiments were conducted on two Christmas tree farms in Alleghany County NC, each with a history of heavy ragweed populations. Preemergence herbicides were applied on March 15 or March 16, 2022. Postemergence herbicides were applied on May 24 or May 25, 2022, when ragweed seedlings were 2 to 4 inches tall and Fraser fir trees had 1 to 4 inches of new growth. Herbicides were applied with a CO_2-pressurized sprayer using drift-reduction TTI 11002 nozzles and calibrated to deliver 15 GPA. All treatments were applied as semi-directed sprays contacting the lower 12 to 18 inches of the Fraser fir trees. Percent ragweed control, as well as injury to Fraser fir trees and white clover ground cover, were visually evaluated. Simazine + pendimethalin provided preemergence control of ragweed at both sites, but this treatment was the most injurious to clover cover. Flumioxazin, indaziflam and oxyfluorfen were each effective at one of the two test sites and did not reduce clover cover. Neither pendimethalin, rimsulfuron, nor flumetsulam controlled ragweed, nor did they injure white clover or Fraser fir. Thus, these three herbicides may have utility for other target weeds. Clopyralid, glufosinate, flumioxazin, and topramezone each controlled ragweed POST but caused unacceptable injury to white clover. Glufosinate and flumioxazin also caused necrosis on Fraser fir branches contacted by the spray. Glyphosate, rimsulfuron, bentazon, and oxyfluorfen did not control ragweed postemergently. Cloransulam @ 0.3 oz/A provided about 50% reduction in ragweed. Where white clover cover is well established, preemergence applications of indaziflam, oxyfluorfen, or flumioxazin may provide control, but results were variable between the two test sites. No POST herbicide tested provided adequate control of ragweed without injury to clover and/or Fraser fir trees. Where maintenance of white clover ground cover is not a management goal, clopyralid or topramezone may be used to control emerged ragweed.

Impact of Temperature on Germination and Growth of Japanese Stiltgrass [*Microstegium vimineum* (Trin.) A. Camus]. Jeffrey Derr*[1], Aman Rana[2]; [1]Virginia Tech, Virginia Beach, VA, [2]Virginia Tech HRAREC, Virginia Beach, VA (382)

Impact of Temperature on Germination and Growth of Japanese Stiltgrass [*Microstegium vimineum* (Trin.) A. Camus]. Jeffrey Derr and Aman Rana Japanese stiltgrass is an invasive summer annual grass native to Asia that grows well in shade, especially in moist sites. It is a weed of turf and landscape ornamentals, as well as in noncrop areas. Japanese stiltgrass has been observed to germinate in early spring prior to southern crabgrass [*Digitaria ciliaris* (Retz.) Koeler]. Information is needed on the temperature needed for Japanese stiltgrass germination, as well as to its tolerance to cold temperatures after emergence. Growth chamber experiments were conducted to address these questions. Germination of Japanese stiltgrass and southern crabgrass was evaluated at 0, 5 and 10 C. There was no germination of Japanese stiltgrass or southern crabgrass at 0 C at 3 weeks after seeding. Japanese stiltgrass germinated at both 5 and 10 C while southern crabgrass germinated at 10 but not at 5 C. Tolerance of 10-cm-tall plants for the two species was evaluated at -5, 0, and 5 C. Different sets of emerged plants were subjected to these temperatures for 1, 3, 5, 7, 9, or 11 days. Neither species survived when exposed to 1 or more days at -5 C. Japanese stiltgrass growth was stunted but plants survived when exposed up to 9 days at 0 C. All southern crabgrass plants died when exposed to 3 or more days at 0 C. Japanese stiltgrass can germinate at lower temperatures than southern crabgrass and emerged Japanese stiltgrass plants can tolerate colder temperatures than southern crabgrass. These findings have implications for the timing of control strategies, including application of preemergence and postemergence herbicides.

Investigating the Genetic Relatedness of Herbicide Resistant *Poa annua* Populations. Claudia A. Rutland*[1], Joseph S. McElroy[2], Jinesh Patel[2]; [1]Auburn University Department of Crop, Soil, and Environmental Sciences, Auburn, AL, [2]Auburn University, Auburn, AL (383)

Poa annua L. (annual bluegrass) is a ubiquitous weed species that has commonly been identified as one of the worst weeds to control in turfgrass systems. The increasing numbers of herbicide resistant populations is a major contributor to the difficulties faced in managing P. annua. With reported resistance to now 12 modes of action, the need to understand how to manage P. annua is at an all time high. Understanding how herbicide resistance is actually spread in P. annua could allow us to update our current management practices. Thus, a population study focusing on the genetic relatedness between resistant P. annua populations was conducted in order to ascertain how these populations were acquiring resistance traits. A selection of 144 individuals resistant to ALS, EPSPS, a-tubulin, and photosystem II herbicides were collected across the southeastern United States to analyze if there was a significant difference between populations. SSR markers were used to produce polymorphic bands, which were scored and used to determine population structure using STRUCTURE. The analysis resulted in a population structure of k=5, with seemingly no correlation between location or resistance type. This begs the question, how are these populations related, and what else do we need in order to understand how herbicide resistance is spread?

Evaluation of Goosegrass (*Eleusine indica*) Populations Suspected Resistant to Revolver and Dismiss.
Bridgette C. Johnson*, Joseph S. McElroy; Auburn University, Auburn, AL (384)

Forty-one goosegrass biotypes from golf courses and country clubs throughout the United States that historically have been treated with foramsulfuron and sulfentrazone were suspected of herbicide resistance. Putative foramsulfuron resistant goosegrass populations were from Grand National Golf Course, Opelika, AL (R-R1)(R- R3), and Waiehu Golf Course, Maui County, Wailuku, HI (R-R2). Estimated sulfentrazone-resistant goosegrass populations were from Grand National Golf Course, Opelika, AL(D-R1), Paris Mountain Country Club, Greenville, SC (D-R2). Resistance was determined by visual injury ratings of plant response to the standard rate of foramsulfuron (28.89 g ai/ha) and sulfentrazone (279.63 g ai/ha) compared to a susceptible standard (PBU). Research was then conducted to assess the level of resistance of the five suspected resistant populations to increasing rates of foramsulfuron (1.81 to 115.56 g ai/ha) and sulfentrazone (26.21 to 1677.78 g ai/ha). Visual injury ratings were taken at 7, 14, 21, and 28 days after treatment (DAT) and aboveground biomass at 28 DAT. Dose-response evaluation showed that each foramsulfuron population R-R1, R-R2, R-R3 had a calculated I50 value of 0.12, 0.12, 0.09, respectively, whereas the susceptible biotype had a 0.02 I50. Each sulfentrazone population D-R1 and D-R2 had a calculated I50 of 0.11 and 0.21, respectively, whereas the susceptible biotype had a 0.03 I50. This data confirms the first E. indica biotypes found resistant to foramsulfuron and sulfentrazone in the world.

Pre-emergent Herbicides and Mulches in Propagation - Rooting and Growth After Transplant. Anthony L. Witcher*[1], Isha Poudel[2]; [1]Tennessee State University, Mcminnville, TN, [2]University of Tennessee, Jackson, TN (385)

Weeds are a major problem in container-grown nursery crops and weed infestation issues commonly originate during propagation. Due to the smaller container sizes used during propagation, there is more competition for resources such as light, nutrients, and space. Manually removing weeds is a time consuming and costly process due to the amount of labor required. The objective of this study was to determine the effect of pre-emergent herbicides and mulches on rooting of stem cuttings and growth after transplant. In April 2020, containers (16.1 cm diameter square) were filled with a 100% pine bark substrate and placed under intermittent mist until saturated. Pre-emergent herbicides (granular formulation - indaziflam, isoxaben + dithiopyr, oxyfluorfen + oxadiazon; sprayable formulation - isoxaben) and mulches (1.3 cm depth; pine pellets and rice hulls) were applied, then cuttings (1 per container) of butterfly bush (Buddleja davidii 'Nanho Blue') and crape myrtle (Lagerstroemia indica 'Catawba') were inserted into each container. After rooting, 17 cuttings were harvested while 8 cuttings were transplanted to 2.4 L containers and grown for 2 (butterfly bush) or 3 (crape myrtle) months. Butterfly bush rooting percentage was 75% or greater for all treatments except pine pellets (68%), isoxaben + dithiopyr (36%) and isoxaben (20%). Cutting root and shoot dry weight was lowest for isoxaben compared to the non-treated control. Due to poor growth or survival in isoxaben and isoxaben + dithiopyr, those cuttings were not transplanted. After transplant, there were no differences in butterfly bush shoot growth or shoot and root dry weight for the remaining treatments compared to the non-treated control. Crape myrtle rooting percentage was 80% or greater for all treatments, while total root length and root dry weight were similar for all treatments compared to the non-treated control. After transplant, there were no differences in crape myrtle shoot growth or shoot and root dry weight. Isoxaben is not labeled for use on butterfly bush, likely causing damage during rooting. Pine pellets caused slight reductions in root biomass for butterfly bush but did not affect crape myrtle. Pre-emergent herbicides and mulches may be viable options for weed control during cutting propagation, but products must be tested on individual plant species to verify crop safety.

Florpyrauxifen-benzyl Resistance in Barnyardgrass (*Echinochloa crus-galli*) Evolved Prior to the Commercialization of Loyant™. Hudson K. Takano*, Scott Greenwalt, David Ouse, Moriah Zielinski, Paul Schmitzer; Corteva Agriscience, Indianapolis, IN (386)

Herbicide options for selective control of monocot weeds in rice have always been limited to a few modes of action such as inhibitors of acetolactate synthase (e.g. penoxsulam). Florpyrauxifen-benzyl (Loyant, Rinskor Active™) is a synthetic auxin molecule introduced to the US rice herbicide market in 2018. It provides broad spectrum of weed control including hard-to-control species such as barnyargrass (*Echinochloa crus-galli*), along with excellent post-emergence rice selectivity at very low use rates. Within a couple of years following its commercialization, field agronomists and academics identified barnyargrass escapes in some areas where Rinskor had been sprayed. Further evaluation under controlled environments confirmed that those plants were able to survive Rinskor application at the label rate. Here, we identify the mechanism of resistance to Rinskor and penoxsulam in two barnyargrass populations from Arkansas (AR-27) and Missouri (MO-18). Using high-resolution mass spectrometry, the two resistant biotypes were compared with known susceptible plants regarding their ability to metabolize florpyrauxifen-benzyl, florpyrauxifen-acid, and penoxulam in planta. We discovered that the resistant plants share a common resistance mechanism to Rinskor and penoxsulam, involving hydrolysis of a methoxy group (likely mediated by a cytochrome P450 monooxygenase) followed by glucose conjugation. Given that penoxsulam has been widely used in rice fields for the past decade, we believe that barnyardgrass may have evolved this resistance mechanism prior to the commercialization of Rinskor.

Molecular Targets of Herbicides and Insecticides - Are There Useful Overlaps? Stephen O. Duke*[1], Amar G. Chittiboyina[1], Thomas C. Sparks[2], Daniel R. Swale[3]; [1]University of Mississippi, Oxford, MS, [2]Agrilucent LLC, Greenfield, IN, [3]University of Florida, Gainesville, FL (387)

New insecticide modes of action are needed for insecticide resistance management strategies. The number of molecular targets of commercial herbicides (ca. 25) is lower than that for insecticides (ca. 35), and only two of the commercial insecticide targets are found in plants. However, ten molecular targets of commercial herbicides are found in insects. For several of these commonly held targets, there are compounds that kill both plants and insects. For example, herbicidal inhibitors of *p*-hydroxyphenylpyruvate dioxygenase (HPPD) such as mesotrione are effective insecticides on several species of blood-fed insects, as HPPD activity is required to reduce toxic levels of tyrosine from accumulating via blood constituents. The glutamine synthetase-inhibiting herbicide glufosinate is insecticidal by the same mechanism of action of action it has as a herbicide, inhibition of glutamine synthetase. These and other examples of shared activities of commercial herbicides with insecticides through the same target site will be discussed. Compounds such as statins with a novel herbicide target (e.g., HMG CoA reductase) shared by insects that is not commercialized will also be discussed. Compounds that are both herbicidal and insecticidal can be used for insect pests not associated with crops or with crops engineered to be resistant to the compounds.

Glyphosate Selection Increases *EPSPS* Copy Number and Glyphosate Resistance Within Individual Plants of *Hordeum glaucum*. Christopher Preston*[1], Patricia Adu-Yeboah[1], Jenna M. Malone[1], Gurjeet Singh Gill[2]; [1]University of Adelaide, Glen Osmond, Australia, [2]University of Adelaide, Adelaide, Australia (388)

Hordeum glaucum Steud. is a widespread weed of pastures and crops in southern Australia. In 2016, failure of glyphosate to control this weed on a field margin in South Australia was reported. The mechanism of resistance to glyphosate was determined as due to gene amplification of the target enzyme 6-enol-pyruvyl-shikimate-s-phosphate synthase (EPSPS). To better understand the stability of EPSPS gene amplification in this population, individual plants were divided into two halves (clones). One clone was treated with glyphosate at 405 g a.e. ha^{-1} and the other clone was left untreated. Seed from each half of each plant was collected separately. Dose response experiments showed a 75% increase in the LD_{50} for progeny from the treated clones compared to progeny from the untreated clones. EPSPS copy number estimates were higher in progeny of treated clones compared to untreated clones for each plant. There was a positive correlation between EPSPS gene copy number and LD_{50} for glyphosate. These results suggest that EPSPS copy number is variable within plants and that glyphosate application is able to increase the average copy number by killing cells with low copy number. For *H. glaucum*, glyphosate selection occurs not only on individuals within the population, but also on cells with an individual. This indicates that repeated glyphosate application can increase the level of resistance in populations of this species.

Herbicides Pose a Unique Challenge: Insights from Omics Analyses of Plants Responding to Sublethal Stresses. Jim Westwood*, Gourav Sharma, Suzanne R. Laliberte, Shawn Askew, Jacob Barney, David C. Haak; Virginia Tech, Blacksburg, VA (389)

Plants may respond to stress by altering their DNA methylation patterns as part of epigenetic mechanisms that help the plant – and perhaps its progeny – to survive the stress. We are interested in weed epigenetic responses to stresses induced by herbicides, and how these may differ from stresses imposed by other weed control methods. In this study, we subjected *Arabidopsis thaliana* to two herbicide stresses (glyphosate and trifloxysulfuron) and two non-herbicidal stresses (clipping and shading), as well as no-stress controls. We measured changes in gene expression, DNA methylation, and small RNA populations following recovery from these stresses in the original plants and in subsequent generations. For stressed plants that had recovered from individual stresses, clear differences were observed in patterns of gene expression, DNA methylation, and small RNAs, demonstrating the uniqueness of each stress on the plants. For the multigenerational aspect, plant phenotypes (e.g., biomass, flowering time, etc.) generally returned to pre-stress levels in the generation following the stress. However, when stresses were repeated in each generation over four generations, plants took longer to adapt, although most phenotypes returned to initial levels after 3 or 4 generations. Plants exposed to the two herbicides responded in a similar way in many cases, differing from non-herbicidal stress. Taken together, results to date suggest that plant epigenetic responses to stress are highly specific to the stress. This work may have implications for herbicide resistance and weed management.

Chromosome-Scale Draft Genomes of Common and Giant Ragweed Reveal a Potential Mechanism of Glyphosate Resistance. Eric Page*[1], Martin Laforest[2], Sara Martin[3], Katherine Bisaillon[4], Brahim Soufiane[4], Sydney Meloche[1], Francois Tardif[5]; [1]Agriculture and Agri-Food Canada, Harrow, ON, Canada, [2]Agriculture and Agri-Food Canada, Saint-jean-sur-richelieu, QC, [3]Agriculture and Agri-Food Canada, Ottawa, ON, [4]Agriculture and Agri-Food Canada, Saint-jean-sur-richelieu, QC, Canada, [5]University of Guelph, Guelph, ON, Canada (390)

Ambrosia artemisiifolia L. (common ragweed) and *Ambrosia trifida* L. (giant ragweed) are two of the most widely distributed and economically important pest species in the world. Across their native North American ranges these species are important weeds of agriculture that cause significant yield losses when left uncontrolled. Both species have developed resistance to many of the most commonly utilized herbicides, including inhibitors of acetolactate synthase (ALS) and enolpyruvyl shikimate 3-phosphate synthase (EPSPS). Although rare, *A. artemisiifolia* and *A. trifida* have also been observed to hybridize, with hybrid individuals reported to contain 30 chromosomes, 18 from *A. artemisiifolia* and 12 from *A. trifida*. In this research, chromosome scale draft genomes of *A. artemisiifolia* and *A. trifida* were produced via a trio binning approach whereby the haploid genomes of hybrid ragweed as well as the diploid genomes of the parental species were sequenced using Pacific Bioscience long read technology. The genomes of the parental species were assembled individually and these results served as reference to sort the sequence reads of the hybrid into 2 groups, referring to the origin of the chromosomes, and therefore permitting the resolution of haplotypes for both species. Genomic assemblies of *A. artemisiifolia* and *A. trifida* facilitated the production of complete sequences of herbicide target site genes, including the three copies of EPSPS. These results revealed a previously unreported Proline to Serine mutation at position 106 (P106S) of EPSP2S in both *A. artemisiifolia* and *A. trifida*. While this mutation has been reported to confer resistance to glyphosate in several other weed species, it is unclear at present what contribution this mutation may make to the level of glyphosate resistance observed in these species.

A Deeper Understanding of Resistance to Glufosinate in *Amaranthus palmeri*. Matheus Machado Noguera*[1], Aimone Porri[2], Eduarda P. Mena Barreto[1], Juan Camilo Velásquez R[1], Martin Penkert[3], James W. Heiser[4], Jens Lerchl[5], Nilda Roma-Burgos[6]; [1]University of Arkansas, Fayetteville, AR, [2]BASF Global Research & Development, Limburgerhof, Germany, [3]BASF SE, Ludwigshafen, Germany, [4]University of Missouri, Portageville, MO, [5]BASF SE, Limburgerhof, Germany, [6]University of Arkansas, Fayetteville, Fayetteville, AR (391)

Dose-Response Characteristics of Global Metabolome as a Potential Avenue to Capture the Secondary Toxicity of Herbicides in Weeds. Shalini priya Etukuri*, Pawanjit Kaur Sandhu, Jisun Lee, Nishanth Tharayil; Clemson University, Clemson, SC (392)

Identification and Characterization of GST Genes Involved in Group 15 Herbicide Detoxification in Resistant Grasses. Evlampia Parcharidou[1], Rebecka Duecker[1], Roland S. Beffa*[2]; [1]Georg-August Universität Göttingen, Göttingen, Germany, [2]Senior Scientist Consultant, Frankfurt, Germany (393)

Herbicides remain one cornerstone technology in Integrated Weed Management. In addition to the right combination of agronomic measures and pesticide applications, the best combinations between pre- and post-emergence herbicides are essential to mitigate resistance evolution. Rye-grass (*Lolium* spp.) and black-grass (*Alopecurus myosuroides* Huds.) have become problematic weeds in cereals not only in Europe but globally. Besides resistance to post-emergent herbicides becoming increasingly widespread, enhanced metabolism of inhibitors of the synthesis of very-long-chain fatty acids (VLCFAs, HRAC Group 15), such as flufenacet, is evolving. Yet, cross-resistance patterns and evolution of this resistance remains poorly understood.The cDNA sequences of glutathione transferases (GSTs) upregulated in flufenacet resistant populations of rye-grass and black-grass were identified and used for protein overexpression. Detoxification of flufenacet was verified for the most overexpressed candidate GSTs. Flufenacet-glutathione conjugate as metabolite was found for most GSTs, except for one black-grass GST where flufenacet-alcohol was produced in presence of reduced glutathione (GSH). Additionally, activity to other VLCFA-inhibitors e.g. acetochlor and pyroxasulfone, and post-emergence herbicides including ACCase- and ALS-inhibitors was verified in vitro. The activity spectrum of the different GSTs showed qualitative and quantitative differences. As several upregulated GSTs detoxified flufenacet in vitro, the decrease in sensitivity observed in flufenacet resistant populations, is likely a result of an additive effect. The polygenic character and the different turnover rate of the individual GSTs may explain the quantitative aspect of at least some metabolism resistance traits and their longer time to evolve than Target Site Resistance traits. The polygenic character might also contribute to the resistance to a broader herbicide spectrum since flufenacet resistance was accompanied by cross-resistance with some, but not all, herbicides of Group 15, and to herbicides with other MoAs. Therefore, not only the rotation of herbicide modes of action, but also of individual active ingredients is important for resistance management. In addition, this suggest that the landscape-scale evolution of NTSR, in that case detoxification of herbicides, results from both parallel and non-parallel patterns of evolution across the genomes. The associated duplication and redundancy in plant genomes means that adaptation may not be mutation-limited and that the repeated evolution of resistance and/or tolerance may rely on neither rare mutational events, nor hard selective sweeps. They also hint that complex adaptations to abiotic and biotic stresses are not constrained by genetic variation and architecture and that convergent phenotypes are shaped by population-specific genome structure and plasticity.

Genetic Factors Influencing Snap Bean Tolerance to Preemergence Herbicides from Three Site-of-Action Groups. Ana I. Saballos*[1], Nicholas Hausman[1], Ed Peachey[2], Patrick Tranel[3], Martin Williams[4]; [1]Global Change and Photosynthesis Research, US Department of Agriculture, Agricultural Research Service, Urbana, IL, [2]Oregon State University, Corvallis, OR, [3]University of Illinois, Urbana, IL, [4]USDA-ARS, Urbana, IL (394)

Weed management is crucial for profitable production of snap bean (*Phaseolus vulgaris* L.), but the challenge posed by *Amaranthus* species is significant. Preemergence herbicides can help the crop gain an advantage over the weeds, but the few preemergence herbicides registered for use on snap bean have poor control of *Amaranthus*. Protoporphyrinogen oxidase (PPO) -inhibiting herbicides, metribuzin (a photosynthesis inhibitor) and pyroxasulfone (a very long chain fatty acid (VLCFA) synthesis inhibitor) are effective in providing early season control of *Amaranthus*, but the extent of tolerance to these herbicides in snap bean is unknown. To quantify tolerance to preemergence herbicides, and determine genomic regions associated with crop tolerance, a snap bean diversity panel was screened for response to application of 6 preemergence herbicides including 4 PPO-inhibitors (flumioxazin, lactofen, saflufenacil, and sulfentrazone), pyroxasulfone and metribuzin. Tolerance to sulfentrazone and saflufenacil was associated with multiple genomic regions including a common region in chromosome 4 that conferred a high proportion of the tolerance. The region contained several candidate genes. Tolerance to flumioxazin was associated with a region on chromosome 2. Tolerance to lactofen is widespread in the panel. Multivariate analysis using the responses of the panel to all 4 PPO herbicides confirmed the roles of loci on chromosomes 2 and 4. Seed weight positively influences the tolerance to PPO-inhibiting herbicides. There is limited evidence of genetic resistance to pyroxasulfone and metribuzin. However, one cultivar presented excellent tolerance to metribuzin across multiple years suggesting that rare genetic tolerance exists in the crop. Our results indicate that genetic tolerance to PPO-inhibitors is available in snap bean. Introgression of 2 genomic regions from tolerant entries may allow the use of PPO-inhibitor herbicides on the crop. The use of metribuzin and pyroxasulfone is less promising, as the margin of crop safety was small and there are limited sources of tolerance genes.

Genome-wide Characterization of MADS-box Transcription Factors Family in Dioecious Amaranth Species. Lucas Kopecky Bobadilla*[1], Patrick Tranel[2]; [1]University of Illinois, Champaign, IL, [2]University of Illinois, Urbana, IL (395)

The MADS-box transcription factor family plays a key role in the development and evolution of flowers in plants. Members of this family are involved in specifying various floral organs, such as sepals, petals, stamens, and carpels, as well as determining flower symmetry and meristem identity. MADS-box has also been implicated in the evolution of novel floral traits and the diversification of flowering plants. _Amaranthus tuberculatus_ and _Amaranthus palmeri_ are the two major US weeds due to the constant evolution of herbicide resistance. Both species are dioecious (e.g., male and female individuals within the species), making them a potential target for sex-ratio manipulation via genetic control. Having the knowledge of the diversification of MADS-box transcription factors families across species and different sexes can provide clues about the evolutionary patterns of dioecy and how we can manipulate it. In this study, utilizing draft genome assemblies and *de novo* transcriptomics, we performed a genome-wide characterization of the MADS-box transcription factor family in two major dioecious amaranth weed species across both sexes. A total of 63 and 37 MADS-box genes were identified in *A. tuberculatus* and *A. palmeri*, respectively. Phylogenetic analysis revealed that these genes could be divided into five major clades, with several members in both species. Expression analysis showed that the multiple MADS-box genes are expressed in at least one floral tissue, with some genes showing sex-specific expression patterns. These results comprehensively overview the MADS-box gene family in dioecious amaranth species. They will be helpful for future functional studies and their roles in the sex-determination mechanism of dioecious amaranths.

Metabolic Resistance Profiles of Palmer Amaranth (*Amaranthus palmeri*) Populations to HPPD-inhibiting Herbicides. Jeanaflor Crystal T. Concepcion*[1], Shiv S. Kaundun[2], James A. Morris[2], Devin Koester[1], Sarah-Jane Hutchings[2], Dean E. Riechers[3]; [1]University of Illinois at Urbana-Champaign, Urbana, IL, [2]Syngenta, Bracknell, United Kingdom, [3]University of Illinois-Urbana, Urbana, IL (396)

Rapid metabolism of 4-hydroxyphenypyruvate dioxygenase (HPPD)-inhibiting herbicides is a widely known mechanism of tolerance in corn and resistance in dicot weed populations. Oxidative (Phase I) metabolism of postemergence HPPD-inhibiting herbicides (mesotrione, tembotrione and topramezone) at least partially confers resistance in waterhemp (*Amaranthus tuberculatus*) and Palmer amaranth (*A. palmeri*) and natural tolerance in corn. Mesotrione resistance in both waterhemp and Palmer amaranth occurs via hydroxylation in the cyclohexanedione ring, forming a 4-hydroxy-mesotrione metabolite also produced in tolerant corn. Topramezone metabolic resistance in multiple herbicide-resistant (MHR) waterhemp is due to Phase I hydroxylation in the isoxazoline ring, which differs from the *N*-demethylation reaction and tolerance mechanism in corn. Tembotrione, on the other hand, is metabolized via reduction and ring hydroxylation in Phase I followed by Phase II glucose conjugation in resistant Palmer amaranth. To date, however, metabolic weed resistance studies based on mass spectral analysis of metabolites have not been reported for mesotrione or topramezone in Palmer amaranth and tembotrione in waterhemp. Our objective was to conduct a liquid chromatography/mass spectrometry (LC/MS)-based comparative investigation of metabolic resistance to mesotrione, tembotrione and topramezone in MHR waterhemp and Palmer amaranth populations from the U.S. An excised leaf assay investigated rates of metabolism and identified putative metabolites of these three herbicides during a 12-hr time-course. Each biological replicate consisted of two leaves per plant treated with 0.15 mM unlabeled herbicide solution for 2-hr. Peak abundances of parent herbicides and metabolite features were derived from LC/MS-based untargeted metabolomics of leaf extracts. Rapid ring hydroxylation

(Phase I) of mesotrione to 4-hydroxy-mesotrione is considered a robust biochemical marker for mesotrione resistance in MHR waterhemp and Palmer amaranth; however, our current results indicate a more complex profile of mesotrione metabolism in MHR Palmer amaranth compared to waterhemp and tolerant corn. Our findings also shed new light on metabolic pathways and putative detoxification mechanisms for three commercial HPPD-inhibiting herbicides in several MHR Palmer amaranth and waterhemp populations. Future work will embark on proteomic and transcriptomic analysis of various control and treated tissues in MHR waterhemp and Palmer amaranth to identify key proteins and genes conferring resistance to HPPD-inhibiting herbicides. Identification of common or unique metabolite profiles when comparing these weedy amaranths and corn will aid in informing and directing new herbicide discovery efforts for improving selective weed management.

Comparative Analysis of *S*-Metolachlor Metabolism Among Palmer Amaranth (*Amaranthus palmeri*) and Waterhemp (*A. tuberculatus*) Populations from the U.S. Jeanaflor Crystal T. Concepcion[1], Noeleen K. Brown[2], James A. Morris[3], Sarah-Jane Hutchings[3], Shiv S. Kaundun[3], Dean E. Riechers*[2]; [1]University of Illinois at Urbana-Champaign, Urbana, IL, [2]University of Illinois-Urbana, Urbana, IL, [3]Syngenta, Bracknell, United Kingdom (397)

Very-long-chain fatty acid (VLCFA) elongase-inhibiting herbicides, including *S*-metolachlor, are widely used for preemergence, residual control of annual grasses and small-seeded broadleaves in crops such as corn, soybean, and cotton. Previous field research investigating two multiple herbicide-resistant (MHR) waterhemp (*Amaranthus tuberculatus*) populations from Illinois demonstrated poor control with *S*-metolachlor, and several Palmer amaranth (*A. palmeri*) populations from the U.S. have shown similar responses in the field. Previous mechanistic research with intact MHR waterhemp seedlings demonstrated that *S*-metolachlor is rapidly metabolized compared to sensitive waterhemp populations; interestingly, the metabolic pathway of *S*-metolachlor in MHR waterhemp is different than tolerant corn and sensitive waterhemp. However, detailed metabolism studies with *S*-metolachlor have not been reported in MHR or sensitive Palmer amaranth. The objective of this research was to conduct a comparative analysis of *S*-metolachlor metabolism in suspected VLCFA inhibitor-resistant Palmer amaranth populations in comparison with sensitive Palmer amaranth and MHR waterhemp populations, as well as tolerant corn seedlings. Metabolism experiments using intact seedlings and radiolabeled *S*-metolachlor combined with thin layer chromatography (TLC) revealed different rates and patterns of *S*-metolachlor metabolism between 2 and 12 hours after treatment (HAT) among several waterhemp and Palmer amaranth populations. Within the 12-hour time course, rapid metabolism of parent *S*-metolachlor to polar compounds (including putative glutathione, dipeptide, and cysteine conjugates) was noted but some metabolites did not match migration distances of these glutathione-derived conjugates, implying that some *S*-metolachlor metabolites may arise from Phase I oxidation and Phase II glucose conjugation reactions in Palmer amaranth seedlings. Current and future research is aimed at quantification and identification of *S*-metolachlor metabolites in Palmer amaranth seedlings (with and without metabolic inhibitor pretreatment) via TLC and LC-MS analysis, investigation of the putative detoxification enzyme(s) and pathways involved, and fold-resistance quantification using dose-response studies with several VLCFA-inhibiting herbicides in the greenhouse. Findings from this research will expand the current knowledge of metabolic detoxification pathways and enzymes specific for weedy dicots and provide metabolic markers for VLCFA-inhibiting herbicide resistance for use in development of new chemistries for selective weed management.

Overcoming Weed Resistance with Discovery of Novel Modes of Action. Karthik Putta*, Paolo Castiglioni, Neville Anthony, David Lauffer, David Vierra, Balaji Aravindhan Pandian, Gary Schaefer, Peter Stchur, Brittany Baisch, Timothy Panosian, Jacqueline Heard, Thomas Meade; ENKO, Mystic, CT (398)

Herbicide resistance is a major factor driving the need for innovation in crop protection chemistry. Enko has built an efficient, proprietary platform (ENKOMPASS™) to bring effective, safe and sustainable crop protection chemistry to address this and other challenges that are facing the industry. ENKOMPASS™ is an integration of multiple tools - DNA-Encoded libraries (DELs), Artificial Intelligence (AI), Machine Learning (ML) and Structure-based Design to discover and develop novel and diverse chemistries for existing and novel targets. An example of the success of this approach is presented here where a DEL screen of more than 120 billion unique molecules was conducted for plant protoporphyrinogen oxidase (PPO), the site of action for commercially successful PPO-inhibiting herbicides. The platform identified molecules that can inhibit the active site of both wild type and commonly selected PPO mutations. Moreover, the platform was able to deliver a molecule with field validated biological efficacy and commercial attributes in 11 months from Hit confirmation. Currently, this molecule has undergone several seasons of field testing and is in advanced development. Our success in developing a PPO-inhibiting herbicide is not an isolated experience. We have screened over 30 targets to date and built a pipeline of novel herbicide, fungicide, insecticide and nematicide chemistries. Over 50% of our targets are new modes of action. Our ENKOMPAS platform continues to evolve, incorporating innovations in DEL, data analysis, and other emerging technologies to continuously accelerate the discovery and development of novel crop protection chemistries.

Which Summer Annual Weeds May Emerge as Important Candidates as Model Species for Research to Combat Climate Change and Global Food Security? Philip Westra*[1], Todd A. Gaines[1], Eric L. Patterson[2],

Jacob S. Montgomery[1], John M. Lemas[1]; [1]Colorado State University, Fort Collins, CO, [2]Michigan State University, East Lansing, MI (399)

Development of the Brazilian Peppertree Biological Control Agent *Pseudophilothrips ichini* (Thysanoptera: Phlaeothripidae) in Florida. Gregory Wheeler*; USDA, Davie, FL (406)

Allelopathic Potential of *Parthenium hysterophorus* Against Selected Weeds of Wheat Crop. Tauseef Anwar*[1], Huma Qureshi[2]; [1]Department of Botany, The Islamia University of Bahawalpur, Bahawalpur-63000, Punjab, Pakistan, Bahawalpur, Pakistan, [2]Institute of Biological Sciences, Gomal University, Dera Ismail Khan-29050, Khyber Pakhunkhwa,Pakistan, Dera Ismail Khan, Pakistan (407)

Heavy doses of synthetic weed control chemicals have facilitated herbicide resistance in weeds as well as predicted to possess toxicities. Natural compounds can be screened as potential herbicides which are more cost effective, efficacious, selective and environmentally safe. Experiments were conducted on selected weeds (Rumex dentatus, Euphorbia helioscopia, Chenopodium album, Avena fatua, Phalaris minor) to study allelopathic potential of Parthenium hysterophorus. Experiments were performed using a medium of 0.75 percent (w / v) agar, filter paper and soil. Parameters studied for assessing allelopathic effects were germination inhibition (percent) along reduction in radicle and plumule length (cm). Results showed that seed germination of R. dentatus, P. minor and C. album is inhibited by P. hysterophorus allelochemicals. Minimum germination for C. album was noted whereas non significant effect on the germination of E. helioscopia, T. aestivum and A. fatua was observed. Leaf extract of P. hysterophorus on the agar, filter paper, and the soil inhibited the radicle and plumule length (cm) of all the test species. Results have indicated that even though radicle length and germination of T. aestivum is not affected by leaf extract of P. hysterophorus, the plumule length was substantially decreased. The retarding effect of growth on wheat seedlings indicates that P. hysterophorus might not be an acceptable candidate for weed control under field conditions.

Dicamba Research Update. Thomas C. Mueller*[1], Larry Steckel[2]; [1]University of Tennessee, Knoxville, TN, [2]University of Tennessee, Jackson, TN (408)

Research focused on two general areas: resistance confirmation and environmental fate of auxinic herbicides. A greenhouse dicamba dose response screen was conducted on 15 Tennessee accessions. Relative resistance factor for dicamba ranged from 1.85-2.49 for several biotypes, upward to 14.25, indicating that this population could no longer be controlled using dicamba. In a separate study dicamba resistance was confirmed, and the addition of malathion did not reverse dicamba resistance from populations collected from Tennessee. This implies that the resistance may not be based upon enhanced herbicide metabolism, but more research is warranted in this area. Experiments were initiated in grower's fields where herbicide failures were previously observed to determine the impact of weed height on Palmer amaranth control following applications of dicamba or 2,4-D. While weed height at the time of application had a significant effect on Palmer amaranth control with auxin herbicides, control was still unacceptable in the field at the labelled rates of dicamba and 2,4-D at <10 cm tall weeds (48% and 53%, respectively). This research confirmed that these Palmer amaranth populations are resistant to both 2,4-D and dicamba. Bare ground experiments were conducted in 2021 and 2022 to determine the best timing and order of sequential applications for 2,4-D or dicamba and glufosinate. Palmer amaranth control increased when the interval between postemergence herbicide applications decreased from 21 to 7 days. At the 7 day interval in a dicamba-based system, the order of herbicides did not affect Palmer amaranth control, but with 2,4-D-tolerant systems greatest control was achieved when 2,4-D was applied first followed by either 2,4-D or glufosinate. Studies examining the off target movement (OTM) of dicamba and 2,4-D continued in 2022. Field studies showed that dicamba emissions were slightly higher than 2,4-D choline, even when labelled adjuvants (VRA) were added to the dicamba. A lab study evaluated the utility of potassium borate (KBr) added to dicamba Diglycolamine formulation to reduce dicamba emissions. A low rate of KBr (0.025 molar) only slightly reduced dicamba emissions, although a KBr dose of 0.1M reduced dicamba emissions equal to the industry standard of Xtendimax+ a Vapor Reducing Agent.

Use of a Spray Drone for Corn Harvest Aid Herbicide Applications. Ryan D. Langemeier*, Livia Pereira, Justin T. McCaghren, Steve Li; Auburn University, Auburn, AL (409)

Pre-harvest desiccation of corn (*Zea maize*) is used to aid in harvest, speeding harvest and reducing operator fatigue. This can be done via ground application, but only if soil conditions allow. In addition, a ground sprayer will knock over crop resulting in lost yield. Airplanes may also be used but can struggle in small and/or oddly shaped fields. Spray drones have made rapid advances in recent years, and are now more capable of covering large acreage, while also alleviating some of the problems with other application methods. However, questions about efficacy remain. With that in mind, an experiment was conducted in early August 2022 at two commercial fields in Autauga and Coffee counties, Alabama. Both experiments included three treatments: 1) glyphosate, 2) glyphosate + carfentrazone, 3) paraquat. All treatments were applied using a DJI Agras T30 using Greenleaf AM110-01 nozzles at 18.7 L ha^{-1}. Plots were ~1.4 hectares in Autauga Co. and 0.6 hectares in Coffee Co. There were three replications of each treatment. Data collection involved visual injury of three species found throughout the plots at each site at 4, 7 and 14 days after treatment (DAT). In addition, multispectral imagery was collected 0, 4, 7, and 14 DAT. By 14 DAT at both sites grass control for all treatments which included glyphosate was >80%. On broadleaves control varied between specie/treatments combinations. In most but not all cases, on broadleaf weeds a mixture of carfentrazone + glyphosate provided the best control. Mean reduction in NDVI from 0 DAT to 14 DAT was highest for glyphosate + carfentrazone, glyphosate, and then paraquat at both sites. These results show that spray drones are a viable method for corn harvest aid applications, and can provide high levels of efficacy.

Investigations of Suspected Weed Resistance to XtendiMax® Herbicide with VaporGrip® Technology.
Aruna V. Varanasi*, Chandrashekar Aradhya, Jeffrey Herrmann, John Willis; Bayer CropScience, Chesterfield, MO (410)

Weeds developing resistance to herbicides pose a great concern to growers in the United States. Introduction of dicamba-tolerant cropping systems in soybean and cotton, stacked with other herbicide tolerant traits such as glyphosate- and/or glufosinate-tolerance, provide farmers the flexibility to use multiple herbicide options to diversify their weed management practices and delay the evolution of resistance. XtendiMax® herbicide with VaporGrip® Technology is a dicamba formulation registered for use in dicamba-tolerant soybean and cotton by the U.S. Environmental Protection Agency (EPA) with certain terms of registration (EPA Reg. No. 264-1210). One of these terms includes investigation of product performance inquiries on lack of XtendiMax herbicide efficacy in the field. Weed populations suspected to have reduced sensitivity to XtendiMax in the field were sampled for testing in controlled environment. Seeds from 62 weed populations, consisting primarily of waterhemp (*Amaranthus tuberculatus*) and Palmer amaranth (*A. palmeri*), were collected in the 2022 crop season and plant bioassays were conducted in the greenhouse at a Bayer CropScience research facility in Chesterfield, Missouri to screen the populations for response to 560g ae ha^{-1} (label rate) and 1120g ae ha^{-1} of XtendiMax. All plants were treated at 7 to 10 cm height and evaluated at 21 days after treatment. Information on the screening process and data on weed efficacy will be presented.

Evaluation of Physical Control Strategies for Kochia (*Bassia scoparia*) Patch Management. Shaun M. Sharpe*, Julia Y. Leeson, Haben Asgedom Telda, Jennifer Town, Tim Dumonceaux, Krista Gill; Agriculture and Agri-Food Canada, Saskatoon, SK, Canada (411)

Kochia is a problematic weed for field crop production on the Great Plains of North America. It is an early emerging, allelopathic, patch forming tumbleweed with resistance to several herbicide modes of action in Canada and the United States. The study objective was to evaluate physical control methods for efficacy in suppressing kochia within established patches on farmland. The control strategies evaluated included: 1) hydro-mulch, 2) field chaff as mulch, 3) plastic mulch, 4) mowing, 5) an untreated control, and 6) an AC Saltlander seeded control. The trial was established in May 2021 at six sites in southern Saskatchewan. Kochia densities and mulch depth were monitored for 9 timings across the 2021 and 2022 growing season. Soil was collected post-harvest in 2021 and 2022 for allelochemical analysis for caffeic acid, chlorogenic acid, ferulic acid, myricetin, and quercetin. The AC Saltlander treatment failed to establish in 2021 or 2022. The hydro-mulch depth was 12 mm at the start of the study and did not decline through the study period. The chaff treatment was 65 mm at the start of the study period, reduced by 15% by August 2021, 40% by May 2022, and 49% by August 2022. Kochia patch densities across sites were between 637 to 928 plants m^{-2} in 2021 and 2,478 to 10,500 plants m^{-2} in 2022. The black plastic mulch treatment was the most consistent, with no kochia emerging in the study period. The chaff treatment was effective, providing 98 to 99% reduction in stand densities through 2021 and 82 to 95% through 2022. The mowing treatment provided between 61 to 98% stand reduction through 2021 and 44 to 70% stand reduction through 2022. No residual allelopathic compounds were detected in the soil above the 1 $\mu mol\ g^{-1}$ soil threshold for the post-harvest timings in 2021 and 2022.

Response of Soybean at Different Growth Stages to Sub-lethal Rates of Imazapyr. Michael W. Marshall*; Clemson University, Blackville, SC (412)

Detection and Site-Specific Treatment of Late-Season Weed Escapes in Rice Using Unmanned Aerial Systems and Deep Learning. Bholuram Gurjar*[1], Bishwa B. Sapkota[1], Isidor Ceperkovic[1], Ubaldo Torres[2], Matthew Kutugata[1], Daniel E. Martin[3], Muthukumar V. Bagavathiannan[1]; [1]Texas A&M university, College Station, TX, [2]Texas Tech University, College Station, TX, [3]United States Department of Agriculture, College Station, TX (413)

Unmanned aircraft systems (UASs), sensors, remote sensing, geographic information systems (GIS), and digital image processing technologies have been transforming agriculture with site-specific, real-time spatial and temporal field data. Significant progress has been made in drone technology and image analysis techniques in conjunction with semi-autonomous aerial and ground vehicle control systems. Weed control is an important part of agriculture, and broadacre production systems depend heavily on herbicides for weed control. In this research, a pipeline involving weed detection in drone-collected multispectral imagery along with a UAS-based spray application were evaluated, targeting weed escapes in a rice field. The research was conducted at the David Wintermann Rice Research and Extension Center, Eagle Lake, TX. The specific objectives of this study were to 1) identify weed patches in a rice crop using image analysis techniques, and 2) compare the efficacy of a drone-based precision herbicide application with the conventional backpack spray application. The weed species targeted in this research were barnyardgrass (*Echinochloa spp.*), amazon sprangletop (*Leptochloa panicoides*), yellow nutsedge (*Cyperus esculentus*), and hemp sesbania (*Sesbania herbacea*). The YOLOv5 model was used to detect late-season weed and find the center pixel coordinate of each weed patch. It determines the position of weeds by converting the pixel coordinates to world coordinates. Results showed that the accuracy of image-based weed detection for hemp sesbania, Amazon sprangletop, yellow nutsedge, and barnyardgrass were 95, 87, 74, and 62 %, respectively. The detection accuracy for barnyardgrass in rice was impacted by crop mimicry, and machine-learning approaches need to be explored to improve detection. The RTK-GPS based site-specific aerial spray application saved up to 45% herbicide volume compared to the broadcast application. Moreover, rice yields were greater with the site-specific drone application compared to the broadcast backpack application since the treatment of weed escapes coincided with the crop reproductive stages. Findings demonstrate the high potential for using deep-learning models and unmanned aerial systems for site-specific management of weed escapes in rice. Future improvements will include real-time weed detection and spraying using an on-board data processing system. Additional experiments are required for improving detection accuracy of barnyardgrass.

Effects of Delaying Rye Cover Crop Termination on Cover Crop Biomass, Weed Suppression, and Corn Yield. Daniel A. Boakye*[1], Parmeshwor Aryal[1], Carlene A. Chase[1], Muthukumar V. Bagavathiannan[2], Andrea Basche[3], Heather Darby[4], Michael L. Flessner[5], Erin Haramoto[6], Ramon G. Leon[7], Steven Brian Mirsky[8], Alison Robertson[9], Matt Ruark[10], Matthew R. Ryan[11], Nicholas J. Seiter[12], Kelley Tilmon[13], Peter Tomlinson[14], John F. Tooker[15], Mark VanGessel[16], John M. Wallace[17], Jared Adam[15], Tauana Ferreira de Almeida[18], Ashley Decker[12], Kristen Loria[11], Rashelle Matthiessen-Anderson[9], Jodie M. McVane[2], Fernando H. Oreja[7], DeAnn Presley[14], Amy Raudenbush[19], Lindsey Ruhl[20], Cynthia Sias[5], Barbara Scott[16], Ethan Sweep[21], Ashley Waggoner[22]; [1]University of Florida, Gainesville, FL, [2]Texas A&M University, College Station, TX, [3]University of Nebraka, Lincoln, NE, [4]The University of Vermont, St. Albans, VT, [5]Virginia Tech, Blacksburg, VA, [6]University of Kentucky, Lexington, KY, [7]North Carolina State University, Raleigh, NC, [8]USDA ARS, Beltsville, MD, [9]Iowa State University, Ames, IA, [10]University of Wisconsin, Madison, WI, [11]Cornell University, Ithaca, NY, [12]University of Illinois, Urbana, IL, [13]Ohio State University, Wooster, OH, [14]Kansas State University, Manhattan, KS, [15]The Pennsylvania State University, University Park, PA, [16]University of Delaware, Georgetown, DE, [17]Penn State University, University Park, PA, [18]University of Nebraska, Lincoln, NE, [19]The Ohio State University, Wooster, OH, [20]University of Vermont, St. Albans, VT, [21]USDA Agricultural Research Service, Beltsville, MD, [22]University of Wisconsin, Mas, WI (414)

Cereal rye (*Secale cereale* L.) is the major cool season cover crop (CC) that is used in no-till corn-soybean production system. Delaying CC termination until after planting the cash crop (planting green) has resulted in adverse effects on the corn crop that are not typically a problem in soybeans. In a multistate, multidisciplinary project aimed at understanding these effects in corn, one of our objectives focused on determining the effect of rye CC termination timing on CC biomass, weed suppression, and corn yield in a no-till system. Results are presented from trials conducted in Pennsylvania, Kansas, Kentucky, Ohio, Texas and Virginia, Delaware, Florida, Iowa, Maryland, New York, Virginia and Vermont. The rye CC was planted immediately after harvesting of soybeans in fall 2020 at seeding rate of 78 kg/ha and the corn was planted in spring 2021. The four treatments were termination of the rye CC at (1) 2-4 weeks before planting (early termination), 3-7 days before planting (late termination), and 0-3 days after planting (post-planting termination), and a no CC treatment (clean fallow) Treatments were arranged in a randomized complete block design with four replications. Glyphosate was used for all the rye terminations. The corn was planted the same day irrespective of the termination time. Total nitrogen (N) application was location specific but a minimum of 50 lb N was applied at planting with a side-dress at around V4-V6. Data were collected on rye CC biomass, weed biomass, and corn yield. Termination after planting resulted in the highest CC biomass accumulation compared to early and late termination regardless of location. Cover crop biomass varied with location with Kansas, Kentucky, Ohio, Texas and Virginia producing more CC biomass than Delaware, Florida, Iowa, Maryland, New York, and Vermont. Weed biomass was low with the rye CC regardless of termination timing. Post-planting termination had the lowest weed biomass accumulation in most of the locations. The rye CC effectively reduced weed biomass through the corn V-5 stage regardless of termination timing. However, a yield penalty was observed with delay of termination of the rye CC that was greater with late termination than with post-planting termination. This is the first year's results from a three-year study that will allow for the development of recommendations for rye termination timing that will allow farmers to plan for specific weed suppression and yield outcomes when utilizing a rye CC.

Maverick™ Corn Herbicide: Application Flexibility with Broad-Spectrum Control. Jonathon Kohrt*[1], Garrison J. Gundy[2], Trevor D. Israel[3], Chad L. Smith[4], John Pawlak[5], Eric Ott[5]; [1]Valent USA LLC, West Des Moines, IA, [2]Valent U.S.A LLC, Mcpherson, KS, [3]Valent USA, Harrisburg, SD, [4]Valent USA, Hallsville, MO, [5]Valent USA LLC, San Ramon, CA (415)

Maverick™ Corn Herbicide is a new corn herbicide offering from Valent USA LLC. *Maverick* is a low-use rate broad-spectrum herbicide premix of clopyralid, mesotrione, and pyroxasulfone that controls problematic annual grasses and broadleaf weeds. Application rates range from 14 fl oz per acre for postemergence applications to 18-32 fl oz per acre for preemergence and burndown applications. Maximum use rates can vary by soil type. Field trials were conducted throughout the Midwestern United Sates to demonstrate that the efficacy of *Maverick* is equivalent to the commercial standards.

Response of North Carolina Palmer Amaranth Accessions to Very-Long-Chain-Fatty-Acid Inhibiting Herbicides. Amanda M. Foderaro*; Syngenta, Raleigh, NC (416)

Investigation of Palmer amaranth Response to VLCFA-Inhibiting Herbicides: Greenhouse dose response studies were conducted to evaluate the differential response of two Palmer amaranth populations from Wake and Martin County, North Carolina, to four Group 15 herbicides, *S*-metolachlor, acetochlor, dimethenamid-P and pyroxasulfone. One susceptible population from Wake County and one tolerant population from Martin County, were evaluated. The results of the study indicated a significant differential response between the two populations to the Group 15 herbicides with the Martin County population requiring 4.1 times more *S*-metolachlor, 5.1 times more acetochlor and 10.5 times more dimethenamid-P to reduce seedling emergence by 50% compared to the more susceptible Wake County population.

Survey and Screening to Evaluate Herbicide Efficacy in Triple Stack Sugarbeet Cropping Systems in SE Montana. Het Samir Desai*[1], Fabian Menalled[1], Lovreet S. Shergill[2]; [1]Montana State University, Bozeman, MT, [2]Montana State University, Huntley, MT (417)

The introduction of glyphosate-resistant (GR) sugarbeet initially reduced the complexity involved in weed management; however, due to the evolution of glyphosate resistance, the need to utilize other weed management strategies that could include alternative herbicide chemistries has emerged. Therefore, genetically engineered triple-stack sugarbeet was developed, having tolerance for three different herbicides: glyphosate, glufosinate, and dicamba. However, the selection of multiple herbicide-resistant biotypes could erode the long-term viability of this technology. To assess the current herbicide-resistance status in sugarbeet growing regions, we conducted a survey prior to the sugarbeet harvest in South-East (SE) Montana and North-West (NW) Wyoming in 2021. Three major weed species: kochia [*Bassia scoparia* (L.) A. J. Scott], redroot pigweed (*Amaranthus retroflexus* L.), and common lambsquarters (*Chenopodium album* L.); three minor weed species: wild oats (*Avena fatua* L.), green foxtail [*Setaria viridis* (L.) P. Beauv], and junglerice [*Echinochloa crus-galli* (L.) P. Beauv.) were collected after visiting 72 sugarbeet fields. Physical screening in greenhouse conditions of glyphosate (840 g ae/ha), glufosinate (590 g ae/ha), and dicamba (280 g ae/ha) was conducted for all samples employing all the populations of each major weed species. Preliminary results show that 97.5% kochia populations [39:40 (survived:total)] survived glyphosate application and 75% survived (30:40) dicamba. For redroot pigweed, 2.3% population (1:44), 2.3% population (1:44), and 61.4% population (27:44) survived glyphosate, glufosinate, and dicamba applications, respectively. For common lambsquarters, 40.7%

population (11:27), 3.7% population (1:27), and 26.0% population (7:27) survived glyphosate, glufosinate, and dicamba application, respectively. All survived populations will be subjected to molecular analysis to identify the potential herbicide resistance mechanisms. Data generated from this study will help understand the eco-evolutionary impacts of triple-stack sugarbeet cultivation in SE Montana and NW Wyoming.

Evaluation of See &Amp; Spray™ Programs for Weed Management in Cotton Production. William L. Patzoldt*[1], Lauren M. Lazaro[1], Michael M. Houston[1], Diego J. Contreras[2], Wesley Everman[2], Tristen H. Avent[3], Jason K. Norsworthy[3]; [1]Blue River Technology, Sunnyvale, CA, [2]North Carolina State University, Raleigh, NC, [3]University of Arkansas, Fayetteville, AR (419)

A prototype See & Spray Ultimate machine was used to access weed control efficacy, input cost savings, and yield protection associated with targeting weeds in cotton (*Gossypium hirsutum L.*) production. Similar to See & Spray Ultimate, the research sprayer was designed with a dual tank delivery system allowing for separate herbicide product mixtures for See & Spray and/or broadcast applications. The dual tank feature has multiple benefits, but the focus of the research was to access 1) different broadcast partners paired with See & Spray applications and 2) the influence of POST-applied residual partners. Field trials were initiated in Greenville, MS, Kinston, NC, and Keiser, AR using cotton with resistance to dicamba, glyphosate, and glufosinate. Data across multiple trials suggests the most consistent weed management programs utilize broadcast applications of a POST-applied residual chemistry (e.g. group 15 herbicide) with See & Spray using the dual tank delivery system. Furthermore, the greatest benefit for See & Spray product savings was when the residual herbicide partner was applied during the first POST timing. In summary, See & Spray Ultimate has the ability to provide equivalent weed control and yield protection as broadcast applications while also reducing herbicide input costs associated with weed management in cotton production.

A23980B: A Step Change for Residual Weed Control in Corn. Sudeep A. Mathew*[1], Scott E. Cully[2], Mark Kitt[3]; [1]Syngenta, Germantown, MD, [2]Syngenta Crop Protection, Marion, IL, [3]Syngenta, Greensboro, NC (420)

A23980B: A Step Change for Residual Weed Control in Corn. Sudeep A. Mathew, Scott E. Cully, Mark J. Kitt, Thomas H. Beckett. A23980B is a new selective herbicide coming soon for weed control in field corn, seed corn, popcorn and sweet corn. A23980B contains ratios of S-metolachlor, pyroxasulfone, mesotrione, bicyclopyrone, and the safener benoxacor that will provide extended residual weed control in corn. Field trials were conducted to evaluate A23980B for residual weed control compared to Acuron®, Acuron Flexi and other corn herbicide premixes in one pass and two pass weed control programs. Results show that A23980B will provide more consistent and longer lasting residual control of difficult to control weeds like *Amaranthus palmeri*, *Amaranthus tuberculatus* and other problematic broadleaf and grass weeds in corn.

What's New in Industry. Sudeep A. Mathew*[1], Marshall Hay[2], Mark Kitt[3]; [1]Syngenta, Germantown, MD, [2]Syngenta Crop Protection, Vero Beach, FL, [3]Syngenta, Greensboro, NC (421)

Estimating Dry-down Performance of Desiccants in Lentils Using Hyperspectral Imaging. Charles M. Geddes[1], Keshav Dev Singh[2], Prabahar Ravichandran*[3]; [1]Agriculture and Agri-Food Canada, Lethbridge, AB, Canada, [2]University of Saskatchewan, Lethbridge, AB, Canada, [3]Agriculture and Agri-Food Canada, Truro, NS, Canada (422)

Will it Dye or Die? Effects of Application Speed on Droplet Distribution and Swath Width of Boomless Nozzles. Kayla L. Broster*, Thomas H. Duncan, John D. Byrd, Jr.; Mississippi State University, Starkville, MS (435)

Boomless nozzles are a convenient technology available for broadcast applications in areas that boom sprayers may not be best for application, due to rough terrain or obstructions. This has made their use more frequent for pasture, forestry, wildlife habitat restoration and roadside and other utility right of way applications. Few studies have evaluated effective spray swath of boomless nozzles at operating pressure and application speeds, in comparison to company advertised swath widths. Two studies were conducted at Mississippi State University (MSU) to evaluate swath width. The first study was conducted in an enclosed facility at the Mississippi Horse Park, while the second study was an unimproved turf area at the R.R. Foil Plant Science Research Center. Both studies were a 4 by 4 factorial arrangement of treatments, with nozzle type and application speed as factors. Nozzles used and their advertised spray swaths included the Boominator 1400ST (no advertised spray swath but static swath of 13 m), Boominator 1870 advertised to cover 11 m, Hamilton #10 advertised to cover 15 m, and Boom Buster 187 advertised to cover 12 m. Swath widths of these nozzles were measured at speeds of 0, 1.6, 3.2, and 6.4 KPH using a 3-point hitch tractor mounted sprayer equipped with Hypro roller pump at engine RPM of 1000. For the indoor study, a mixture of water and FD&C Blue 1 food dye (Flavors and Color) solution was applied over six rows of 21.5 X 28 cm Kromekote (CutCardstock) sheets, placed every 0.3 m out to 0.6 m past the advertised swath of each, for each nozzle and speed. The sheets were scanned at 600 dot per inch (dpi) and processed through the USDA Automatic Paper Analysis tool in the DepositScan software add-on for Image J (Version 1.38X) to obtain coverage data, which was used to develop a relative deposition curve to estimate total and effective swaths of each nozzle at each speed. The field study was conducted on a flat grassy area, with mixed grass and clover, with applications of MSMA 6 Plus (monosodiums acid methanearsonate) at a 10.5% v/v solution mixed with FD&C Blue 1 food dye. These applications were made in manner described above minus Kromekote paper on August 8, 2022. Swath widths were measured immediately after application and 3 days after treatment (DAT), when visual injury was obvious. Data were analyzed with RStudio (Version RStudio 2022.07.1) and two-way ANOVA, with means separated using Fisher's Protected LSD (alpha = 0.05). Data were transformed to percent difference of effective kill swath to total swath to allow for more equal comparison between nozzles and the effect of speed on spray distribution. Nozzle pattern uniformity determined by Christiansen's Uniformity Coefficient, for the indoor study. Nozzles were analyzed separately across speed as their responses differed. The Hamilton #10 significantly decreased in swath width when accelerating from 0 to 1.6 KPH, all other nozzles did not. However, as speed increased past 1.6 KPH the relative effective swath of each nozzle decreased, except for the Boom Buster 187, where the swath increased between 3.2 KPH and 6.4 KPH. There were no differences in uniformity between each nozzle at any speed when using the measured effective swath, based on the uniformity coefficient obtained from each nozzle and speed combination. Comparing relative effective swath and uniformity coefficient to determine the widest and consistent effective swath pattern overlap, the Boom Buster 187 provided the most efficient distribution of spray solution, potentially requiring fewer in field passes to uniformly cover large areas.

Chemical Control of Terrestrial Alligatorweed in Mississippi. Gray Turnage*; Mississippi State University, Starkville, MS (436)

Alligatorweed (*Alternanthera philoxeroides*) is the most widespread aquatic weed in Mississippi. Alligatorweed is capable of growing in aquatic, wetland, and terrestrial sites and switches growth form from a rhizomatous mat forming plant (aquatic) to a prostrate growth form with tap roots (terrestrial) which allows it to survive stressors such as drawdown or drought. Much is known regarding the impacts and control of alligatorweed in aquatic sites, however, less is known regarding control of alligatorweed in terrestrial sites. Muscadine Farms Wildlife Management Area (WMA) is a series of 90 ponds (each 6.1-8.1 hectares in size) that covers approximately 809 ha in the Mississippi flyway in western Mississippi; the WMA is primarily managed for waterfowl habitat. The terrestrial form of alligatorweed has infested many ponds in the WMA and displaced desirable vegetation utilized as waterfowl forage. In 2022, an herbicide trial was conducted and repeated two weeks later to assess the reduction of alligatorweed at Muscadine Farms by herbicides utilized for operational control of alligatorweed as well as newer herbicides labeled for use in aquatics in the last two decades. Herbicides were applied to 83.6 m^{-2} (9.1 by 9.1 m) plots at Muscadine Farms. High and low rates of imazapyr (8.2 and 4.1 kg a.e. ha^{-1}), bispyribac-sodium (0.11 and 0.055 kg a.i. ha^{-1}), topramezone (11.4 and 5.7 kg a.e. ha^{-1}), florpyrauxifen-benzyl (0.02 and 0.01 kg a.i. ha^{-1}), and fluridone (10.9 and 5.4 kg a.i./ha^{-1}) were tested as stand-alone foliar herbicide treatments. Glyphosate (11.0 kg a.i. ha^{-1}), triclopyr (11.7 kg a.e. ha^{-1}), metsulfuron-methyl (0.036 kg a.i. ha^{-1}), and two-way tank mixes of the three were also tested as foliar treatments. All herbicide applications included a 0.5% v:v MSO surfactant and were applied at a 467.7 L ha^{-1} (50 gal ac^{-1}) diluent rate. Each treatment was replicated four times. This is the first report of alligatorweed control by bispyribac-sodium, topramezone, or fluridone suggesting that these herbicides may provide another alligatorweed control option for resource managers.

Assessment of Reverse Cut Stem Treatments for Difficult to Control Tropical Vines. Stephen F. Enloe*; University of Florida, Gainesville, FL (437)

Arrowhead vine (*Syngonium podophyllum*) is an invasive vine in the Araceae family that is native to Latin America. Although a popular houseplant for decades, it has escaped cultivation and is now problematic in south and central Florida forested wetlands. Syngonium forms dense cover and climbs into trees up to 10 m. Management has proven problematic as climbing vines are extremely difficult to manually remove from trees. Additionally, they survive after cutting via adventitious aerial roots that regrow to the ground. To address this, a field study was conducted on the Loxahatchee River floodplain near Jupiter, Florida in 2020 and 2021 to assess the effectiveness of manual removal and herbicide foliar and reverse cut stem treatments. Plots were 10 by 10 m and replicated four times per treatment. Foliar treatments were applied to arrowhead vine ground cover and included imazamox, a tank mix of imazamox + carfentrazone, or no herbicide. Reverse cut stem treatments included glyphosate at 50% v/v or no herbicide, applied with an herbicide foamer or a two second dip method. Hand removal included the complete removal of all vines covering the ground and trees. Syngonium ground and tree cover were recorded at multiple times from 0 to 360 days after treatment (DAT). Imazamox alone and the imazamox + carfentrazone tank mix did not effectively reduce Syngonium ground cover, while hand removal significantly reduced ground cover compared to the untreated control. Reverse cut stem treatments with glyphosate applied with a foamer applicator or a dip method reduced climbing vine cover to near zero and were generally more effective than cutting alone. Surviving climbing vines that had been cut with no herbicide treatment had regrown aerial roots 120 cm to the ground. This study indicates that a combination of manual removal and reverse cut stem treatment techniques may be the most effective approach for managing Syngonium in forest wetlands. Future work should address additional herbicides for foliar ground treatments.

Aquatic Weed Control Research Update. Benjamin P. Sperry*; US Army Corps of Engineers, Gainesville, FL (438)

The US Army Corps of Engineers has been involved in aquatic plant control research and operations since the 1800s after the introductions of exotic waterhyacinth (*Eichhornia crassipes*) and alligatorweed (*Alternanthera philoxeroides*). Over 100 years later, these agency efforts have continued now focusing on the development of environmentally compatible management strategies for new and existing invasive plant species. This presentation will introduce the US Army Engineer Research and Development Center's Aquatic Plant Management Team, its members, research capabilities, and a brief overview of current research projects.

Smutgrass (Sporobolus Indicus) Identification and Management with Unmanned Aerial Systems.
Zachary S. Howard*[1], Scott A. Nolte[2], Daniel E. Martin[3], Chenghai Yang[3]; [1]Texas A&M University, College Station, TX, [2]Texas A&M AgriLife Extension, College Station, TX, [3]United States Department of Agriculture, College Station, TX (439)

Perennial grass weeds are often of the most difficult to control weeds in a pasture setting. Smutgrass (Sporobolus indicus) can significantly impact forage quantity, and therefore lead to required reductions in livestock stocking rate to protect from forage overgrazing, ultimately impacting profitability. Traditionally this plant is difficult to control and requires high inputs of either time or monetary resources. Site specific weed management (SSWM) could not only lead to effective utilization of non-selective herbicide glyphosate to control smutgrass but can spare the forage such that it is sprayed only on the weed. In this study, pastureland infested with smutgrass was characterized by red, green, blue, near-infrared, and a canopy height model and further processed for the identification of smutgrass by a machine learning algorithm. 25 points where smutgrass was identified in two locations where then utilized for UAS (unmanned aerial system) spraying. Additionally, 25 plants were sprayed by individual plant treatment method (IPT) to compare the time and productivity of each event. On target percentage and control measures were taken to evaluate the effectiveness of the UAS spray application. Various markers of productivity indicate UAS mapping and spraying can be a viable option for SSWM, and that there is room for improvement.

How Much Pre-plant Weed Suppression is Practical for Riparian Forest Buffer Establishment? Arthur E. Gover*, Emily N. Rojik; Penn State, University Park, PA (440)

Evaluation of Factors That Influence Herbicide Spray Retention on Floating and Emergent Aquatic Vegetation. Christopher R. Mudge*[1], Benjamin P. Sperry[2], Michael W. Durham[3], Kurt D. Getsinger[4]; [1]U.S. Army Engineer Research & Development Center, Baton Rouge, LA, [2]US Army Corps of Engineers, Gainesville, FL, [3]University of Florida, Gainesville, FL, [4]U.S. Army Engineer Research and Development Center, Vicksburg, MS (441)

Management of invasive floating and emergent plants in the U.S. has been accomplished through the use if foliar applied aquatic herbicides. For decades, small-scale research trials and operational use has developed and improved use patterns and efficacy to selectively manage unwanted plants for most of the active ingredients registered in aquatic settings. However, limited information exists on what happens to the spray solution after being applied. One of the goals of plant management is to maximize herbicide retention on the target plant to aid in efficacy, but some deposition (or spray loss) reaches the water column when foliar applications are administered. As result of this data gap, several research trials were conducted in 2020 and 2021 in Baton Rouge, LA and Gainesville, FL to evaluate the influence of factors including spray trajectory angle, spray pattern type, and broadcast vs spot spray application methods on foliar applied spray retention vs. spray loss. Foliar applications of the fluorescent tracer dye rhodamine WT was used as a cost-effective herbicide surrogate to evaluate retention vs. spray loss under these various situations. Water hyacinth, water lettuce, and giant salvinia were the test species.

Control of Wolftail Sedge (*Carex cherokeensis*) in Forages. David P. Russell*; Auburn University, Madison, AL (442)

Wolftail sedge, or Cherokee sedge (*Carex cherokeensis*), is a cool-season perennial that resides among permanent forage pastures and hayfields across its native range in the Southeastern U.S. It is considered a weed that spreads by rhizomes, displaces desirable forage species, and lowers forage nutritive quality. Mechanical mowing has proven impractical, but 2,4-D has shown activity on established populations in Black Belt soils. The objective of this study was to evaluate the effect of 2,4-D when combined with herbicides with a history of sedge control. Field studies were conducted in north Alabama from 2020 to 2022 in wolftail sedge-infested tall fescue hayfields to determine proper herbicide selection and application timing to minimize the presence of sedge. Four ALS-inhibiting herbicide active ingredients (imazethapyr, sulfosulfuron, halosulfuron, and metsulfuron) were combined with 2,4-D ester and 2,4-D amine and compared to 2,4-D (ester and amine) alone, 2,4-D + florpyrauxifen-benzyl, and a non-treated check. These twelve herbicide treatments were arranged in a randomized complete block design and applied in the fall (late-November or early-December) and spring (March) to established wolftail sedge populations using a CO2-pressurized backpack sprayer in a carrier volume of 140 L ha^{-1} using AIXR11002 nozzles. Data collection included visual estimates of control, along with percent sedge occurrence using square meter quadrats at 6 and 2.5 months after the fall and spring application, respectively. All data were subjected to ANOVA using PROC GLIMMIX and means separated using Fisher's Protected LSD at α = 0.05. Application timing and herbicide treatment had a significant effect on percent occurrence. Collectively, spring treatments resulted in at least twice the level of control compared to fall applications. Data suggest that 2,4-D ester + imazethapyr (Pursuit) or 2,4-D ester + halosulfuron (Permit) should be applied during spring. Each of these spring treatments provided greater than 70% visual control three months following treatment. These treatments may also be applied during the fall for at least 59% visual sedge control which may provide greater forage safety especially in tall fescue pastures or hayfields.

Preventing Preventable Problems on the Farm. Eric P. Prostko*; University of Georgia, Tifton, GA (443)

Growers are under a tremendous amount of pressure to plant, spray, harvest, and pay the banker on-time. These factors coupled with weather, equipment, and personnel issues often lead to unwanted mishaps on the farm. One of the roles of an extension weed specialist is to help troubleshoot problems with the goal of improving the current situation (if possible) and preventing those problems from happening again in the future. Most weed science related problems can be traced back to specific weather events, sprayer cleanout, mixing errors, sprayer plumbing, inadequate record keeping, and label misinterpretation. Top 10 tips for preventing preventable problems on the farm include the following: 1) Phone/text a colleague before mixing/spraying; 2) Only have available on hand the products that need to be applied on the day of application; 3) Never use chemicals from unlabeled pesticide containers. Dispose of pesticides in unlabeled containers according to local requirements; 4) Before mixing/spraying, make sure spray tank is empty and that the sprayer has been properly cleaned; 5) Consider pesticide container color coding or painting. Some manufactures have different colored labels for different types of pesticides; 6) When applying multiple pesticides and/or fertilizers with 1 applicator, make sure spray lines are plumbed correctly; 7) Maintain adequate sprayer and field records to prevent misapplications; 8) Thoroughly review all pesticide labels before application. Label reform is needed to make it easier to find important topics of interest such as rain-free period or crop rotation restrictions; 9) Avoid solving every insect/weed/disease/fertility problem with a single application. It is more likely for undesirable mixing issues and crop injury to occur when multiple products are applied in a single application. Mixing sequence is important. Adequate research to address all possible tank-mixtures/application conditions is impossible; and 10) Pay close attention to weather conditions and nearby sensitive crops. The use of all off-target mitigation practices (nozzle type, GPA, boom height, tractor speed, pressure, drift retardants, etc.) does not guarantee 100% safety for neighboring sensitive crops.

And You Think Pigweed Control is a Challenge!! John D. Byrd, Jr.*, Kayla L. Broster, Thomas H. Duncan; Mississippi State University, Starkville, MS (444)

Extension Weed Scientists (Specialists), like County Extension Agents, are faced with a diverse array of questions posed by clientele. In 2022, an email from an architectural firm requested assistance with a weed management issue on a historical preservation project funded by the Mississippi Department of Archives and History. Windsor mansion was an antebellum Greek Revival mansion built in 1861 south of Port Gibson, MS for Smith Coffee Daniell II, a cotton planter. The 3-story 20 m by 20 m 1,600 m^2 home was supported by 29 Corinthian columns built of brick handcast by plantation labor with an external stucco covering. Columns sit atop 3 m tall by 1.5 m square plinths and are topped with a cast iron capital. All that remain today are 23 mostly complete and 5 partial 12.2 m tall fluted columns as fire destroyed the mansion in 1890. The property was designated a Historic Place on the National Register in 1971 and a Mississippi Landmark in 1985. In the 1970's, the hollow columns were encapped with cement to prevent seed and other organic matter accumulation. Vandalism and weathering over time have caused chunks of stucco as well as cement bonding of bricks in the columns to deteriorate. A diverse spectrum of weedy vegetation has crawled into or emerged from stucco or cement gaps as well as from the remaining cast-iron capitals. Several species of sedges (*Cyperus* spp. and *Carex* spp.), including purple and yellow nutsedges (*Cyperus rotundus* and *C. esculentus*), dandelion (*Taraxacum officinale*), bermudagrass (*Cynodon dactylon*), poison ivy (*Toxicodendron radicans*), Japanese climbing fern (*Lygodium japonicum*), dogfennel (*Eupatorium capillifolium*), smooth and southern crabgrass (*Digitaria ischaemum* and *D. ciliaris*), green ash (*Fraxinus pennsylvanica*), poplar (*Populus alba*), marestail (*Conyza canadensis*), ferns, white clover (*Trifolium* spp.), hyssopleaf sandmat (*Chamaesyce hyssopifolia*), American sycamore (*Platanus occidentalis*), trumpet creeper (*Campsis radicans*), green alder (*Alnus viridis*), Virginia pepperweed (*Lepidium virginicum*), hop clover (*Trifolium campestre*), Pennsylvania pellitory (*Parietaria pensylvanica*), wild violet (*Viola* spp.), and Virginia creeper *(Parthenocissus quinquefolia*) could be identified from images provided by the architects and a site visit. The challenge presented by these restoration architects is to control these plants without the use of any salt, acid or other chemical compound that could damage the soft brick, cement, or stucco manufactured in the mid-1800's.

US-Herbicide Resistance Action Committee (HRAC): Who We Are and What We Do? Zahoor A. Ganie[1], Greg Elmore[2], Lowell Sandell[3], Chandrashekar Aradhya*[4]; [1]FMC, Newark, DE, [2]Bayer Crop Science, Saint Louis, MO, [3]Corteva, Ashland, NE, [4]Bayer Crop Science, Chesterfield, MO (445)

United States-Herbicide Resistance Action Committee (US-HRAC) is an industry group operated by technical members of the companies represented in CropLife America. US-HRAC has a mission to protect crop yields and quality in the US by supporting fight against herbicide resistant weeds, and to promote proper product stewardship required to delay the evolution of new herbicide resistance issues. Diverse weed management approaches can help growers to mitigate the evolution and spread of herbicide-resistant weeds. We support development and distribution of research based technical information on diverse weed management strategies that are effective, reliable, practical, and economical. Our primary audience are Grower Influencers. These are industry experts (personnel, ag advisors, extension experts, retail agronomists etc.) who work directly with farmers. To achieve its goals USHRAC has three strategic focus areas that includes regulatory engagement, stakeholder engagement and communication.

Public Parks, Family Farms, and Cover Crops: Implementing Ecological Weed Management with Diverse Stakeholders in Oregon Hazelnuts. Steven Haring*, Marissa Lane-Massee, Cal Penkauskas, Alejandro Brambila, Ari Brown, Lauren Hallett; University of Oregon, Eugene, OR (446)

Cover crops can be a foundational tool for integrated pest management, but researchers face many challenges introducing cultural management practices to new cropping systems. These challenges are compounded when combining cover crops with other complex management goals such as ecological restoration, as well as when working outside of traditional agricultural institutions. Through a series of experiments using a translational ecology framework and adaptive management in hazelnut orchards, particularly using Willamette Valley native wildflower species as cover crops, we explored different modes of research with, by, and for various stakeholders. These experiments underscored the importance of practicing intentional community engagement in agricultural research, including different dimensions of stakeholder-created research questions, farmer research review, and reciprocity. By prioritizing these dimensions of community engagement, scientists both within and outside of traditional agricultural institutions can improve the robustness and practicality of their research.

Challenges with Co-existence of Multiple Herbicide-Resistant Soybean Traits. Amit J. Jhala*; University of Nebraska Lincoln, Lincoln, NE (447)

A number of herbicide-resistant soybean traits are available commercially in the market, making it complicated for growers to understand which soybean trait is resistant to what herbicide(s). The objectives of this project were (1) To provide information on seven herbicide-resistant soybean traits available in the marketplace and discuss which herbicides they are and resistant to, and (2) To provide information on factors to consider when applying post-emergence herbicides in multiple herbicide-resistant soybean, and (3) To create awareness about miss-application of post-emergence herbicide in soybean. Post-emergence herbicides must be selected depending on which soybean trait you have planted. For example, when Roundup Ready 2 Xtend® soybean is planted, dicamba-based herbicides such as XtendiMax®, Engenia®, or Tavium® can be applied. Several growers rely on certified pesticide applicators for their herbicide application. Therefore, good communication is needed between the owners or leaseholder of the field and the certified applicators to make sure the applicator knows which soybean trait is planted in which field and what herbicide(s) to spray. There are some conventional/organic soybean growers in Nebraska. Care must be taken when conventional and organic soybean fields are in proximity to avoid off-target injury from post-emergence herbicides applied in multiple herbicide-resistant soybean. Organic and specialty crop growers should register on Drift Watch (https://ne.driftwatch.org/). This website is a voluntary communication tool that enables organic and specialty crop producers and pesticide applicators to work together to protect from off-target injury through use of mapping programs.

Climate Change and Weed Management: Synthesizing Information for New York Farmers. Caroline Marschner*[1], Sharon Bachman[2], Catalina Ferreria-Dias[3], Isabella Colucci[1], Rebecca Stup[1], Claire Liu[1], Antonio DiTommaso[1]; [1]Cornell University, Ithaca, NY, [2]Cornell Cooperative Extension, East Aurora, NY, [3]Cornell University, Santa Ana, CA (448)

Climate change may increase the threats weeds pose to agricultural productivity by altering weed distributions, weed competitiveness, and management efficacy. To minimize agronomic impact, farmers require accurate predictions about how climate change will affect local weed communities. Factors associated with climate change have different effects on different weeds, climate impacts vary by region, and the messaging around weed management can be complex. We are conducting scoping reviews of current literature on three relevant questions for New York field crop and dairy farmers: 1) how does climate change impact herbicide efficacy, 2) what is known about range-shifting weeds that are likely to impact our farms, and 3) how will weed competitiveness change with the predicted changes to New York's climate? We have extracted data from 54peer-reviewed articles that focus or summarize research on herbicide efficacy and climate change. The articles included field studies, most of which were conducted in North America, as well as laboratory studies, simulations, and review articles. Of the 54 articles, 31 articles assessed the impacts of climate change on glyphosate efficacy. The most commonly studied aspects of climate change were carbon dioxide concentration, soil moisture, and temperature. The direction and magnitude of climate effects on herbicide efficacy varied based on specific herbicide, weed species, and even susceptible vs. herbicide-resistant biotypes of the same species. Qualitatively, elevated carbon dioxide tended to reduce herbicide efficacy for herbicide-resistant species and low soil moisture tended to reduce herbicide efficacy for C4 species. High temperatures tended to reduce herbicide efficacy in annual grasses and herbicide-resistant C4 species. The next steps of our project will be to connect the species and herbicide-specific information from our three scoping reviews with the 2023 climate projections for the Northeast, and work with New York weed scientists and extension professionals to develop practical weed management resources for New York farmers.

Advancing Physical Weed Control: Farmers Love Field Days, But On-farm Research is Very Challenging.
Eric R. Gallandt*[1], Daniel C. Brainard[2]; [1]University of Maine, Orono, ME, [2]Michigan State University, East Lansing, MI (449)

Physical weed control (PWC) remains a foundational part of a diverse weed management program on organic farms, with many farmers relying on time-tested, familiar tools. Unfortunately, most weeding tools offer relatively poor efficacy and selectivity; e.g., outcomes of 50% weed control and 10% crop stand reduction are not unusual. Because efficacy is typically density independent, the consequences of poor efficacy are greater with increasing initial weed densities as there are simply more surviving weeds. Increased weed survival, concomitant seed rain, high seeding densities, and poor efficacy often result in and increasing weed problem for these farmers. We are completing a four-year project in northern New England and the midwestern U.S. to test the hypothesis that improved PWC, as may be achieved with improved tool selection, adjustment, and perhaps stacking of tools, combined with proven methods to reduce the germinable weed seedbank, could establish a virtuous cycle of improved PWC and reduced weed seed rain resulting in a declining weed population. To encourage farmer adoption and testing of improved PWC tools and seedbank management we set up an ambitious program in which farmers would be provided funding to support an on-farm research intern, PWC tools that they considered promising improvements over their existing tools, as well as data and networking with other participating farmers. Despite the investment in farm personnel to encourage engagement and improve the quality of on-farm data, and our development of a professional Data Dashboard to aggregate and share participants' data, it was the rare case that could be considered a real success as we envisioned. Participating farmers were highly engaged when we delivered tools and worked with them to adjust and optimize settings, but the stresses of daily operations on these diverse organic vegetable farms ultimately overwhelmed their capacity to successfully establish, maintain, collect and post data and experiences from the replicated field experiments we designed based on their interests. The project could be considered a success as several participating farms either purchased, or decided not to purchase, new equipment based on this project. However, contrary to the data-driven decision-making we aimed to promote, field testing and observation were of far greater interest to the majority of our participating farmers.

A Story of Success and Challenges in Eradicating Palmer Amaranth (*Amaranthus palmeri*) in Minnesota. Debalin Sarangi*[1], Eric Y. Yu[1], Anthony Brusa[2], Shane Blair[3], Monika Chandler[3], Denise Thiede[3], Anthony Cortilet[3], Melissa Wilson[1], Jeffrey Gunsolus[1], Roger Becker[1], Thomas J. Peters[4]; [1]University of Minnesota, St. Paul, MN, [2]University of Minnesota, Saint Paul, MN, [3]Minnesota Department of Agriculture, St. Paul, MN, [4]North Dakota State University, Fargo, ND (450)

Palmer amaranth (*Amaranthus palmeri* S. Watson), a native to the southwestern US, is the most problematic weed in row-crop production systems in the US. It was first reported in Minnesota in 2016 and currently, Palmer amaranth is a Minnesota Prohibited Noxious Weed, Eradicate Species, meaning that the weed must be eradicated whenever sighted. Since 2016, the Minnesota Department of Agriculture (MDA) investigated potential sources for all Palmer amaranth infestations in the state, and contaminated commercial seed lots, sunflower screenings as livestock feed, and contaminated manure spreading on the fields were identified as the major pathways for Palmer amaranth introduction in the state. Therefore, MDA, University of Minnesota (UMN) Extension, Conservation Crops Minnesota and Iowa (CCMI), farmers, and other partners started working together to eradicate these infestations before this weed becomes a major problem in the state. Communications between these partners are still occurring with bimonthly meetings throughout the year. In 2016, Palmer amaranth-contaminated seed mixes were sown in 35 sites in Lyon and Yellow Medicine Counties as a part of the Conservation Reserve Program (CRP), but with the continuous monitoring and eradication efforts, no Palmer amaranth was found at any of these sites by 2018. MDA and the Minnesota Environment and Natural Resources Trust Fund, as recommended by the Legislative-Citizens Commission on Minnesota Resources (LCCMR) offered financial support to the UMN Research and Extension groups for leading the research and outreach efforts. Cost-effective genetic testing methods were developed to detect Palmer amaranth seeds in commercial seed lots. Several research projects are ongoing at UMN to determine how to detect, extract, and kill Palmer amaranth seeds in manure. Finally, we have learned that communication between the Department of Agriculture, legislators, university Extension personnel, researchers, farmers, and other stakeholders is the key to success in an effort like this. Our success is more likely to persist if the surrounding jurisdictions collaborate with us. dsarangi@umn.edu

POSTER

WSSA Public Awareness Committee Update. Theresa Piskackova*[1], Eric Gustafson[2]; [1]Czech University of Life Sciences Prague, Prague 6 - Suchdol, Czechia (Czech Republic), [2]Weed Science Society of America, Westminster, CO

Taxonomic Pattern in Seed Dormancy and its Breaking Methods in Globally Important Weed Species. Aniruddha Maity*, Akashdeep Singh; Auburn University, Auburn, AL (1)

Seed dormancy is the inability of a viable seed to germinate even under favourable conditions. It can be of a variety of forms, such as innate, induced, and enforced dormancy. Seed dormancy is an evolutionary critical trait that sustains species persistence by restricting seed germination under unfavourable ecological conditions that would normally allow a poor likelihood of seedling survival. Owing to the variable magnitude of seed dormancy, weed seeds hibernate in the soil seedbank and emerge periodically over many years, jeopardising the application timing of weed control strategies across global cropping systems. Understanding weed seed dormancy cycle is critical as to optimise the execution timing of available weed control strategies without interfering with the cash crop growth and development. To this effect, we reviewed 98 research studies conducted across five continents that included 50 globally important weed species from 18 taxonomic families. In this preliminary data mining, *Portulaceae*, *Geraniaceae*, *Ambrosiceae*, *Polygonaceae*, and *Asteraceae* family weed seeds were found to be more dormant than seeds from other families, with most weed species having a fresh seed dormancy of 90-100%. *Panicum dichotomiflorum*, a summer annual *Poaceae* weed, was found to show the greatest seed dormancy (93-100%). In laboratory condition, seed dormancy release by external treatments substantially varied according to the taxonomic family. The *Poaceae* weeds mostly released seed dormancy in exposure to incubation at high temperatures and/or with red light irradiation and imbibition or scarification using a number of chemicals such as potassium nitrate, gibberellic acid, and sulfuric acid. Whereas cold stratification (4?) and gibberellic acid under dark conditions were commonly used methods for breaking seed dormancy in the *Amaranthaceae* family. Such a data base with critical interpretation, once finalized, will help devising and incorporating appropriate seed dormancy breaking treatments into an integrated weed management strategy, which will aim to stimulate the weed seeds to germinate during pre-planting operations and be killed using pre-emergence herbicides. This will help in effective weed management by exhausting the weed seedbank and reducing the number of post-emergence herbicide applications, ultimately lessening the financial burden on farmers.

The Interaction of Pyroxasulfone and Flumioxazin Applied Preemergence for the Control of Multiple-herbicide-resistant Waterhemp [Amaranthus Tuberculatus (Moq) J.D. Sauer] in Soybean. James C. Ferrier, Nader Soltani*, Peter H. Sikkema; University of Guelph, Ridgetown, ON, Canada (2)

Six field experiments were conducted to investigate any interaction between pyroxasulfone and flumioxazin on soybean tolerance and control of multiple-herbicide-resistant (MHR) waterhemp in soybean during 2016 and 2017 in Ontario, Canada. There was a synergistic increase in soybean injury with the co-application of pyroxasulfone and flumioxazin at all rates evaluated at 2 weeks after emergence (WAE), the two highest rates evaluated (134/106 and 268/211 g ai ha^{-1}) at 4 WAE, and the highest rate (268/211 g ai ha^{-1}) evaluated at 8 WAE. Soybean injury with all pyroxasulfone and flumioxazin treatments was transient and had no adverse effect on soybean grain yield. Pyroxasulfone applied preemergence (PRE) at 45, 89, 134, and 268 g ai ha^{-1} controlled MHR waterhemp up to 72, 89, 92, and 95%, respectively. Flumioxazin applied PRE at 35, 70, 106, and 211 g ai ha^{-1} controlled MHR waterhemp up to 78, 90, 93, and 96%, respectively. Pyroxasulfone/flumioxazin applied PRE at 45/35, 89/70, 134/106, and 268/211 g ai ha^{-1} controlled MHR waterhemp up to 92, 96, 98 and 100%, respectively. There were no significant antagonistic or synergistic interactions for the control of MHR waterhemp with pyroxasulfone/flumioxazin at rates evaluated except at 268/211 g ai ha^{-1} which provided a synergistic increase in MHR waterhemp control at 4 WAE. The MHR waterhemp biomass and density reductions followed a similar trend as visible control. Pyroxasulfone/flumioxazin at 268/211 g ai ha^{-1} caused a synergistic response in biomass reduction (9% difference). Based on these results, there is an additive increase in MHR waterhemp control and potential for synergistic soybean injury with the co-application of pyroxasulfone plus flumioxazin.

Planting Green in Soybean: Evaluation of Residual Herbicide Interaction, Weed Control Programs, and Soybean Yield. Trey P. Stephens*, Amit J. Jhala; University of Nebraska-Lincoln, Lincoln, NE (3)

Control of Larger Herbicide-Resistant Common Ragweed (*Ambrosia artemisiifilia*) in Soybean. Kurt M. Vollmer*; University of Maryland, Queenstown, MD (4)

Populations of common ragweed (*Ambrosia artemisiifolia* L.) in Maryland have developed resistance to herbicide groups 2, 9, and 14. This limits postemergence control options in soybean, in particular when common ragweed has exceeded 10 cm in height. Two studies were conducted to evaluate single and/or sequential applications of one or more herbicides for controlling ragweed common ragweed greater than 15 cm tall. In the first study, single applications containing 2,4-D, combinations of 2,4-D, fomesafen, and/or glufosinate, or sequential applications of one or more of these herbicides controlled common ragweed at least 92% when applied to 15 cm to 30 cm tall plants compared to a single application of fomesafen (79%) or glufosinate (83%) 3 wk after treatment (WAT). In the second study, 2,4-D, 2,4-D + glyphosate, glufosinate + glyphosate, 2,4-D + glufosinate + glyphosate controlled common ragweed at least 93% when applications were made to 15 cm to 30 cm tall common ragweed. However, when these same treatments were applied to 36 to 46 cm tall common ragweed there was an average 24% decrease in control. These results show tank-mixing herbicides or sequential applications are needed to provide optimal control of herbicide-resistant common ragweed once it has reached 15 cm to 30 cm in height; however, control is likely to decrease if applications are further delayed.

Urea Ammonium Nitrate as the Carrier for Preplant Burndown Herbicides. Kaone Mookodi*[1], Jared Spackman[2], Albert T. Adjesiwor[1]; [1]University of Idaho, Kimberly, ID, [2]University of Idaho, Aberdeen, ID (5)

Pre-plant weed control is a common practice for the Pacific northwest. The timing of the spring burndown herbicide application often coincides with starter nitrogen (N) application. Co-application of the herbicide and nitrogen fertilizer such as UAN, can reduce the number of trips to the farm, the labor costs and the costs of N and herbicide applications. However, there is a dearth of information on the effect of herbicide-N fertilizer mixtures on herbicide efficacy. Field studies was were conducted in the summer of 2021 and 2022 to evaluate the effect of urea ammonium nitrate (UAN; 32-0-0) rate (0, 25, 50, 75, and 100% of carrier volume) on the efficacy of three non-selective herbicides [glyphosate (1260 g ae ha-1), paraquat (560 g ai ha-1), and tiafenacil (74 g ai ha-1)]. There was no statistically significant effect of UAN volume on herbicide efficacy. Glyphosate, paraquat, and tiafenacil did not significantly reduce weed control efficacy when applied with UAN as a carrier. At 3 weeks after herbicide application, glyphosate efficacy ranged from 92 to 94% (broadleaved weeds) and 97 to 97% (grassy weed). Paraquat efficacy ranged from 72 to 78% (broadleaved weeds) and 63 to 87% (grassy weed). Tiafenacil efficacy ranged from 74 to 75% (broadleaved weeds) and 51 to 80% (grassy weed). Higher application volume may be needed to increase the efficacy of contact herbicides such as paraquat and tiafenacil.

Brake, Valor, and Their Tank-Mixture Combinations with Residual Herbicides for Broad- Spectrum Weed Management Programs in Mississippi Peanut. Taghi Bararpour*; Mississippi State University, Stoneville, MS (6)

Weed control is one of the most limiting factors facing peanut (*Arachis hypogaea*) producers in Mississippi. A field study was conducted in 2022 at the Delta Research and Extension Center, in Stoneville, Mississippi, to evaluate Brake, Valor, and their tank-mixture combinations with residual herbicides for broad- spectrum weed management programs in Mississippi peanut. Peanut (Georgia-06G) was planted at a seeding rate of eight seeds ft^{-1} on May 24, 2022 and emerged on May 30. Plot size was 13 ft wide by 20 ft long. The plot area contained Palmer amaranth (*Amaranthus palmeri*), entireleaf morningglory (*Ipomoea hederacea var. integriuscula*), prickly sida (*Sida spinosa*), broadleaf signalgrass (*Urochloa platyphylla*), and hemp sesbania (*Sesbania herbacea*). The study was designed as a randomized complete block with 18 treatments and four replications. Herbicide treatments were as follows (rate in oz/a): 1) Brake at 16; 2) Brake at 32; 3) Valor (flumioxazin) at 1.5; 4) Valor at 3; 5) Brake at 16 + Valor at 1.5; 6) Brake at 16 + Valor at 3; 7) Brake at 32 + Valor at 1.5; 8) Brake at 32 + Valor at 3; 9) Valor at 3 + Dual Magnum (*S*-metolachlor) at 32; 10) Strongarm (diclosulam) at 0.45 + Dual Magnum; 11) Brake at 16 + Dual Magnum; 12) Valor at 3 + Brake at 16 + Dual Magnum; 13) Valor at 3 + Dual Magnum + Strongarm; 14) Valor at 3 + Dual Magnum + Strongarm + Brake at 16; 15) Brake at 16 + Strongarm + Dual Magnum; 16) Brake at 16 + Warrant (acetochlor) at 48; 17) Brake at 16 + Warrant + Valor at 1.5: and 18) nontreated check. All treatments were applied preemergence (PRE). Peanut injury was 11 and 2; 23 and 13; 4 and 1; 10 and 5; 6 and 3; 11 and 9; 7 and 3; 8 and 5; 11 and 11; 0 and 3; 10 and 2; 1 and 1; 0 and 0; 0 and 0; 2 and 0; 2 and 0; 0 and 0 from treatment 1 through 17 at 1- and 2- weeks after emergence (WAE), respectively. There was no peanut injury after 4 WAE. Palmer amaranth control was > 95% for treatment 3, 4, 6, 9, 12, and 15 at 6 WAE. Entireleaf morningglory control was 95, 99, 95, 100, 100, and 100% from treatment 5, 9, 10, 13, 14, and 15 at 6 WAE, respectively. The application of treatment 16 and 17 provided only 68 and 74% entireleaf morningglory control, respectively. All herbicide treatments provided 95 to 100% control of prickly sida. Hemp sesbania control was 98 to 100% from all herbicide applications. Broadleaf signalgrass was a difficult weed to control. Treatment 1 through 17 provided 49, 66, 64, 67, 81, 84, 91, 85, 98, 95, 91, 95, 100, 100, 100, 64, and 69% control of broadleaf signalgrass at 6 WAE, respectively. Based on this study, treatments 13 (Valor at 3 + Dual Magnum + Strongarm) and 14 (Valor at 3 + Dual Magnum + Strongarm + Brake at 16) were the best applications in terms of longer residual activity and broad-spectrum weed control.

Weed Control Efficacy and Dry Bean Response to Preemergence Herbicide Activation Timing. Prayusha Bhattarai*[1], Albert T. Adjesiwor[2]; [1]University of Idaho, Moscow, ID, [2]University of Idaho, Kimberly, ID (7)

Field experiments were conducted in 2021 and 2022 to evaluate the effect of irrigation timing on EPTC, flumioxazin, pyroxasulfone, and flumioxazin plus pyroxasulfone weed control efficacy and safety in dry beans. There were 14 treatments arranged in a randomized complete block with four replications. Treatments comprised of a weedy and hand-weeded check, and EPTC (3,430 g ai ha-1), flumioxazin (53.6 g ai ha-1), pyroxasulfone (119 g ai ha-1), and flumioxazin + pyroxasulfone (70.4 + 89.3 g ai ha-1) incorporated with overhead irrigation at 1, 4, and 8 days after herbicide treatment (DAT). Activation timing did not affect the phytotoxicity of flumioxazin. Delaying activation until 8 days after herbicide application resulted in 11 and 19% phytotoxicity in the pyroxasulfone and flumioxazin + pyroxasulfone treatments, respectively, but the crop recovered within 5 weeks after treatment. Activation timing did not affect the efficacy of flumioxazin, pyroxasulfone and flumioxazin + pyroxasulfone. However, flumioxazin + pyroxasulfone provided better weed control compared to flumioxazin or pyroxasulfone applied alone, irrespective of the activation timing. Delaying EPTC activation until 8 days after treatment reduced weed control by 48 to 59% compared to 1 day after treatment activation timing. Herbicide treatment and activation timing did not reduce dry bean yield. Seed yield was 223 kg ha-1 in the weedy check and 1,420 to 2,657 kg ha-1 in herbicide-treated plots which were statistically similar to the hand-weeded check. Uncontrolled weeds reduced dry bean yield by 91% compared to the hand-weeded check.

Postharvest Control of Russian Thistle (*Salsola kali*) in Spring Wheat Stubble. Fernando H. Oreja*[1], Drew J. Lyon[2], Jennifer A. Gourlie[3], Henry C. Wetzel[2], Judit Barroso[1]; [1]Oregon State University, Adams, OR, [2]Washington State University, Pullman, WA, [3]Oregon State University, Pendleton, OR (8)

In the rainfed cropping region of the US Inland Pacific Northwest (PNW), water consumption by crops and weeds is an important component of the water balance in agroecosystems. In these systems, the period after harvest is important to accumulate enough water for the next crop to achieve an economically viable yield. *Salsola tragus* L. is one of the dominant broadleaf weeds in semi-arid regions of the PNW, especially in no-till fallow. After harvest Russian thistle increases root growth and consumes up to 60% of the water it will use during its entire life cycle. Chemical weed control is the most important weed management tool in no-tillage systems. The size and maturity of Russian thistle post-harvest and increasing herbicide resistance may make it more difficult to control with herbicides. This may also be influenced by low water status in the plants, since this is typically a very dry period in the climate of the

PNW. This research was to look at the optimum timing of glyphosate as well as to determine if there were other modes of action that could be rotated for control of Russian thistle. The objective of this research was to determine the efficacy of different post-harvest application timings (24 h after harvest, 1 week after harvest (WAH), 2 WAH, and 3 WAH) of three herbicide treatments at different heights. In a split-plot block design with four replicates, two experiments were conducted in Adams, Oregon and Lind, Washington, repeated in two consecutive years. The herbicides applied were paraquat, glyphosate, and either a pre-mix of bromoxynil + pyrasulfotole at Lind or bromoxynil + metribuzin at Adams. Efficacy was evaluated at three, six, and nine weeks after treatment (WAT) at Adams, and one, two, three and four WAT at Lind. Paraquat provided the greatest control in all scenarios (100%). The pre-mixtures with bromoxynil performed better than glyphosate, in general. Treatments with glyphosate and pre-mixtures with bromoxynil performed better on short stubble than on tall stubble (94% vs 83%). At Lind, the greatest control with glyphosate was achieved applying it 1 WAH (91% in 2020 and 74% in 2021). The best option to control Russian thistle and reduce water consumption for the next season is the application of paraquat or glyphosate/pre-mixtures with bromoxynil as soon as possible after harvest on short stubble.

Identification of Palmer Amaranth Using Machine Learning. Akhilesh Sharma*[1], Vipan Kumar[2], Ajay Sharda[1], William Hsu[1]; [1]Kansas State University, Manhattan, KS, [2]Kansas State University, Hays, KS (9)

Rapid evolution of herbicide-resistant (HR) Palmer amaranth (*Amaranthus palmeri* S. Wats.) populations and increasing costs of their control build a strong rationale for developing site-specific Palmer amaranth management strategies. Site-specific Palmer amaranth control warrants its rapid and accurate detection in various cropping situations. The main objectives of this research were (1) to identify and detect Palmer amaranth plants using two open-source object detection algorithms (Detectron2 and YOLOv7x), and (2) compare the performance of both object detection algorithms for speed and accuracy in Palmer amaranth detection. A soybean field site with natural infestation of Palmer amaranth population was established in the 2022 growing season at Kansas State University Agricultural Research Center near Hays, KS. Images (total 372) of Palmer amaranth plants at various vegetative growth stages (cotyledons to 76 cm tall) were manually collected using a GoPro Hero camera at ground level. All collected images were pre-processed to the size of 640 x 640 pixels and were manually annotated with bounding boxes with 650 total annotations. The annotated images of Palmer amaranth were used to train open-source Meta's Detectron2 with fast regional convolutional (Fast R-CNN) configuration and YOLO (you only look once) v7x algorithms. We used mean average precision (mAP) and inference speed metrics to evaluate the accuracy and speed of both object detection algorithms in identifying and locating Palmer amaranth. Results indicated that both Detectron2 and YOLO v7x algorithms performed well in detecting Palmer amaranth. The YOLOv7x algorithm achieved the mAP score of 78.4% @0.5 and 51.9 % @ 0.75 and an inference speed of 30.8 milliseconds. However, the Detectron2 algorithm with Fast R-CNN configuration had the mAP score of 80.76% @0.5 and 45.58% @0.75 and an inference time of 0.2 seconds. These results suggest that the YOLOv7x algorithm can play a critical role in site-specific weed management by detecting Palmer amaranth plants more accurately and about 6.5 times faster than Detectron2 in agricultural fields. Contact E-mail: akhileshsharma@ksu.edu

Drill-Spacing And Bed Width Influence on Canopy Coverage, Weed Control and Yield in Furrow-Irrigated Rice. Koffi Badou-Jeremie Kouame*[1], Thomas R. Butts[1], Noah H. Reed[2], Brad M. Davis[1], Leah M. Collie[3], Troy W. Dillon[1], Jason K. Norsworthy[2], Tom Barber[1], Jarrod Hardke[4], Jason A. Bond[5], Hunter D. Bowman[6]; [1]University of Arkansas System Division of Agriculture, Lonoke, AR, [2]University of Arkansas, Fayetteville, AR, [3]University of Arkansas, Lonoke, AR, [4]University of Arkansas System Division of Agriculture, Stuttgart, AR, [5]Mississippi State University, Stoneville, MS, [6]Mississippi State University, Starkville, MS (10)

Furrow-irrigated rice is a relatively new production system in the Mid-South for which the critical cultural practice of flooding is removed. Thus, a viable option for an integrated weed management program is lacking, negatively impacting weed management efforts. Therefore, alternative production practices and management strategies are needed in this production system. Field experiments were conducted in 2021 and 2022 in Lonoke and Pine Tree (Arkansas) to investigate the impact of drill row spacing and bed width on weed control, canopy coverage, and rough rice yield in furrow-irrigated rice. A randomized complete block split-plot design was used with four replications. Treatments consisted of three bed widths (whole plot factor) (76-cm, 97-cm, and 152-cm) and four drill row spacings (subplot factor) (13-, 19-, 25-, and 38-cm). Hybrid rice cultivar RT7521 FP was drill-seeded and standard practices for fertility and irrigation were followed. Barnyardgrass density was assessed at the 5- to 6-leaf rice stage and preharvest. Aerial digital images from small unmanned aircraft systems (sUAS) were collected at the 3-6 leaf and panicle differentiation stages and processed with Field Analyzer software to estimate canopy coverage. Rice was harvested with a plot combine at maturity, and rough rice yield was adjusted to 13% moisture. Results showed that across locations and years, barnyardgrass density at both the 5- to 6-leaf rice stage and preharvest stage was influenced by bed width and drill row spacing. In general, as drill row spacing increased, barnyardgrass density also increased. However, as bed width increased, the beneficial effects observed of narrower drill row spacings (13- and 19-cm) were minimized or negated completely. In 2021, rice yield was greater for 76-cm bed widths in Pine Tree while no differences were detected between bed widths in Lonoke. In contrast, in 2022, 76-cm bed widths had the lowest yields at both locations. Rough rice yield was greatest for the 13-cm drill row spacing in three of four site-years. Averaged across bed widths at the 3-6 leaf stage, the 13- and 19-cm drill row spacings had the greatest canopy coverage in Lonoke and Pine Tree, respectively. At the panicle differentiation stage, averaged across locations, canopy coverage was numerically greater for smaller drill row spacings (13- and 19-cm), but no statistical differences were detected. Despite large variability observed in the data, probably due to furrow-irrigated planting practices that affected crop stand in some plots, this research demonstrated that greater rice canopy coverage and weed suppression could be provided by narrower drill row spacings (13- and 19-cm) while optimizing rough rice yield. Keywords: drill-spacing, bed width, yield, barnyardgrass, canopy coverage.

Evaluating the Metabolic Response of Cloquintocet-Mexyl in Winter Wheat. William Kramer*, Todd A. Gaines, Franck E. Dayan; Colorado State University, Fort Collins, CO (11)

Cloquintocet-mexyl is a popular herbicide safener that selectively boosts crop herbicide metabolism while not compromising weed control. Enhanced crop safety is commonly achieved through increased expression of genes involved in herbicide metabolism. Previous literature connects cloquintocet-mexyl to transcriptional activation of genes that encode for various detoxifying enzymes, particularly glutathione-S-transferase (GST). In this research, cloquintocet-mexyl is used as a tool to modulate herbicide metabolism in winter wheat to identify genes that are involved in enhanced herbicide tolerance. To investigate the transcriptomic response associated with this potential differential metabolism, wheat was sprayed with 10 g ha^{-1} of cloquintocet-mexyl and leaf-tissue was sampled 12-hours after application. Total RNA was extracted and used to generate paired-end cDNA Illumina libraries for sequencing. Differential gene expression results indicate that various metabolic enzymes are responding to cloquintocet-mexyl applications. Many upregulated genes are directly associated with GST enzyme induction and have previously been reported to respond to cloquintocet-mexyl applications. More specifically, these genes have been previously identified as the mechanism of non-target site resistance in various weed species. The selected GST genes will serve as targets to further understand the molecular mechanism associated with cloquintocet-mexyl applications in wheat. Ultimately, inducible GSTs genes involved in herbicide metabolism can be used as molecular markers for future wheat breeding programs.

Confirmation of Glyphosate Resistance in New York Palmer Amaranth (*Amaranthus palmeri*) Populations and Response to Alternative Chemistries. Aleah L. Butler-Jones*[1], Elizabeth C. Maloney[2], Lynn M. Sosnoskie[2]; [1]Cornell University, Ithaca, NY, [2]Cornell University, Geneva, NY (12)

Chemical weed control has been the primary approach to managing weeds in production settings due to its high efficacy and relatively low cost. Glyphosate has been extensively used, particularly on acreage planted to herbicide-tolerant agronomic crops. The repeated use of glyphosate over space and time has facilitated the spread of glyphosate-resistant weeds, including Palmer amaranth. Palmer amaranth (Amaranthus palmeri) is one of the most troublesome weeds due to its fast growth, high fecundity, potential to severely reduce crop yields, and the evolution of herbicide resistance. Though it is primarily found in the South and Midwest, Palmer amaranth was recently identified in a limited number of soybean fields in New York (NY). In 2021 and 2022, dose-response studies were conducted under greenhouse conditions to evaluate the response of three populations of Palmer amaranth (NY-Orange County, NY-Steuben County, Nebraska (known sensitive)) to glyphosate. Seed were planted in 7.6 cm diameter pots filled with a commercial potting mix and placed in a greenhouse set to a constant temperature of 25°C with a 16:8 light:dark cycle. Seedlings were hand-thinned to one plant per pot and treated with glyphosate at the 2 to 4 leaf stage at doses of: 0.07, 0.14, 0.28, 0.56 (1X rate), 1.12, 2.24, and 4.48 kg ae ha-1 of Roundup PowerMax (Bayer Crop Science). Applications were made using a cabinet sprayer with a single nozzle (8002E, Teejet Technologies) boom set to deliver 187 L ha-1. Each herbicide by rate combination was replicated up to 12 times for each of the four horseweed populations and the study was repeated in time. At 21 days after treatment, surviving plants were harvested and weighed to assess biomass accumulation. Data was analyzed with mixed models ANOVA and log-logistic regression using R statistical software. Results demonstrated significant ($P < 0.05$) differences among the populations with respect to herbicide injury responses. The sensitive Nebraska population was effectively controlled at all rates tested. New York Palmer amaranth populations had plants surviving glyphosate applications up to 2.24 kg ae ha-1, although resulting biomass was significantly reduced (-65% Orange County, -56% Steuben County) under greenhouse conditions. Additional screens indicate that all three Palmer populations were also resistant to ALS-inhibiting herbicides used in soybean production, but sensitive to other modes of action including WSSA 10 (glufosinate), WSSA 14 (acifluorfen, fomesafen), and WSSA 27 (mesotrione). 2023 studies will describe the mechanisms of glyphosate resistance in the NY populations.

Does Row Spacing Matter for Weed Suppression in Corn and Soybean Cropping Systems? A Meta-analysis. Mandeep Singh*[1], Resham Thapa[2], Navdeep Singh[3], Steven Brian Mirsky[4], Amit J. Jhala[1]; [1]University of Nebraska-Lincoln, Lincoln, NE, [2]USDA-ARS Beltsville Agricultural Research Center, Beltsville, NC, [3]Punjab Agricultural University, Ludhiana, India, [4]USDA ARS, Beltsville, MD (13)

Narrow row spacing (<76 cm) is generally advocated for weed suppression and yield benefits in row crops, especially soybean. Many studies have assessed the effect of narrow rows on weed suppression in corn and soybean. However, the results from these studies have not been compiled for quantitative synthesis. Therefore, we are conducting a meta-analysis to estimate the overall effect of narrow rows on weed suppression in corn and soybean in the United States. We searched the relevant keywords (row spacing OR row width AND corn soybean OR maize OR weed control) in google scholar, Scopus, and weed science journals. This gave us 35 relevant studies. Weed control and crop yield data for row spacing treatments were extracted from these studies to calculate overall effect sizes. Preliminary analysis indicates that narrow rows were usually beneficial in suppressing weeds in soybean, but not so in corn. The detailed results of this meta-analysis will be presented at the conference.

Response of Glyphosate-Resistant Kochia (*Bassia scoparia*) to Dichlorprop-P, 2,4-D, Dicamba Alone or Their Premixture. Sachin Dhanda*[1], Vipan Kumar[1], Augustine Obour[1], Elizabeth Yeager[2], John Holman[3], Anita Dille[2]; [1]Kansas State University, Hays, KS, [2]Kansas State University, Manhattan, KS, [3]Kansas State University, Garden City, KS (14)

The widespread evolution of multiple herbicide-resistant (MHR) kochia has become a serious management challenge for producers across the North American Great Plains. Main objective of this research was to investigate the response of MHR kochia populations to POST applied 2,4-D, dichlorprop-p, and dicamba alone or in a premixture (2,4-D/dichlorprop-p/dicamba). Dose-response experiments were conducted in a greenhouse and in a fallow field at Kansas State University Agricultural Research Center in Hays (KSU-ARCH), KS. Two previously confirmed MHR kochia populations (resistant to glyphosate/dicamba/fluroxypyr/atrazine/metribuzin) from Garden City, KS along with one susceptible population from Hays, KS were tested in greenhouse study. The field site had natural infestation of glyphosate-resistant (GR) kochia. Both field and greenhouse experiments were conducted in a randomized complete block design with 4 and 12 replications, respectively. The tested doses for 2,4-D were 0, 269, 538, 1076, 2152, 4304, and 8608 g ae ha^{-1}. Dichlorprop-p doses including 0, 280, 560, 840,1120, 1400, 1680, and 3360 g ae ha^{-1} were tested. Dicamba doses were 0, 140, 280, 560, 840, and 1120 g ae ha^{-1}. A premixture of 2,4-D/dichlorprop-p/dicamba was tested at 0, 373, 746, 1119, 1492, and 2238 g ae ha^{-1}. Data on percent visual control of MHR or GR kochia at 7, 14, and 28 days after treatment (DAT) and biomass reduction (%) at 28 DAT were recorded. In field study, the predicted 2,4-D, dichlorprop-p, dicamba, and 2,4-D/dichlorprop-p/dicamba doses that provided 90% control of GR kochia were 23177, 2222, 1272, and 772 g ha^{-1} at 28 DAT, respectively. In greenhouse study, the predicted 2,4-D, dichlorprop-p, dicamba, and 2,4-D/dichlorprop-p/dicamba doses that provided 90% control of MHR kochia populations were 56200 to 117843, 2484 to 8832, 3501 to 5323, and 817 to 1015 g ha^{-1} at 28 DAT, respectively. In contrast, only 471 g ha^{-1} dose of 2,4-D/dichlorprop-p/dicamba premixture was needed for 90% control of the susceptible kochia population. Altogether, these results suggest that 2,4-D, dichlorprop-p, and dicamba in a premixture may possibly act synergistically for control of GR, MHR, or susceptible kochia populations.

Impacts of Florpyrauxifen-benzyl on Rice and Smallflower Umbrella Sedge, *Cyperus difformis*, at Different Growth Stages. Deniz Inci*, Kassim Al-Khatib; University of California, Davis, Davis, CA (15)

Cyperus difformis L. (CYPDI), smallflower umbrella sedge, is a troublesome sedge weed in California (CA) rice agroecosystem. The repeatedly use of the same modes of action herbicides and the lack of crop rotation have resulted in herbicide-resistant CYPDI. Florpyrauxifen-benzyl (FB), Loyant, is a new synthetic auxin-type herbicide recently registered for CA rice. The application timing of FB is from the two-leaf rice growth stage until 60 days before harvest. Two field research experiments were conducted during the 2021 and 2022 growing seasons at the California Rice Experiment Station in Biggs, CA, to elucidate rice and CYPDI response to FB when applied at different growth stages. In the first study, clomazone at 622 g ai/ha was applied to all plots except untreated control (UTC), and FB was applied at 40 g ai/ha to 1-leaf, 10, 15, 20, and 25 cm height CYPDI growth stages. In the second study, FB was applied at 40 and 80 g ai/ha after the rice panicle initiation. All FB treatments included methylated seed oil at a 584 ml/ha rate. Both studies were conducted as a randomized complete block design, where experimental units were 3X6 m treatment plots. All plots were evaluated for weed control and crop injury at 7, 14, 21, 28, 35, and 42 days after treatments (DAT). Weeds were counted at 28 DAT within two randomly selected areas in each plot, and plots were harvested at the end of the season, and the yield was measured. FB at the 1-leaf stage was the most effective treatment for CYPDI at 28 DAT with 98% control and for *Echinochloa* species (1ECHG) at 42 DAT with 100% control. The highest yield at harvest was achieved as 12,432 kg/ha at 25 cm CYPDI growth stage application. In the second study, after rice panicle initiation applications, the highest necrosis, 32%, was observed at 7 DAT of FB at 80 g ai/ha treatment. Rice gradually recovered from all the injuries by the 42 DAT. FB at 80 g ai/ha was the most effective treatment. The highest CYPDI and 1ECHG control were observed at 42 DAT at 96% and 93%, respectively. Nevertheless, the highest yield at harvest was achieved as 9,620 kg/ha at FB at 40 g ai/ha treatment. While 40 and 80 g ai/ha FB treatments caused 8% grain blanking in rice panicles, UTC showed 14% blanking. Correspondence: inci@ucdavis.edu

Late-Season Control of Glyphosate-Resistant Palmer Amaranth with Glufosinate in XtendFlex Soybean. Vipan Kumar*[1], Sachin Dhanda[1], Rui Liu[2], Amit J. Jhala[3]; [1]Kansas State University, Hays, KS, [2]Washington State University, Prosser, WA, [3]University of Nebraska Lincoln, Lincoln, NE (16)

Glyphosate-resistant (GR) Palmer amaranth (*Amaranthus palmeri* L.) has rapidly spread in the Central Great Plains, including Kansas. The commercialization of newly developed glyphosate/dicamba/glufosinate (GDG)-resistant soybean (*Glycine max* L.) varieties allows POST applications of dicamba and glufosinate for in-season control of GR Palmer amaranth. However, the cutoff date for in-season dicamba applications in GDG-resistant soybeans is June 30 to prevent dicamba drift to sensitive crops, leaving glufosinate as sole herbicide option for late-season control of GR Palmer amaranth. The main objectives of this research were to (1) determine the effectiveness of late-season applications of glufosinate for GR Palmer amaranth control, and (2) the impact of those glufosinate applications on grain yields in GDG-resistant soybeans. Field experiments were conducted in 2022 growing season at Kansas State University Agricultural Research Center in Hays, KS. A GDG-resistant soybean variety 'AG37XF1' was planted on May 25, 2022. Study site had a natural infestation of GR Palmer amaranth. Total ten treatments comprising various glufosinate rates (594, 655, and 737 g ha^{-1}), timings (single or sequential), and tank-mix combinations were tested on 70 to 90 cm tall Palmer amaranth. A nontreated weedy check and a sequential treatment of glyphosate followed by (fb) glyphosate were also included. Study was conducted in a randomized complete block design with four replications. Data on percent visual control of GR Palmer amaranth were collected at 7, 14, and 28 days after treatment (DAT) and shoot dry weight reduction (% of nontreated) at soybean maturity were collected. Results indicated that single application of glufosinate at 655 or 737 g ha^{-1} and all sequential applications of glufosinate provided an excellent (87 to 93%) control of GR Palmer amaranth 28 DAT, which did not differ for glufosinate + acifluorfen and glufosinate + pyroxasulfone + fluthiacet-methyl treatments. GR Palmer amaranth control with single application of glufosinate at 594 g ha^{-1} or glufosinate + s-metolachlor (655 + 1337 g ha^{-1}) was moderate (82 to 84%). The least control (11%) was observed with a sequential treatment of glyphosate *fb* glyphosate. Majority of the tested treatments reduced shoot biomass of GR Palmer amaranth by 83 to 91% and the least reduction (33%) was observed with glyphosate *fb* glyphosate treatment. Majority of the tested glufosinate-based treatments resulted in soybean grain yield of 626 to 701 kg ha^{-1}. The least soybean grain yield (270 kg ha^{-1}) was observed with sequential treatment of glyphosate *fb* glyphosate. These results suggest that single applications of glufosinate at 655 or 737 g ha^{-1} or sequential applications in late-season can provide effective control of tall GR Palmer amaranth in GDG-resistant soybeans.

Weed Survey in Colorado Sugarbeet Farms Reveals Resistance to Dicamba and Glyphosate. André Lucas Simões Araujo*, Eric P. Westra, Todd A. Gaines; Colorado State University, Fort Collins, CO (17)

Colorado is part of the Central Great Plains where sugarbeet is produced. Weed competition causes severe sugarbeet yield losses. Glyphosate is one of the most used post-emergence herbicides to control weeds in transgenic glyphosate-resistant sugarbeet. Resistance cases in weeds to this mode of action are a current problem in sugarbeet farms. A new genetically engineered trait for sugarbeets is under development that will confer resistance to glufosinate, glyphosate, and dicamba. Since these three active ingredients have a history of use in fallow and crop rotations in the region, growers are concerned that resistance for those molecules may already exist. The research objective was to perform a resistance survey to evaluate the trait efficacy for Colorado. Thirty-seven sugarbeet fields in Colorado were visited in the Fall of 2021, and three weed species were surveyed: kochia (*Bassia scoparia*), common lambsquarters (*Chenopodium album*), and Palmer amaranth (*Amaranthus palmeri).* The collection was created by selecting and combining seeds from 10-15 weed plants present in each sugarbeet field. From each collection site, 32 seeds were planted and screened with glyphosate (840 g ae ha^{-1} + AMS 20g/L), glufosinate (590 g ae ha^{-1} + AMS 2.5lb/a), and dicamba (280 g ae ha^{-1} + NIS 0.25% v/v) at the spray volume of 187 L/ha. Populations were graded as susceptible (0% to 2% survival), low resistance (2% to 20% survival) and resistant (20% to 100% survival). Resistance mechanisms in the kochia populations were investigated. Increased EPSPS gene copy number assays were conducted in accessions that survived glyphosate. For dicamba-resistant populations, Sanger sequencing was performed to verify the occurrence of a mutation in the *IAA16* degron. Twenty-six kochia populations, one common lambsquarters population, and two Palmer amaranth populations were identified with a resistance frequency of over 20% survival to glyphosate. Three kochia populations were identified with over 20% survival to dicamba. Populations resistant to glufosinate were not identified. Resistance to dicamba was not identified in common lambsquarters and Palmer amaranth populations. Resistance in the majority of kochia populations tested so far is likely due increased EPSPS copy number. A mutation, gly to asn, in the *IAA16* degron was identified in one population that possessed low resistance frequency. This mutation in the *IAA16* degron was not identified in populations classified as resistant, and the resistant mechanism in those populations remains unknown. Although the new sugarbeet trait will facilitate growers in managing weeds, concerns about the long-term utility of this technology should be considered based on resistance occurrence before the trait launches. Integrated pest-management approaches must be associated with this technology to mitigate resistance evolution in the Central Great Plains.

Impact of Soil-Applied Herbicides on Early Planted Soybean. Matthew S. Goddard*, Christy L. Sprague; Michigan State University, East Lansing, MI (18)

The number of days that a farmer has to conduct field operations in the spring are declining, due to more variable higher rainfall events. Early soybean planting is one way that growers are adapting to overcome these challenges. Currently, very little information is available on the effectiveness and crop safety of soil-applied preemergence (PRE) for early planted soybean. A field study was conducted in 2022 at two locations, East Lansing (EL) and Frankenmuth, MI (SVREC) to: 1) compare PRE and delayed PRE (DPRE) herbicide applications in early planted soybean, and 2) examine weed control and crop safety of several PRE herbicides in early versus normal planted soybean. Soybean was planted on April 20 and May 20 (~4 weeks later) for the early and normal planted soybean, respectively. The treatments of: metribuzin, flumioxazin, s-metolachlor, pyroxasulfone, saflufenacil + metribuzin + pyroxasulfone and metribuzin + flumioxazin were applied immediately after planting for each planting date. Metribuzin, flumioxazin, s-metolachlor, and pyroxasulfone were also applied as a DPRE (~80 GDD) in the early planted soybean. A weed-free and untreated control were included in each planting time. When weeds were ~10 cm tall 2,4-D choline + glyphosate + ammonium sulfate was applied POST. Averaged across all herbicides, the DPRE application reduced soybean stand 10% and the DPRE of flumioxazin at EL caused 20% injury, 21 day after emergence (DAE). The DPRE of pyroxasulfone also resulted in 92% more weed biomass at POST compared with pyroxasulfone PRE. However, there was no differences in weed biomass between the PRE and DPRE for the other herbicide treatments. Yield was also 12% lower from the DPRE treatments compared with the PRE treatments for early planted soybean at EL. At SVREC, there were no differences between the PRE and DPRE treatments. At EL, there was a 25% stand reduction between the early and normal planted soybean, regardless of herbicide treatment. Even though stand was lower in the early planted soybean, none of the soil-applied treatments caused more injury than the normal planted soybean, 21 DAE. At the time of POST herbicide application, all PRE treatments provided similar reductions in weed biomass with the exception of metribuzin and s-metolachlor in the early planted soybean. Differences in biomass between the early and normal planting were 98 and 91% for metribuzin and s-metolachlor, respectively. At this location, soybean yield was 15% higher for the normal planted soybean compared with the early planting, much of this may have been due to overall reduced stands from crusting early in the season. At SVREC, there were no differences in weed biomass between the planting times or herbicide treatments, or in soybean yield across both planting dates and weed control treatments. Overall, more work is needed to examine weed control systems in early planted soybean.

Estimating Harvest Loss of Corn from Growers' Fields in Nebraska and Implication for Volunteerism.
Trey P. Stephens[1], Vipin Kumar*[2], Jenny Rees[3], Amit J. Jhala[1]; [1]University of Nebraska-Lincoln, Lincoln, NE, [2]Virginia Polytechnic Institute and State University, Painter, VA, [3]University of Nebraska-Lincoln, York, NE (19)

Corn is one of the most important commodities for Nebraska agriculture, grown over 3.9 million ha in 2022. Most of the corn cultivars/hybrids grown in Nebraska are resistant to herbicides like 2,4-D, glyphosate, glufosinate, and quizalofop which help in diversifying herbicide program for weed management in corn. However, harvest loss of herbicide-resistant (HR) corn results in infestation of HR volunteer corn in rotated and corn-corn cropping systems, which is a serious concern. To estimate the extent of corn harvest loss and its implication for volunteer corn infestations in successive years, 26 and 21 corn growers' fields were sampled for corn harvest loss in six counties across southeastern and south-central Nebraska in 2020 and 2021, respectively. From each field, 16 samples were collected for loose kernels and ears using 0.5 m^2 quadrat, following a W pattern. Loose kernels and kernels separated from ears were dried and adjusted to 15.5% moisture content before taking 100 count seed weight. Germination test was conducted to estimate the seed bank addition of volunteer corn. Average kernel loss was 68 ± 8.8 and 33 ± 3 m^{-2} which equaled to yield loss of 209 ± 28.1 and 111 ± 10.9 kg ha^{-1} in 2020 and 2021, respectively. Harvest loss equates to 2% and 1% of statewide average corn yield for Nebraska in 2020 (10893 kg ha^{-1}) and 2021 (12177 kg ha^{-1}), respectively. Germination percentage for harvest loss kernels was found to be 51%, which result in addition of 35 ± 4.5 and 17 ± 1.5 germinable volunteer seed m^{-2} area.

Molecular Basis for Resistance to ACCase-inhibiting Herbicides in Downy Brome (*Bromus tectorum*) Populations. Victor Ribeiro*[1], Caio A. Brunharo[2], Carol Mallory-Smith[1], Darrin Walenta[3], Judit Barroso[4]; [1]Oregon State University, Corvallis, OR, [2]Penn State University, University Park, PA, [3]Oregon State University, La Grande, OR, [4]Oregon State University, Adams, OR (20)

The prevalent use of acetyl-coenzyme A carboxylase (ACCase)-inhibiting herbicides for downy brome (*Bromus tectorum* L.) control in fine fescue (*Festuca* L. spp.) grown for seed has selected ACCase inhibitor-resistant downy brome populations. The objectives of this study were to (1) evaluate the response of nine downy brome populations to ACCase-inhibiting herbicides and the acetolactate synthase (ALS) inhibitor sulfosulfuron and (2) characterize the resistance mechanisms. Greenhouse dose-response studies were conducted in a completely randomized design with six replications. Herbicide rates tested ranged from 0 to 8X for sethoxydim (X = 525 g a.i ha-1), clethodim (X = 136 g a.i ha-1), fluazifop (X = 140 g a.i ha-1), quizalofop (X = 92 g a.i ha-1), and sulfosulfuron (X = 35 g a.i ha-1), where X is the labeled herbicide rate. Downy brome biomass was collected 14 days after treatment, dried, and weighed. Three and four-parameter log-logistic models were fitted to the relative biomass (% of untreated control) data to estimate the rates to reduce growth by 50% and resistance ratios (RR). All populations were resistant to fluazifop (RR =2.7), clethodim (=5.0), quizalofop (=14.2), and sethoxydim (=15.5) when compared to a susceptible population. All populations were susceptible to sulfosulfuron. ACCase gene sequencing revealed a single nucleotide polymorphism that encodes amino acid position 2096, resulting in a Gly2096Ala substitution in 7 of the 9 resistant populations. Herbicides with different modes of action are needed to control downy brome in fine fescue fields and to reduce the selection pressure of ACCase inhibitors in this system.

Expression Levels of Selected GSTs in Palmer Amaranth Resistant to Trifluralin. Fidel Gonzalez Torralva*, Jason K. Norsworthy; UNIVERSITY OF ARKANSAS, Fayetteville, AR (21)

Glutathione *S*-transferases (GSTs) are soluble enzymes that play crucial roles in different detoxification processes. Such enzymes protect organisms from biotic and abiotic stresses such as herbicide damage. In a trifluralin-resistant Palmer amaranth (*Amaranthus palmeri* S. Watson) accession, GST-mediated metabolism was suggested as contributing to trifluralin resistance. In this work, we described the basal expression levels of specific GSTs in a trifluralin-resistant and -susceptible Palmer amaranth accessions. For that, RNA and complementary DNA synthesis were carried out following standard procedures. GST specific primers were designed using Primer-BLAST software. Quantitative Polymerase Chain Reactions (qPCR) were used to estimate the basal gene expression levels in both the resistant and susceptible

Palmer amaranth accessions relative to two reference genes. Basal expression values differed among GSTs and accessions. Ratios obtained (resistant divided by susceptible values) were in the range of 0.2 to 6-fold. Results suggest that those GSTs would contribute to trifluralin resistance on this trifluralin-resistant Palmer amaranth accession from eastern Arkansas. In addition, results obtained in this research may aid developing new herbicides for control of this resistant biotype.

Impact of Spring-Applied Herbicides on Cover Crops Planted After Wheat. Claudia R. Walz*, Christy L. Sprague; Michigan State University, East Lansing, MI (22)

Wheat acres account for six percent of Michigan farmland. Planting cover crops following wheat harvest allows growers to keep their land covered throughout the year. Many herbicide labels lack basic information on the rotation restriction regarding cover crops or suggest growers conduct a bioassay. Therefore, the objective of this study is to evaluate the effects of commonly used wheat herbicides on cover crop establishment and growth following winter wheat harvest. Field experiments were established at three locations for two years for a total of six site-years. The sites varied in climatic conditions as well as soil properties. Soils were a sandy clay loam with 3.3% organic matter (OM), a loam soil with 1.8% OM, and a clay loam soil with 3.1% OM. Winter wheat was drilled in the fall prior to herbicide treatment establishment. In the spring when wheat ranged between Feekes stage 4 and 5, four replications of 10 herbicide treatments, including an untreated control, were applied in 3 by 27 m strips across the field. The herbicides evaluated included: pyrasulfotole + bromoxynil, bicyclopryrone + bromoxynil, thifensulfuron + tribenuron, halauxifen + florasulam, mesosulfuron, mesosulfuron + thiencarbazone, pyroxsulam, pinoxaden + fenoxaprop, and clopyralid. Two to three weeks after wheat harvest, nine different cover crop species were drilled perpendicular to the herbicide treatments, directly into the wheat stubble. The cover crops included: annual ryegrass, dwarf Essex rapeseed, crimson clover, mustard Caliente, red clover, cereal rye, oat, oilseed radish, and Austrian winter pea. One month after planting, the covers were evaluated for injury and stand, 2-0.25 m^2 quadrats per plot. Approximately 3 months after planting, prior to the first frost, covers were evaluated again for injury, aboveground biomass was sampled, and dry weights were determined. From the time of herbicide application in the spring to cover crop planting growing degree days (GDD) ranged from 2901-3158 with 25-79 cm of rainfall in 2021, and 1583-1866 GDD with 12-21 cm of rainfall in 2022. In 2021, the Austrian winter pea did not establish at any location. At the initial evaluation, pyrasulfotole + bromoxynil caused bleaching of the leaf margins for red clover and crimson clover in 5 of 6 and 2 of 6 site years, respectively. Bicyclopyrone + bromoxynil also caused leaf margin bleaching in 3 sites in 2021 for red clover. However, both red and crimson clover outgrew these symptoms by the end of the fall and these treatments did not impact clover biomass. Halauxifen + florasulam caused injury to oilseed radish in 2 of 6 site years. In general, it appears that 2022, planting pea after thifensulfuron + tribenuron showed higher injury than other herbicides. Overall, for our initial research, annual ryegrass, dwarf Essex rapeseed, crimson clover, mustard Caliente, red clover, cereal rye, oat, oilseed radish, and Austrian winter pea established and were able to produce significant amounts of biomass following spring applications of herbicides commonly used in winter wheat.

PSII Inhibitor Resistance in Italian Ryegrass (Lolium Multiflorum) Populations in Louisiana Sugarcane.
Alice A. Wright*[1], Gerald A. McCollam[1], Albert J. Orgeron[2]; [1]USDA-ARS, Houma, LA, [2]LSU AgCenter, Baton Rouge, LA (23)

Herbicide resistance poses a threat to the continued efficacy of herbicides, particularly in crops like sugarcane that have a limited number of registered herbicides. Photosystem II (PSII) inhibitors comprise a large percentage of herbicides used in sugarcane. Recently, populations of Italian ryegrass in Louisiana sugarcane fields were identified as resistant to PSII inhibitors. Seeds were collected from these populations for pre-emergence and post-emergence dose response studies. Atrazine, diuron, metribuzin, and terbacil were selected as these herbicides are both registered for use in sugarcane and are representative of different PSII inhibitor chemistries. In post-emergence applications, the White Castle population exhibited resistance to both metribuzin and diuron with an ED_{50} for dry weight of 5769 g ai/ha and 4221.9 g ai/ha, respectively, and for injury 6120 g ai/ha and 7210 g ai/ha, respectively. Terbacil controlled all populations at the field rate. Dose responses are ongoing for pre-emergence applications. Sequencing of *psb*A to identify target site mutations is ongoing and metabolic inhibitors will be used to determine if a metabolic resistance mechanism is present. Other herbicides were investigated as alternatives; PSII inhibitor resistant plants are controlled by glyphosate and clethodim, which are applied when soybean is rotated with sugarcane. Clomazone, applied pre-emergence, and S-metolachlor provide control at the field rate. Alternative control options that can be implemented immediately are important as ryegrass outcrosses and can easily spread resistance traits within and between populations. This work will provide insight into the resistance mechanisms in these populations and provide growers with alternative control strategies. Alice.wright@usda.gov

Optimizing Dicamba and Glufosinate Application Timings for Weed Control in XtendFlex Soybeans in Central Kansas. Monica R. Marrs*[1], Vipan Kumar[2], Anita Dille[1], Rui Liu[3], Augustine Obour[2]; [1]Kansas State University, Manhattan, KS, [2]Kansas State University, Hays, KS, [3]Washington State University, Prosser, WA (24)

Glyphosate-resistant (GR) Palmer amaranth (Amaranthus palmeri S. Wats.) is one of the most troublesome weeds for soybean producers across the United States. Xtendflex soybeans are a traited variety that is glyphosate/dicamba/glufosinate-resistant (GDG-R) that allows growers to use POST applications of dicamba and glufosinate for in-season control of GR Palmer amaranth. Palmer amaranths extended emergence can make management decisions harder to decide on an optimum time to apply POST application treatments. The main objective of this research was to find the optimum timings of POST applied dicamba and glufosinate for GR Palmer amaranth control in GDG-R soybeans under no-tillage dryland conditions. In 2021 two separate field experiments, two row spacings, were conducted at Kansas State University Agricultural Research Center (ARCH) in Hays, KS. This location had a natural seedbank of GR Palmer amaranth population. In 2022 one field experiment was conducted at two locations, ARCH and at a grower field in Great Bend, KS, including both row spacings in one experiment. Both fields were overseeded with GR Palmer amaranth. For each experiment, the GDG-R soybean variety 'AG37XF1' seeded at a population of 156,900 seeds ha-1 for all site years. Four POST herbicide treatments: a) nontreated, b) dicamba applied at 560 g ae ha-1 alone, c) dicamba followed by (fb) glufosinate at 603 g ae ha-1 plus acetochlor at 1260 g ae ha-1 (sequential 7 days apart), and d) dicamba plus acetochlor fb glufosinate (sequential 7 days apart) applied at three different timings at 14, 21, and 28 days after soybean drilling. Each experiment was conducted in a split-plot randomized complete block design, the split being 25 cm row and 76 cm row, and factorial arrangement of treatments (herbicide program and timings) with 4 replications. Percent visual control of GR Palmer amaranth collected at 7-day intervals after first POST application. Two 0.25 m2 quadrats were used to estimate Palmer amaranth density and aboveground biomass at each evaluation timings. All data were subjected to analysis of variance (ANOVA) using PROC Mixed in SAS. In 2021 and 2022 results indicated that sequential POST treatments of dicamba or dicamba + soil residual herbicide fb a glufosinate alone or glufosinate + soil residual herbicide were more effective in controlling weeds in XtendFlex soybeans than dicamba alone. In 2022 weed control using dicamba alone had greater control in 38 cm row spacing than 76 cm row spacing

.

Optimizing Cover Crop Termination and Herbicide Application Timings to Manage Herbicide-Resistant Waterhemp in Soybean. Ramawatar Yadav*[1], Edward S. Dearden[2], Ryan Hamberg[1], Alexis L. Meadows[1], Austin H. Schleich[1], Prashant Jha[1]; [1]Iowa State University, Ames, IA, [2]Iowa State University, Crop Production & Physiology, Ames, IA (25)

Field experiments were conducted in 2022 in Ames, IA to quantify optimum timing of cover crop termination and herbicide application to manage herbicide-resistant waterhemp (*Amaranthus tuberculatus* [Moq.] J.D. Sauer) in soybean. A three-factor split-split plot design was used with three replications. The first factor consisted cereal rye (*Secale cereale* L.) cover crop vs. no cover crop. The second factor consisted two timings of cover crop termination; two weeks before or one week after soybean planting (WAP). The third factor consisted four herbicide programs, which included 1) glyphosate for cover crop termination, 2) glyphosate followed by (fb) glufosinate, 3) glyphosate plus S-metolachlor and fomesafen fb glufosinate, or 4) glyphosate fb glufosinate plus S-metolachlor and fomesafen. Preemergence (PRE) herbicide application was made at cover crop termination whereas postemergence (POST) herbicide application was made at 4 WAP. Cereal rye cover crop had a biomass of 2,100 kg ha^{-1} at early termination timing compared with 5,300 kg ha^{-1} at late termination timing. Cover crop reduced waterhemp aboveground biomass by more than 75% compared with no cover crop at 4 and 8 WAP. Herbicide programs had a significant effect on waterhemp biomass regardless of the presence of cover crop and cover crop termination timings. Waterhemp biomass was lowest in the plots treated with glyphosate plus a residual herbicide (S-metolachlor and fomesafen) at cover crop termination compared with no residual herbicide at cover crop termination, or a residual herbicide applied at the time of POST application. Addition of a residual herbicide with POST application of glufosinate did not reduce waterhemp biomass compared to POST application of glufosinate only. In conclusion, a residual herbicide should be included at the time of cover crop termination to suppress herbicide-resistant waterhemp biomass and potential seed production.

Palmer Amaranth (*Amaranthus palmeri*) Control Options in the Absence of Dicamba. Tyler S. Soignier*, Brian K. Pieralisi, Darrin M. Dodds, Jason Krutz, Dave Spencer, Whitney S. Crow, Bradley J. Norris, William J. Rutland, Eli B. Hobbs; Mississippi State University, Starkville, MS (26)

Investigation of Fluroxypyr Resistance in a Kochia Population from Colorado. Eduardo Souza de Amorim*, Olivia E. Todd, Eric P. Westra, André Lucas Simões Araujo, Todd A. Gaines; Colorado State University, Fort Collins, CO (27)

Kochia (*Bassia scoparia*) is an economically problematic annual broadleaf weed in crop systems in the western US. Cases of resistance to synthetic auxinic herbicides are of concern for the maintenance of control technology for this species. The objective of this work was to characterize resistence to fluroxapyr in a kochia population (Flur-R) from the State of Colorado. A dose response showed that the Flu-R population was 40x more resistant than the susceptible population (S). To understand if this population would be controlled by dicamba, a Kompetitive allele specific PCR (KASP) assay was used to determine if a known resistance mechanism (G127N mutation of *IAA16*) is present in Flur-R. For this KASP assay, DNA was extracted from plant tissue collected before fluroxypyr application at 157 g ae ha^{-1}. Thirty five Flur-R kochia plants were used for the assay. Twelve plants from a known dicamba-resistant population (9425) were used as a positive control, and twelve plants from a known herbicide-sensitive population (S) were used as a negative control. Out of 35 samples, seven amplified the wildtype allele, seven amplified the mutante allele and 21 amplified for heterozygous. Future work will focus on repeating the fluroxypyr dose response experiment to confirm our initial findings and to test for dicamba resistance associated with the *IAA16* mutation found in Flur-R and 9425.

Fitness of Different Ploidy Plants of F1 Progeny Derived from an Interspecific Cross Between Cultivated Sorghum and Johnsongrass. Usha R. Pedireddi*, Rana Afshang, Maximillian Chung, Kayla Buenaventura, Aleena Jiby, Nithya K. Subramanian, Muthukumar V. Bagavathiannan; Texas A&M University, College Station, TX (28)

Differential Control of ALS- and PPO-inhibiting Herbicide-resistant Amaranthus Retroflexus Populations with Other Postemergence Herbicides. Eric A. Jones*, Diego J. Contreras, Charlie W. Cahoon, Ramon G. Leon, Wesley Everman; North Carolina State University, Raleigh, NC (29)

Recently, two distinct ALS- and PPO-inhibiting herbicide-resistant redroot pigweed (*Amaranthus retroflexus*) populations were confirmed in North Carolina. The Camden County population carried a Trp_{574}Leu mutation in the *ALS* gene and Arg_{98}Gly mutation in the *PPX2* gene. The Pasquotank County population carried a Pro_{197}His mutation in the *ALS* gene and no mutation in the *PPX2* gene. Farmers in proximity to the fields where these populations were identified asked what herbicide(s) were still effective on these two redroot pigweed populations. Atrazine, glyphosate, glufosinate, mesotrione, 2,4-D, and dicamba were applied postemergence to plants from both populations with a discriminating rate. No plants from either population survived glufosinate treatment. 2,4-D and dicamba-treated plants exhibited survival 21 d after treatment but were controlled 28 d after treatment. The Pasquotank County population exhibited 17 and 37% survival when treated with atrazine and mesotrione, respectively. The Camden County population exhibited 0% and 14% survival when treated with atrazine and mesotrione, respectively. The Camden and Pasquotank County population were effectively controlled with glyphosate but a population from Yadkin County exhibited 13% survival. Subsequent dose-response assays were conducted with glyphosate and mesotrione to determine if the Yadkin and Pasquotank County populations were resistant, respectively. The lethal glyphosate dose to control 50% of the plants from Yadkin County was significantly higher (approximately two-fold) than the other tested populations. The lethal mesotrione dose to control 50% of the plants from Pasquotank County was significantly higher (approximately ten-fold) than the other tested populations. The results of the experiment provide effective herbicides to control ALS- and PPO-inhibiting herbicide-resistant redroot pigweed found in North Carolina, but two additional glyphosate- and mesotrione-resistant populations have been confirmed in the state. More research is needed to determine if the Pasquotank County population is atrazine-resistant.

Evaluation of Fall-Applied Residual Herbicides for Weed Management in Chickpeas(Cicer Arietinum). Akamjot S. Brar*[1], Qasim Khan[1], Fabian Menalled[1], Zach Miller[1], Clint Beiermann[1], Kent McVay[1], Lovreet S. Shergill[2]; [1]Montana State University, Bozeman, MT, [2]Montana State University, Huntley, MT (30)

Weed management at earlier crop stages is crucial for successful establishment of pulse crops. Since pulse crops such as chickpea are poor competitors with weeds due to slow germination and early growth, fall application of herbicides can aid in early season weed suppression and improve crop establishment. Field experiments were conducted in 2022 at two sites in Montana (Huntley and Corvallis) to evaluate crop safety and weed control by soil active herbicides before ground freeze in fall followed by a POST application in spring planted chickpeas. Ethalfluralin at 1.05 kg ha-1 + triallate at 1.68 kg ha-1 ,and pyroxasulfone at 131 g ha-1 + flumioxazin at 60.6 g ha-1 provided 80-90% control before POST application for Kochia scoparia and Amaranthus retroflexus at Huntley. Metribuzin at 420 g ha-1, Dimethamid at 950 g ha-1 + pendimethalin at 2.13 kg ha-1 and pyroxasulfone at 131 g ha-1 + flumioxazin at 60.6 g ha-1 provided 80-90% control before POST application of Chenopodium album at Corvallis. The follow-up POST application of clethodim at 280 g ha-1 and bentazon at 1.12 kg ha-1 helped in ensuring season long control by increasing control from 90 to 100% by eliminating escaped weeds. The treatments of dimethamid, simazine, flumioxazin, carfentrazone + sulfentrazone and pyroxasulfone provided poorer weed control than other treatments. There was no significant visual injury of any herbicide application and yield reductions in chickpea. These herbicide programs can be used in integration with other weed management tactics in pulse crops for effective weed management.

What Are We Missing? Undetected Changes to Fiber Quality from Low Rates of 2,4-D Using AFIS. Kyle R. Russell*[1], Peter A. Dotray[2], Brendan R. Kelly[1]; [1]Texas Tech University, Lubbock, TX, [2]Texas Tech University, Texas A&M AgriLife Research, Texas A&M AgriLife Extension, Lubbock, TX (31)

An increase in the number of preplant, at-plant, and post-plant applications of improved low volatility formulations of dicamba and 2,4-D choline to aid in the control of troublesome broadleaf weeds including glyphosate-resistant Palmer amaranth has increased with the adoption of auxin-tolerant cotton. With this increase in the number of applications, the risk of off-target movement of auxinic herbicides also has increased. Field studies were conducted at the Texas Tech University New Deal Research Farm in 2019, 2020, and 2021 to evaluate dicamba-tolerant cotton response to various rates of 2,4-D choline when applied at four growth stages [first square (FS) + 2 weeks (wk), first bloom (FB), FB + 2 wk, and FB + 4 wk]. Applications of 2,4-D choline were applied at 1.06 (1x), 0.106 (1/10x), 0.021 (1/50x), 0.0106 (1/100x), 0.0021 (1/500x), and 0.00106 (1/1000x) kg ae ha^{-1} to Deltapine 1822 XF cotton. Yield loses were observed in all years when rates of 2,4-D choline = 1/100x were applied at FS + 2 wk and FB. Only the 1x rate of 2,4-D choline resulted in yield loses at the FB + 4 wk timing in all three years. Short fiber content, neps, and seed coat neps, which are not tested using the High Volume Instrument, increased where micronaire, fiber length, and uniformity were negatively impacted.

Developer of Novel Adjuvant Systems for Herbicides. Jim T. Daniel[1], Chase T. Boman*[2]; [1]Daniel Ag Consulting, Keenesburg, CO, [2]AgraSyst, Spokane, WA (33)

Seed-Shattering Phenology of Grass and Broadleaf Weeds at Soybean (*Glycine max*) and Corn (*Zea mays*) Harvest in SE Montana. Het Samir Desai*[1], Fabian Menalled[2], Lovreet S. Shergill[3]; [1]Southern Agricultural Research Center, Montana State University, Huntley, MT, [2]Montana State University, Bozeman, MT, [3]Montana State University, Huntley, MT (34)

Due to escalating cases of multiple herbicide resistance in weeds, there is a need to develop effective non-chemical weed management tactics. Concurrent maturation of crops and infesting weeds presents an opportunity to use harvest weed seed control (HWSC) systems. HWSC systems disrupt the redistribution of weed seed at harvest to minimize the weed seedbank replenishment. Weed seed retention at crop harvest determines the successful implementation of HWSC. A field experiment was conducted to assess the seed retention and seed-shattering phenology of kochia [*Bassia scoparia* (L.) A. J. Scott], common lambsquarters (*Chenopodium album* L.), redroot pigweed (*Amaranthus retroflexus* L.), and green foxtail [*Setaria viridis* (L.) P. Beauv.] in soybean and corn. Early anthesis was observed in corn than in soybean for all tested weed species. All weed species retained 85-90% at corn and soybean physiological maturity except *C. album*. *Chenopodium album* retained 75% and 80% of seeds at corn and soybean physiological maturity, respectively. At 28 days after corn and soybean maturity, plants of all weed species retained 40-50% seeds. A greater proportion of seed retention at corn and soybean physiological maturity suggests that all tested weed species are suitable candidates for HWSC. Data generated from this project will aid in further research for assessing different HWSC methods such as chaff-lining, narrow windrow burning, impact mill, etc.

Confirmation of a Three-way (Thifensulfuron, Atrazine and Glyphosate) Herbicide Resistant Population of Palmer Amaranth (*Amaranthus palmeri*) in North Carolina. Ronel J. Argueta*, Eric A. Jones, Diego J. Contreras, Charlie W. Cahoon, Wesley Everman; North Carolina State University, Raleigh, NC (35)

Atrazine has historically been efficacious herbicide in North Carolina, where almost every hectare of corn receives a pre- and postemergence application. In 2016, a farmer in Washington Co. North Carolina reported a control failure on *Amaranthus palmeri* S. Watson (Palmer amaranth) in a corn field treated with atrazine. The objectives were to evaluate the response of the putative atrazine-resistant population from Washington Co. (R) at different atrazine rates and determine effective herbicides to control the putative atrazine resistant population. Two atrazine-susceptible populations from Edgecombe Co. (S) and Johnston Co. (S) were used susceptible controls. The experimental design was completely randomized with four replications. Atrazine was applied at 0, 56, 177, 560, 1770, 5600 and 17,700 g ai ha^{-1} and included crop oil concentration (1% v v^{-1}). Treatments were applied to plants 7.6 to 9 cm in height. Plant survival was evaluated 21 days after treatment. The LD_{50} for each population were calculated from a three-parameter log-logistic model. A resistance ratio was calculated for each atrazine-susceptible population. The LD_{50} value was 1354 g ai ha^{-1} for R population. The R/S were 6.8 to 8.7 compared to the S populations. The Washington Co. population has evolved atrazine resistance. Subsequently, these *A. palmeri* populations were treated with 7 herbicide modes of action for further characterization under similar conditions and evaluations. All populations exhibited resistance to thifensulfuron and glyphosate with survival =70% and =30%, respectively. 2,4-D and dicamba provided variable control on all populations. The population is resistant to thifensulfuron, atrazine and glyphosate.

Axant™ Flex Cotton Response and Residual Weed Management Using HPPD Inhibiting Herbicides. Megan Mills*[1], Peter A. Dotray[2], Gregory B. Baldwin[3], Scott Asher[4], Adam C. Hixson[4], Maxwell E. Smith[1], Blaine A. Patton[1], Kyle R. Russell[1], Bobby Rodriguez[5]; [1]Texas Tech University, Lubbock, TX, [2]Texas Tech University and Texas A&M AgriLife Research & Extension Service, Lubbock, TX, [3]BASF, Research Triangle Park, NC, [4]BASF, Lubbock, TX, [5]Texas A&M AgriLife Extension, Lubbock, TX (36)

Palmer amaranth (*Amaranthus palmeri* S. Wats) is a summer annual weed native to the southwestern United States and is regarded as one of the most problematic weeds in agronomic crops. The surge of herbicide resistant Palmer amaranth across the Cotton Belt contributes to its troublesome status. In 2022, three field studies were conducted at New Deal and Halfway, Texas to evaluate Axant™ Flex cotton response to topramezone applied at two early season growth stages, as well as to determine the efficacy of Alite™27 (isoxaflutole) when used preplant incorporated (PPI) or preemergence (PRE). Crop response studies at New Deal consisted of isoxaflutole or prometryn applied preemergence followed by (fb) topramezone applied alone, or tank mixed with isoxaflutole, glufosinate, glyphosate, or dicamba applied early-postemergence (EPOST) or mid-postemergence (MPOST). At both growth stages, crop response was =5% at 28 days after the EPOST and MPOST applications. Lint yield was not adversely affected in either study. The non-crop study at Halfway consisted of pendimethalin applied alone or in a tank mix with isoxaflutole applied PPI, followed by (fb) PRE applications of prometryn, fluometuron, diuron, fluridone, fomesafen, acetochlor, or S-metolachlor applied alone or tank mixed with isoxaflutole. No treatment contained isoxaflutole at more than one application timing. Palmer amaranth was completely controlled 15 days after PRE applications (DAA) for all treatments except isoxaflutole and pendimethalin PPI fb no PRE or prometryn PRE, and pendimethalin PPI fb isoxaflutole with fluometuron PRE. At 50 DAA, Palmer amaranth control was = 95% with applications of isoxaflutole and pendimethalin PPI fb fluridone, acetochlor, or S-metolachlor PRE, and pendimethalin PPI fb isoxaflutole and acetochlor PRE. These results support the use of isoxaflutole and topramezone in Axant™ Flex cotton production to help manage herbicide resistant weeds with no adverse effects on cotton yield and quality.

Hybrid Rice Tolerance with Various Fenclorim Seed Treatments and Microencapsulated Acetochlor.

Tristen H. Avent*, Jason K. Norsworthy, Leonard B. Piveta, Casey H. Arnold, Pamela Carvalho-Moore; University of Arkansas, Fayetteville, AR (37)

Rice (*Oryza sativa* L.) producers in the U.S. need effective herbicides to control problematic weeds. Previous research has demonstrated that acetochlor can provide in-season weed control in rice; however, undesirable injury is common. Thus, trials were initiated on a silt loam soil at the Rice Research and Extension Center near Stuttgart, AR, in 2020 and 2021 to evaluate a dose-response of a fenclorim seed treatment ranging from 0 to 5 g ai kg^{-1} of seed with and without a microencapsulated acetochlor formulation. For both years, acetochlor was applied delayed-preemergence at 1,260 g ai ha^{-1} (4 to 7 days after planting). The two hybrid rice cultivars RT 7231 FP and RT 7521 FP did not differ in response to either the seed treatment or acetochlor. In the absence of acetochlor, the fenclorim seed treatment caused less than 10 and 5% injury in 2020 and 2021, respectively. In the presence of acetochlor, the fenclorim seed treatment rate of 2.5 g ai kg^{-1} reduced rice injury and increased rice plant heights and shoot numbers relative to acetochlor without fenclorim. Additionally, the same fenclorim rate with acetochlor provided comparable heights and shoots to the nontreated control in all evaluations. Acetochlor or fenclorim did not influence rough rice yield for both cultivars. Based on the results of these studies, the fenclorim seed treatment rate at 2.5 g ai kg^{-1} of seed provided the most consistent safening response for hybrid rice tolerance to the microencapsulated formulation of acetochlor.

Transgenes Flow from Corn Confer Resistance to Glyphosate and Glufosinate-Ammonium in Teosinte (Zea Mays Subsp. Mexicana). Luan Cutti*[1], Paula Sinigaglia Angonese[1], Guilherme Menegol Turra[1], Estéfani Sulzbach[1], Catarine Markus[1], Todd A. Gaines[2], Aldo Merotto[1]; [1]Federal University of Rio Grande do Sul, Porto Alegre, Brazil, [2]Colorado State University, Fort Collins, CO (38)

The corn's wild ancestor teosinte has emerged as a new weed in summer crops as a consequence of pasture use for dairy production in southern Brazil. The objective of this study was to investigate the occurrence and mechanism of resistance to glyphosate and glufosinate in teosinte. Seeds from teosinte plants that survived to glyphosate applied at 1110 g ha^{-1}, one susceptible teosinte biotype, and two corn hybrids resistant to glyphosate (*cp4-EPSPS* event and *mEPSPS* with TIPS mutation event) were used for a dose-response assay, with doses from 0 to 71040 g ha^{-1}. The resistance factor to glyphosate of the resistant in comparison with the susceptible teosinte biotype was 196. Survival of the resistant teosinte plants was similar to both transgenic corn. The resistant teosinte had two *EPSPS* fragments, one is the wild-type *EPSPS* (bigger fragment), and the second one the *mEPSPS* (smaller fragment), observed by PCR amplification. The *mEPSPS* transgene from resistant biotype showed TIPS mutation (Thr102Ile/Pro106Ser). The same results were found for positive control transgenic hybrid corn. In addition, the *cp4-EPSPS* gene was identified in the resistant teosinte biotype and in positive control transgenic hybrid corn. The *cp4-EPSPS* sequence homology between resistant biotype and maize was close to 100%. The resistant teosinte carrying *mEPSPS* had 10 times more *mEPSPS* copies than transgenic corn also carrying *mEPSPS*. We found three different genotypes among the resistant teosinte plants: plants with only the *cp4-EPSPS* transgene, plants with only the *mEPSPS*, and plants with both. Results also showed that some teosinte plants resistant to glyphosate are also resistant to glufosinate-ammonium carrying the *pat* transgene. We conclude that glyphosate resistance in the teosinte was originated due to gene flow from the *mEPSPS* and the *cp4-EPSPS* transgenic corn. This is the first report of transgene escape from corn in field conditions, which is a challenge for the sustainability of herbicide resistance technologies in this crop. E-mail: luancutti@hotmail.com

Comparing Remotely-Sensed Data for Estimating Palmer Amaranth and Soybean Canopy Cover. Byron Evers[1], Hannah Buessing[2], Christopher Weber[2], Sarah Lancaster*[2]; [1]Bayer CropScience, Manhattan, KS, [2]Kansas State University, Manhattan, KS (39)

Palmer amaranth is among the most troublesome weeds in Kansas soybean fields. Cultural management practices, such as narrow row spacing, that achieve rapid canopy closure are recommended to suppress Palmer amaranth growth later in the growing season. Remote sensing has been investigated as a tool to increase the efficiency of Palmer amaranth management . To further evaluate the potential of remote sensing technology to detect weed and soybean canopy cover percent, an Unoccupied Aerial Vehicle (UAV) equipped with a multi spectral sensor was used to capture reflectance at two locations in Kansas. Data were collected from an experiment designed to investigate the interaction of planting date and row spacing on weed management in Kansas soybeans. The dominant weed species at both locations was Palmer amaranth. Canopy classifications from the spectral data were delineated through two methods: (1) by using the excess green index and (2) through maximum likelihood supervised classification. Weed control was estimated by visual assessment throughout the growing season. Both methods were capable of distinguishing weeds and soybean and were correlated with visual estimates of weed control. Supervised classification resulted in a stronger relationship with weed control.

A Metabolization System Affected by Temperature is Associated with Florpyrauxifen-benzyl Resistance in Barnyardgrass (Echinochloa Crus-galli). Luan Cutti*[1], Crystal D. Sparks[2], André Lucas Simões Araujo[2], Angela Da Cas Bundt[3], Catarine Markus[1], Todd A. Gaines[2], Aldo Merotto[1]; [1]Federal University of Rio Grande do Sul, Porto Alegre, Brazil, [2]Colorado State University, Fort Collins, CO, [3]Corteva Agriscience, Santa Maria, Brazil (40)

The herbicide florpyrauxifen-benzyl was recently released aiming to control quinclorac, ALS, and ACCase-resistant biotypes of *Echinochloa*. The objective was to identify biotype of barnyardgrass resistant to florpyrauxifen-benzyl and its mechanism of resistance. 23 biotypes of barnyardgrass from Brazil that were never sprayed with florpyrauxifen-benzyl were screened with the label rate 30 g ha^{-1}. The most resistant biotype was selected with florpyrauxifen-benzyl for three generations. The barnyardgrass resistant biotype and one susceptible were submitted to dose-response curves evaluating cytP450 (malathion) metabolic inhibitor in greenhouse (28ºC ± 3). The doses sprayed ranged from 0 to 240 g ha^{-1}. It was also performed dose-response curves for both biotypes under two different temperatures 35ºC/30ºC and 25ºC/20ºC day/night in growth chamber. The fresh shoot weight was evaluated, and the data fitted to three parameters log-logistic curve. RNA sequencing was performed with samples collected at 24 hours after florpyrauxifen-benzyl treatment at 35ºC/30ºC. The resistant biotype showed a resistance factor (RF) of 3.57 to florpyrauxifen-benzyl when compared to susceptible. The previous spray with cytP450 inhibitor reduced the RF to 1.5. LC-MS/MS analysis showed the resistant biotype accumulates less florpyrauxifen-acid at 8 and 24h after herbicide spray. When the dose-response curves were performed at 35ºC/30ºC the RF increased to 5.21, while for 25ºC/20ºC the RF was 2.37. RNAseq data analysis showed 551 genes up-regulated in the resistant biotype and among them genes that codify four *CYPs*, seven *GSTs*, six *UGTs*, five *AKRs*, three *PERs*, and three *ABCs*. It was identified a biotype with innate resistance to florpyrauxifen-benzyl by enhanced metabolization. Metabolic resistance is a huge concern, due to its ability to evolve to resistance to multiple herbicides, but lower temperatures can be an opportunity to manage this mechanism of resistance. E-mail: luancutti@hotmail.com

Evaluation of the Axant™ Flex Cotton System. Zachary R. Treadway*[1], Jennifer Dudak[1], Todd A. Baughman[1], Adam C. Hixson[2], Gregory B. Baldwin[3]; [1]Oklahoma State University, Ardmore, OK, [2]BASF, Lubbock, TX, [3]BASF, Research Triangle Park, NC (41)

Use of New Sorghum Herbicide Technologies to Control Annual Grass Weeds. Blaine A. Patton*[1], Peter A. Dotray[2]; [1]Texas Tech University, Lubbock, TX, [2]Texas Tech University, Texas A&M AgriLife Research, Texas A&M AgriLife Extension, Lubbock, TX (42)

Sorghum [*Sorghum bicolor* (L.) Moench] hectares may increase with the development of new insect and drought-resistant traits. Grassy weeds continue to be a problem in sorghum, but improved control may be accomplished with the introduction of new herbicide technologies. Non-crop field studies were conducted in 2021 and 2022 at the Texas Tech New Deal Research Farm to evaluate the effectiveness of imazamox, nicosulfuron, and quizalofop to control barnyardgrass [*Echinochloa crus-galli* (L.) P. Beauv], large crabgrass [*Digitaria sanguinalis* (L.) Scop], and Texas millet [*Urochloa texana* (Buckley) R. Webster]. Imazamox (0.0526 and 0.079 kg ai ha^{-1}), nicosulfuron (0.0352 and 0.07 kg ai ha^{-1}), and quizalofop (0.0463 and 0.0695 kg ai ha^{-1}) were applied in separate trials seeded to each weed species when plants reached 10 cm and 20 cm in height. Plots, two 102-cm rows by 7.62 m, were replicated four times for each grass species. In 2021, quizalofop at 0.0695 kg ai ha^{-1} provided >70% control in all species when evaluated 14 days after treatment. Nicosulfuron at 0.07 kg ai ha^{-1} controlled Texas millet 70% 14 days after application. Grass control following imazamox did not exceed 50%. In 2022, quizalofop at 0.0695 kg ai ha^{-1} controlled 20 cm barnyardgrass and 10 cm Texas millet 92% and 80%, respectively, when evaluated 14 days after treatment. Quizalofop controlled 20 cm large crabgrass 77% when evaluated 28 days after treatment. In all studies, quizalofop seemed to provide the most consistent control of each weed species. Further research is needed to evaluate control of other grass and broadleaf weeds in these new sorghum technologies.

Distribution of Target Site Mutations in PPO2 Gene in Waterhemp (Amaranthus Tuberculatus) and Palmer Amaranth (A. Palmeri) Populations in the USA Conferring Resistance to PPO Inhibiting Herbicides. Daljit Singh[1], Jacob S. Montgomery[2], Jaishree Chitoor[3], Chandrashekar Aradhya[3], Alejandro Perez-Jones*[3]; [1]Bayer, Chesterfield, MO, [2]Colorado State University, Fort Collins, CO, [3]Bayer Crop Science, Chesterfield, MO (43)

Palmer Amaranth (*Amaranthus palmeri*) Management in Kansas Cotton (*Gossypium hirsutum*). Tyler P. Meyeres*[1], Lucas Haag[2], Gaylon Morgan[3], Sarah Lancaster[1]; [1]Kansas State University, Manhattan, KS, [2]Kansas State University, Colby, KS, [3]Cotton Incorporated, Cary, NC (44)

Residual herbicides are needed to control herbicide-resistant weeds in cotton. Multiple applications of residual herbicides offer overlapping weed control, which is critical for weeds like Palmer amaranth (*Amaranthus palmeri* S. Watson) that have extended windows of emergence. Experiments were established to compare weed control with currently available residual herbicides applied preemergence (PRE) or postemergence (POST) in Enlist (2,4-D resistant = EN) and Xtendflex (dicamba resistant = XT) cotton systems in Kansas. Field experiments were conducted in Hutchinson, KS in 2021 and 2022, where cotton was grown under overhead irrigation. In 2021, treatments consisted of PRE applications of pendimethalin followed by (*fb*) early POST (EPOST) applications of acetochlor, dimethenamid-P, or *S*-metolachlor + trait herbicide (EN = 2,4-D choline; XT = dicamba) or trait premix (EN = glyphosate + 2,4-D choline; XT = dicamba + s-metolachlor) *fb* late POST (LPOST) applications of glyphosate + trait herbicide. In 2022, treatments consisted of PRE applications of fluometuron alone or fluometuron + acetochlor, dimethenamid-p, s-metolachlor, or pendimethalin *fb* EPOST applications of glyphosate + trait herbicide alone or a second application of the residual herbicides. The dominant weed species was Palmer amaranth. Weed control was recorded prior to LPOST (2021) and EPOST (2022) applications. Weed control, weed biomass, and weed density were recorded at harvest. No yield data was measured in 2021. In 2022, yield components including fruiting nodes, total bolls, and harvestable bolls plant^{-1} were measured. All data were subjected to ANOVA and means were separated with Tukey's HSD test (a= 0.05). Prior to LPOST applications in 2021, there was a significant interaction of herbicide trait system and residual herbicide. The addition of EPOST applications resulted in greater control than PRE only systems, which resulted in 5% or less control. Prior the LPOST in 2021, there were no differences in control between the herbicide trait systems regardless of residual herbicide utilized, except in the case of dimethenamid-P. Applications of dimethenamid-P resulted in 70% and 31% control in XT and EN systems, respectively, which indicates less control was observed when combined with 2,4-D choline. Prior to EPOST applications in 2022, PRE applications of fluometuron resulted in 56% control while combinations of fluometuron with dimethenamid-P or *S*-metolachlor resulted in greater than 83% control. When POST applications were utilized, control at harvest in 2021 ranged from 48 to 71% regardless of residual herbicide or herbicide trait system. In 2022 at harvest, XT systems resulted in 79% control and EN systems resulted in 68% control. Treatments with two applications of acetochlor, dimethenamid-P, or *S*-metolachlor resulted in 74% or greater Palmer amaranth control at harvest. In 2022, there were no differences among residual herbicides for all yield components measured. However, when *S*-metolachlor was applied PRE and EPOST, fruiting nodes, total bolls, and harvestable bolls plant^{-1} were also similar to the non-treated check. Results of this trial indicate that the cotton herbicide trait system may influence Palmer amaranth control; but, residual herbicide selection, the inclusion of POST applications, and layered residual herbicides are of greater importance.

Changes in Nutrient Partitioning Following Low Rates of 2,4-D. Kyle R. Russell*[1], Peter A. Dotray[2], Irish L. B. Pabuayon[3], Glen L. Ritchie[1]; [1]Texas Tech University, Lubbock, TX, [2]Texas Tech University, Texas A&M AgriLife Research, Texas A&M AgriLife Extension, Lubbock, TX, [3]LSU AgCenter, Rayne, LA (45)

With the use of auxinic herbicides increasing in recent years, the risk of off-target movement to susceptible cotton has also increased. Field trials were conducted in 2020 and 2021 at the Texas Tech University New Deal Research Farm in New Deal, TX to evaluate changes in nutrient partitioning following applications of 2,4-D applied to different cotton maturity stages. Applications of 2,4-D at 0.107 (1/10×) and 0.0107 (1/100×) kg ae ha^{-1} were applied using a CO_2-pressurized backpack calibrated to deliver 140 l ha^{-1} to DP 1822 XF at first square + 2 weeks, first bloom, first bloom + 2 weeks, and first bloom + 4 weeks. In both 2020 and 2021, lint yield decreased following both rates of 2,4-D when applied at first square + 2 weeks and first bloom. The 1/100× rate of 2,4-D did not decrease lint production when applied at first bloom + 2 weeks and first bloom + 4 weeks. Nitrogen, phosphorus, and potassium accumulation in the leaves, stems, bracts, and the seed increased while at the same time lint yield decreased. When no differences in yield were observed, little to no differences in nutrient accumulation was observed relative to the nontreated control. While differences in nutrient accumulation were observed, further research is needed to determine differences in plant function following exposure to 2,4-D.

First Report of Multiple-Herbicide Resistance in Walter's Barnyardgrass in Texas Rice. Nithya K. Subramanian*, Tamara T. Mundt, Isidor Ceperkovic, Gabriella Elizarraras, Muthukumar V. Bagavathiannan; Texas A&M University, College Station, TX (46)

First Report of Multiple-Herbicide Resistance in Walter's Barnyardgrass in Texas Rice Nithya K Subramanian*, Tamara Mundt, Isidor Ceperkovic, Gaby Elizarraras and Muthukumar V BagavathiannanTexas A&M University, College Station, TX- 77845 *Contact: nithya.subramanian@ag.tamu.edu The genus *Echinochloa* (Poaceae) is one of the most important weed species in rice production systems worldwide. Walter's barnyardgrass (*E. walteri*) is a less common weed species in Texas rice, but a suspected herbicide-resistant population of this species was identified in a rice field near Beaumont, TX during the 2021 field season. The objective of this study was to evaluate the sensitivity of this population to a number of commonly used rice herbicides. A total of 10 herbicides belonging to five different modes of action were tested at the recommended label rate (1X rate). These include cyhalofop, fenoxaprop, imazethapyr, bispyribac-sodium, penoxsulam, nicosulfuron, quinclorac, propanil, florpyrauxafen-benzyl, and glyphosate. Four to six seedlings per treatment were sprayed at the three-to-four leaf stage using a track sprayer, and the experiment was repeated twice. The survival rate (0 to 100%) was recorded at 28 days after the herbicide treatment. Following this, a confirmative study was conducted using the 2.5X rate. Among the herbicides tested, resistance was documented for six different herbicides belonging to three different herbicide modes of action [acetolactate synthase (ALS)-inhibitors (imazethapyr, penoxsulam, bispyribac-sodium, nicosulfuron), auxin analog (quinclorac), and PSII-inhibitor (propanil)]. Thus, results show the occurrence of multiple resistance, as well as cross-resistance (to different herbicide families within ALS-inhibitors) in this population. Molecular mechanisms and dose-response studies are currently underway to further characterize resistance. This is the first report of herbicide resistance in this species globally, and the development of alternative management options is imperative.

Preemergence Control of *Amaranthus palmeri* and Axant™ Flex Cotton Weed Management Systems Using Isoxaflutole. Maxwell E. Smith*[1], Peter A. Dotray[2], Adam C. Hixson[3]; [1]Texas Tech University, Lubbock, TX, [2]Texas Tech University, Texas A&M AgriLife Research, Texas A&M AgriLife Extension, Lubbock, TX, [3]BASF, Lubbock, TX (47)

Amaranthus palmeri S. Wats. (Palmer amaranth) is one of the most common and most troublesome weeds in cotton (*Gossypium hirsutum* L.) production on the High Plains of Texas. Glyphosate-tolerant *Amaranthus palmeri* presents challenges to producers when it comes to effective season-long control. Palmer amaranth resistance to other modes of action have been reported, which will add to the complexity of controlling this weed. Axant™ Flex cotton from BASF may soon be available, which will allow producers to use a new mode of action for use in cotton. In 2021 and 2022, field studies were conducted at the Texas Tech University New Deal Research Farm to evaluate herbicide programs in Axant™ Flex cotton using isoxaflutole, an HPPD-inhibiting herbicide. Studies were conducted in 2021 and 2022 in Halfway, TX to evaluate tank-mix partners with isoxaflutole. The objective of these studies was to evaluate crop response and the control of *Amaranthus palmeri* in HPPD-tolerant cotton systems using isoxaflutole, and to evaluate length of residual control of *Amaranthus palmeri* with isoxaflutole plus other herbicides applied in tank-mix. In studies at New Deal, *Amaranthus palmeri* control and crop response were evaluated after preemergence and early-postemergence applications in different herbicide programs. The herbicide programs that included isoxaflutole were compared to standard local programs. Cotton crop response 21 days after the preemergence application was <12% for all treatments in 2021 and no response was observed in 2022. Control of *Amaranthus palmeri* 29 days after the preemergence application was >92% and >88% for all treatments that included isoxaflutole in 2021 and 2022, respectively. In 2021 and 2022, control of *Amaranthus palmeri* 28 days after the early postemergence application was =88% for all treatments that included a 0.105 kg ai ha $^{-1}$ rate of isoxaflutole in the preemergence application. In 2021 at Halfway, TX, *Amaranthus palmeri* was controlled >45% 28 days after application for all treatments that included isoxaflutole. In 2022, control of *Amaranthus palmeri* for all treatments that included isoxaflutole was >68% 28 days after application. Axant™ Flex cotton offers a new mode of action to become available for use in cotton and can provide residual control of *Amaranthus palmeri* to cotton producers.

Evaluation of Axant Flex Systems in Mid-South Cotton. Bradley J. Norris*, Brian K. Pieralisi, Darrin M. Dodds, Tyler S. Soignier, William J. Rutland; Mississippi State University, Starkville, MS (48)

Crabgrass and Texas Panicum Control with Group 15 Herbicides Applied Preemergence. Livia Pereira*, Steve Li, Ryan D. Langemeier, Justin T. McCaghren; Auburn University, Auburn, AL (49)

Gramineous weeds continue to be challenging to control in crops, especially in crops where herbicide tolerance traits are not available. The objective of this study was to compare the efficacy of four Very Long-chain Fatty Acid Inhibiting herbicides for preemergence control on two grass species with different seed sizes. The study was conducted in two sites in central and south Alabama during the 2022 growing season. The seeds of *Digitaria anguinalis* and *Panicum texanum* were spread onto tilled 1.83 x 1.83 m plots and incorporated with a rotary tiller, and then treated with herbicides immediately after. Plots were hand weeded to remove all other weed species as needed. Data collection involved weed counts and visual rating at 14, 28, and 42 days after treatment (DAT). Additionally, a handheld greenseeker and a multispectral imaging camera on an unmanned aerial vehicle were used to assess growth plus biomass collected at 42 DAT. Imagery was analyzed in QGIS 3.22 and statistics were conducted in SAS 9.4. The results showed a control over 94% for all products on *Digitaria*, and overall *Panicum* presented more difficult to control. The treatment that consisted of pyroxasulfone plus carfentrazone had the highest level of control for *Panicum*, while all treatments had same level of control for *Digitaria* when compared to non-treated control (NTC). Both *Panicum* and *Digitaria* presented no significant difference for stand counts across products even at 42 DAT, but pyroxasulfone plus carfentrazone was numerically lower for both species. Similar to stand count, no significant difference for dry biomass for *Digitaria*. However, pyroxasulfone plus carfentrazone was the only treatment significantly lower than NTC for *Panicum*. The research highlights the species-specific differences, indicating the need for site-specific recommendations depending on the weed species.

Effect of Grassy Weed Herbicides on Cereal Rye. Ethan B. Whitmoyer*, Dwight Lingenfelter, John M. Wallace; Penn State University, University Park, PA (50)

Cereal rye (*Secale cereale* L.) is a commonly used cover crop species and one of the least expensive seed options. Since cereal rye is a useful cover crop species, some farmers are interested in growing rye not only as a winter cover but also to harvest the seed for sowing in subsequent years. Currently, cereal rye has the fewest number of herbicides registered compared to other commonly planted small grains; and furthermore, none have activity on weedy grasses. Therefore, field studies were conducted in 2019, 2020, 2021, and 2022 to determine the safety of various small grain herbicides on cereal rye with an emphasis on those that control grassy weeds. Cereal rye (var. 'Aroostook') was planted each year in late September at 67 or 101 kg ha-1. Studies were arranged in a randomized complete block design with three replications. Herbicides were applied with a small-plot, CO2-backpack sprayer system that delivered 140 L ha-1 thru TeeJet AIXR110015 nozzles. Treatments were applied to the rye in the fall (2-leaf stage; 5-10 cm tall) and in the spring (23-33 cm tall). Delayed preemergence treatments were added to the 2022 study. Treatments included: flufenacet + metribuzin (381 g ai ha-1), pendimethalin (1064 g), pyroxasulfone (119 g), metribuzin (157 g), mesosulfuron (15 g), pyroxsulam (18 g), halauxifen + florasulam (11 g), thifensulfuron + tribenuron (26 g), chlorsulfuron + metsulfuron (21 g, 2020 and 2021 studies). All herbicides were applied in the fall and spring except for flufenacet + metribuzin, pendimethalin, and pyroxasulfone which were only fall-applied. Appropriate adjuvants were included where necessary. Visual crop injury ratings were collected. Results across all studies for visual crop injury data from any application timing indicate that all herbicides on average cause no more than 5% injury, except metribuzin (14-35 % injury) and mesosulfuron (13-57% injury). Mesosulfuron showed the highest level of injury when applied either in the fall or spring, however, rye was injured most when applied during the early stages of growth. Metribuzin and flufenacet + metribuzin showed variable injury (2-35%), however more injury was observed from the fall application. In summary, the results of these crop safety studies of herbicides currently not labeled for use on cereal rye show there is potential for some to receive registration for this crop. Especially those that provide control of weedy grass species such as pyroxsulam and thifensulfuron + tribenuron, a frequently used premix in other small grains, for control of many broadleaf weeds. Future research will further explore cereal rye tolerance to delayed preemergence treatments. Furthermore, herbicide tolerance of other small grain cover crop species such as triticale (*Triticale hexaploide*), hybrid rye, and spelt (*Triticum spelta*), should be examined.

Evaluating Weed Control and Crop Response of Pyridate in Field Pea. Adam Leise*[1], Amit J. Jhala[1], Dennis Long[2]; [1]University of Nebraska Lincoln, Lincoln, NE, [2]Belchim, Manchester, TN (51)

Controlling weeds throughout the growing season is one of the biggest problems growers in NE are dealing with annually. Integrated Weed Management is the control of weeds throughout the growing season through chemical, biological, and mechanical methods. Multiple modes of action chemically in cash crop fields have become suggested in order to best control weeds through the growing season. Field Peas are increasing in popularity in Nebraska due to their nitrogen fixation and efficient water usage. Efficient weed control in field peas is needed in order to maximize yield while also minimizing plant competition. Traditionally, group 2 and group 5 herbicides have been used to control weeds in field pea rows. Tough 5EC (pyridate) is a group 6 herbicide that is being tested within field pea rows. In order to determine Tough 5EC effectiveness, this study: (i) Assessed control ratings when using Tough 5EC herbicide in field pea trials, (ii) compared crop damage in field peas when using various rates of Tough 5EC herbicide, and (iii) determined the overall efficiency of a new mode of action control in field peas. In this study, we compared two different rates of Tough 5EC. The two different rates were 6.9 fl.oz./ a against 13.8 fl.oz./a. Our results showed that an increase in the rate of Tough, with a combination of Basagran or Metribuzin provided the most effective weed control. Crop damage was assessed by visual ratings of chlorosis, plant vigor, and phytotoxicity. Tough 5EC with a rate of 13.8 fl.oz./a had the highest crop damage with significant chlorosis. A balance of effective weed control with minimal crop damage will be the goal of our future research.

Evaluation of Biological Nitrification Inhibition (BNI) Potential in Cover Crops. Nithya K. Subramanian*, Nithya Rajan, Sanjay Antony-Babu, Tamara T. Mundt, Megan L. Schill, Dinesh Phuyal, Muthukumar V. Bagavathiannan; Texas A&M University, College Station, TX (52)

Long-term Palmer Amaranth (*Amaranthus palmeri*) Management in Dicamba-Resistant Cotton Using Herbicide Programs, Crop Rotations, and Tillage. Rohith Vulchi*[1], Joshua A. McGinty[2], Scott A. Nolte[3]; [1]Texas A&M University, College Station, TX, [2]Texas A&M AgriLife Extension, Corpus Christi, TX, [3]Texas A&M AgriLife Extension, College Station, TX (53)

Dicamba-resistant cotton technology was introduced in the US to manage glyphosate-resistant (GR) Palmer amaranth. However, with sole reliance on herbicides for weed control, GR Palmer amaranth evolved resistance to dicamba also in Tennessee and Kansas. Previous research shows tillage, crop rotation, and using in-season residual herbicides provide significant GR Palmer amaranth control. Unfortunately, research integrating all these three weed management strategies in a system was not done before. Therefore, a three-year field experiment was conducted at College Station and Thrall, TX to evaluate the efficacy of dicamba-based herbicide programs under cotton-cotton-cotton and cotton-sorghum-cotton cropping sequences in no-till cover cropping, strip tillage, conventional tillage for GR Palmer amaranth control. Herbicide programs include a high-input herbicide program (HI) with in-season residual herbicides and a low-input herbicide program (LI) without residual herbicides. HI in conventional tillage controlled all weeds =95% at both locations. Using HI under cotton-sorghum-cotton provided >90% Palmer amaranth control in no-till cover cropping and strip tillage up to 28 DA second POST application from 2019-21 at College Station. In strip tillage at Thrall, the efficacy of LI reduced from 97% in 2019 to 85% by 2021 under cotton-sorghum-cotton rotation at 28 DA second POST application timing. Similarly, the efficacy of LI reduced from 96% in 2019 to 80% by 2021 under continuous cotton production. In no-till cover cropping, the efficacy of LI remained around 99-100% under cotton-sorghum-cotton rotation from 2019-21 and did not change significantly. However, the efficacy of LI reduced from 96% in 2019 to 88% by 2021 under continuous cotton production. These results indicate using in-season residuals and rotating with other crops improve or sustain GR Palmer management in different tillage types in Dicamba-resistant cotton.

Spot Treatment of Clethodim on Weedy Rice. Rasim Unan*, Liberty B. Galvin, Aaron Becerra-Alvarez; University of California, Davis, Davis, CA (54)

Utilizing Novel Application Methods to Mitigate Auxin + Graminicide Herbicide Antagonism. Jake A. Patterson*[1], Darrin M. Dodds[1], Antonio Augusto Correa Tavares[2], Gresham A. Stephens[1], Caleb K. Meyer[1], Sydney C. Baker[1]; [1]Mississippi State University, Starkville, MS, [2]University of Nebraska-Lincoln, North Platte, NE (55)

Herbicide antagonism is an economically and environmentally troublesome issue that many growers across the United States are experiencing. Published literature indicates that certain herbicides, when mixed together in solution, antagonize one another resulting in decreased weed control compared to each herbicide applied independently. While herbicide antagonism is well documented, it is less known if methods exist which can overcome antagonism without increasing herbicide rates or making repeated applications. Field experiments were conducted in 2022 at the Black Belt Branch Experiment Station near Brooksville, MS to evaluate grass weed control in response to the application of dicamba, 2,4-D, glyphosate, and clethodim using a commercially available direct-injection system compared to tank-mix combinations of these herbicides. A sprayer equipped with a commercially available direct-injection system was used to apply treatments as follows 1) the two herbicides tank-mixed and sprayed through a single boom, 2) the auxin herbicide applied from the bulk tank and the grass herbicide direct-injected into the spray boom, 3) the grass herbicide applied from the bulk tank and the auxin herbicide direct-injected into the spray boom, and 4) the grass herbicide applied alone. Each herbicide combination was applied with the application methods listed above. At 14 and 28 days after treatment, regardless of which auxin herbicide was included, glyphosate provided greater barnyardgrass [*Echinochloa crus-galli* (L.) Beauv.] control than clethodim. At 14 and 28 days after treatment, when dicamba was used, no differences in barnyardgrass control were observed within glyphosate- nor clethodim-containing treatments. At 14 days after treatment, when clethodim was injected into 2,4-D, increased control was observed compared to when 2,4-D was injected into clethodim or the two herbicides were tank-mixed. These findings suggest that, in general, auxin + graminicide treatments applied via direct-injection do not provide greater barnyardgrass control compared to tank-mix treatments. However, these results do indicate that, regardless of application method, glyphosate is more efficacious in controlling barnyardgrass than clethodim. Although no differences in barnyardgrass control were observed with dicamba, there is still value in utilizing this application method to potentially alleviate tank-contamination.

Early Season Weed Control in Rice on Organic Soils with Preemergence Herbicides. D Calvin Odero*; University of Florida, Belle Glade, FL (56)

Impact of Cereal Rye Cover Crop Seeding Rate and Termination Timing on Weed Management in Soybean. Eric Y. Yu*[1], Axel Garcia y Garcia[2], Lizabeth Stahl[3], Ryan P. Miller[4], Gregg A. Johnson[1], Ce Yang[1], William Lazarus[1], Debalin Sarangi[1]; [1]University of Minnesota, St. Paul, MN, [2]University of Minnesota,

Lamberton, MN, [3]University of Minnesota Extension, Worthington, MN, [4]University of Minnesota, Rochester, MN (57)

Fall-planted cover crops can outcompete weeds in the spring and provide early-season weed control. However, weed suppression is dependent on cover crop biomass, termination timing, and weed emergence periodicity. Although cover crops are gaining traction in Minnesota, data on critical cover crop biomass production for weed suppression is limited in the Upper Midwest. The objective of this study was to evaluate the effect of cereal rye (CR) cover crop seeding rate and termination timing on CR biomass production, weed suppression, and soybean yield in Minnesota. Field studies were conducted in 2021-2022 at a farmer's field in Rochester, MN. Soybean was planted on two different dates (early and late) and CR seeding rate treatments included 0, 67, 101, and 135 kg ha^{-1}. Termination timing treatments were: 7 days before soybean planting (T1); at soybean planting (T2); at soybean planting with PRE (T2+PRE); and 7 days after soybean planting (T3). The seeding rate did not affect CR biomass production and weed control; however, differences were observed between soybean planting date and termination timing. At 42 d after planting (DAP), common lambsquarters (*Chenopodium album*) control was highest (89 to 99%) in termination treatments T2, T2+PRE, and T3 in late-planted soybean, which was similar to T2+PRE (91%) and T3 (87%) in early-planted soybean. Similarly, yellow foxtail (*Setaria pumila*) control was highest (95 to 98%) in T2, T2+PRE, and T3 in late-planted soybean, which was slightly better than T2+PRE (81%) and T3 (85%) in early-planted soybean. Despite greater weed control, soybean yield was negatively impacted in the late soybean planting date. Therefore, CR at 67 kg ha^{-1} terminated at T3 in an earlier soybean planting date is recommended. yu015348@umn.edu

Intercropping Soybean and Corn into the New Winter Oilseed Crop Golden Pennycress (*Thlaspi arvense*). Brent S. Heaton*, Tad L. Wesley, Winthrop B. Phippen, Mark L. Bernards; Western Illinois University, Macomb, IL (58)

The use of cover crops has become increasingly important for reducing nutrient loss, suppressing weeds, and enhancing soil health. Golden pennycress (*Thlaspi arvense*) is a winter-annual oil-seed domesticated using gene editing that will be used as a biofuel, livestock feed, and as specialty food oil. Fitting pennycress into a corn-soybean rotation will likely require earlier than normal harvest of one crop, and later than normal planting of the second crop. Intercropping into pennycress may improve corn or soybean yield, but may also negatively affect pennycress production. Our objectives were to: 1) measure pennycress yield as affected by intercropping corn or soybean prior to mid-bolting in pennycress, and 2) measure corn and soybean yield as effected by intercropping into pennycress. Pennycress was planted September 2021. Corn and soybean were planted into pennycress and into bare-ground plots on March 21 and April 12, 2022. Pennycress was harvested June 5, 2022, and corn and soybean were planted immediately after harvest into pennycress stubble and into bare-ground plots. All crops were harvested with small-plot combines. Disturbance from planting equipment (tires, row openers) did not affect pennycress yield, and whole-plot pennycress yield was not affected by intercropping prior to the mid-bolt stage in pennycress. Soybean and corn demonstrated season-long negative responses to intercropping, as evidenced by delayed development, delayed maturity, and reduced canopy cover. Corn yields were lower when intercropped into pennycress than when planted after pennycress harvest, and were lower when planted into pennycress stubble than into bare-ground. Soybean yields were the same for both intercropped and post-harvest planting, but intercropped soybean yields were less than bare-ground soybean yields. Cropping techniques that optimize yield of pennycress and the subsequent corn or soybean crop need to be developed to ensure farmer profitability.

Quantifying the Impact of Herbicide Drift on Peanuts Using Unmanned Aerial System. Pratap Devkota*, Navjot Singh, Joseph E. Iboyi; University of Florida, Jay, FL (59)

Peanut growth and yield can be negatively impacted by herbicide drift. The off-target movement of herbicides applied in other agronomic crops such as cotton, soybean, and corn can result in a significant yield reduction in peanuts The research was conducted to evaluate the impact of herbicide-drift on peanuts using an unmanned aerial system (UAS) and field-based measurements. The study evaluated the impact of 25% of the label rate of glufosinate, glyphosate, dicamba, paraquat, and lactofen applied at 25 and 60 days after planting (DAP) peanuts. NDVI, peanut injury, yield, height, width, and leaf area index (LAI) were measured at 4 weeks after herbicide application. The Normalized Difference Vegetation Index (NDVI) was calculated from images collected with a multispectral camera mounted on a UAS. Among different herbicides, the 25% label rate of dicamba, glufosinate, and glyphosate applied at 25 DAP resulted in the highest peanut injury. These herbicides caused 42 to 52% peanut injury and 25 to 57% NDVI reduction. Glyphosate (25% of the label rate) applied at 25 and 60 DAP reduced the peanut yield by >65%. Likewise, dicamba (25% label rate) applied at 60 DAP reduced the peanut yield by 51%. In this study, NDVI showed a strong correlation with peanut injury, canopy height, canopy width, and yield with R^2 of 0.71 to 0.85, 0.76 to 0.94, 0.89 to 0.94, and 0.66 to 0.69, respectively. Therefore, UAS-derived NDVI can be a helpful parameter to determine peanut injury and yield reduction caused by herbicide drift.

Weed Seed Fate at Corn Harvest. Haylee E. Schreier*[1], Jacob E. Vaughn[1], Kevin W. Bradley[1], Bryan G. Young[2], Jason K. Norsworthy[3], Prashant Jha[4], Travis Legleiter[5], Bill Johnson[6]; [1]University of Missouri, Columbia, MO, [2]Purdue University, Brookston, IN, [3]University of Arkansas, Fayetteville, AR, [4]Iowa State University, Ames, IA, [5]University of Kentucky, Princeton, KY, [6]Purdue University, Brooksone, IN (60)

The increasing problem of herbicide resistance in weeds such as waterhemp (Amaranthus tuberculatus) and Palmer amaranth (Amaranthus palmeri) has resulted in a greater need for a more integrated approach to weed management, especially in U.S. corn, soybean, and cotton production systems. Previous research in Australia has shown that harvest weed seed destruction (HWSD) can be a successful method to reduce weed seed from returning to the soil. One form of HWSD is the use of impact mills that destroy weed seed as they pass through the combine, and two of the most common HWSD implements that are starting to be commercially available to U.S. growers are the Redekop™ and Seed Terminator™. Although a variety of research has been conducted with these devices when harvesting soybean, very little research has been conducted to determine the utility of these devices for use in U.S. corn. This may be a significant oversight, as it is likely that substantial differences may exist with respect to the performance of these implements in corn versus soybean. Additionally, the extent to which weed seed will enter the combine with a traditional corn header compared to what occurs with a soybean header is unknown. A field experiment was conducted at the time of corn harvest in Arkansas, Kentucky, Missouri, Indiana, and Iowa in 2022 to: 1) quantify the amount of weed seed that enters and exits the combine during typical corn harvest operations in the U.S., and 2) determine the effectiveness of on-combine HWSD implements at destroying weed seed and reducing weed seedbanks following corn harvest. To determine the amount of weed seed that physically entered and exited the combine, a series of collections were taken to determine weed seed fractions lost at the header, deposited in the grain tank, and discharged out of the rear of the combine with chaff. The combined data from all sites that contained Amaranthus species revealed that there was an average of 668 Amaranthus seed per m^2 lost at the header as a result of seedheads shattering after contact with the combine reel, while an average of 194 Amaranthus seed per m^2 were discharged out of the rear of the combine with the chaff fraction. Additionally, there was an average of 533 Amaranthus seed diverted to the grain tank per bushel of corn grain harvested. Three of the five locations contained an on-combine seed destruction device. Data from the Arkansas location indicated that harvesting corn with a combine equipped with an impact mill resulted in a 74% reduction in the total amount of weed seed deposited out of the rear of the combine and back onto the soil. Overall, the preliminary results from this research collected thus far indicate that a significant fraction of weed seed are lost at the corn header and diverted into the grain tank, but that combines equipped with impact mills can contribute to a reduction in weed seed return to the soil seedbank.

Seed Germination Responses of Branched Broomrape (*Phelipanche ramosa*) to Different Sanitation Chemicals and Exposure Durations. Pershang Hosseini*; Brad Hanson, University of California, Davis, Davis, CA (61)

Branched broomrape is a root parasitic weed that recently re-emerged in California tomato fields in several counties. It is currently classified as an "A" pest and subjected to state-enforced action like eradication. Preventive measures must be taken to stop or significantly reduce the spread of branched broomrape seeds to other areas. This parasitic weed can produce thousands of tiny seeds, easily spread by farm machinery over short and long distances. To prevent branched broomrape seed dispersal, disinfection of farm machinery is necessary before entering a new farm. We, therefore, tested the effectiveness of various ammonium compounds, including didecyl dimethyl ammonium chloride (DDAC), alkyl dimethyl benzyl ammonium chloride (ADBC), didecyl dimethyl ammonium bromide (DDAB), ammonium bromide (AB), and ammonium chloride (AC) on prevention of branched broomrape seed germination in different exposure duration. For all chemicals, we tested seven concentrations: 0 (distilled water), 0.05, 0.1, 0.2, 0.5, 1, and 2.5 % (w/v) of distilled water. The exposure times tested were 1, 3, 5, and 10 minutes. All treatments were replicated four times in 2022 at the Contained Research Facility of UC Davis. Dose-response analysis showed that three quaternary ammonium compounds, ADAC, DDAB, and DDAC, effectively prevented broomrape seeds' germination. The effective dose for 50% reduction in germination varied significantly across the above three chemistries, and exposure durations ranged from 0.00002 (SE=0.00) in DDAC at 1 minute to 0.402 (SE=0.035) (w/v) in ADAC at 10 minutes. The germination responses of branched broomrape seeds to DDAB and DDAC were very similar. The decline occurred sharply, and complete prevention was achieved below 0.5 concentration at minimum exposure duration, i.e., 1 minute. However, the declining response to ADAC was slower than other compounds. AC and AB had no sanitation effects on branched broomrape seeds, even at the highest concentration rate and longest exposure duration. Optimum concentrations can provide similar inhibition effects when combined with exposure times.

Sensitivity of Selected mid-Atlantic Vegetables to Simulated Dicamba Drift Rates. Maggie H. Wasacz[1], Baylee L. Carr[2], Mark VanGessel[3], Thierry E. Besancon*[4]; [1]Rutgers University, Wall Township, NJ, [2]Rutgers University - Weed Science, Chatsworth, NJ, [3]University of Delaware, Georgetown, DE, [4]Rutgers University, Chatsworth, NJ (62)

Dicamba is a synthetic auxin herbicide that may be applied over the top of transgenic dicamba-tolerant soybeans. The increasing prevalence of herbicide-resistant weeds such as Palmer amaranth, common ragweed, and horseweed has resulted in dicamba-based herbicides to become important tool in soybean production systems. Because dicamba is prone to off-target movement, there has been increased concern regarding dicamba drift onto nearby specialty vegetable crops. The present study evaluates twelve selected mid-Atlantic vegetable crops species for sensitivity to sub-lethal rates of dicamba. Soybean, snap bean, lima bean, tomato, eggplant, bell pepper, cucumber, summer squash, watermelon, pumpkin, sweet basil, lettuce, and kale were grown in the greenhouse and exposed to dicamba at 0, 0.056, 0.11, 0.28, 0.56, 1.12, 2.24 g ae ha^{-1}, which is respectively 0, 1/10,000, 1/5,000, 1/2,000, 1/1,000, 1/500, and 1/250 of the maximum recommended label rate for soybean application (560 g ae ha^{-1}). Plant foliage was evaluated weekly up to 4 weeks after treatment using visual rating methods and leaf deformation index (LDI) measurements. Overall, snap bean was the most sensitive crop species with dicamba rates as low as 0.11 g ae ha^{-1} resulting in significantly higher leaf deformation levels compared to the nontreated control. Other Fabaceae and Solanaceae species also demonstrated high sensitivity to sub-lethal rates of dicamba with rates ranging 0.28 to 0.56 g ae ha^{-1} causing significantly higher leaf deformation compared to the nontreated control. Cucurbitaceae species demonstrated low to moderate sensitivity, while sweet basil, lettuce and kale demonstrated tolerance to dicamba rates up to 2.24 g ha^{-1}.

Integrating Pre-emergence Herbicides with Potential Safeners for Purple Nutsedge (*Cyperus rotundus* L.) Management in Florida's Tomato Plasticulture Production. Ruby Tiwari*[1], Nathan Boyd[2], Pamela Roberts[1], Samira Daroub[3], Ramdas Kanissery[1]; [1]University of Florida, Immokalee, FL, [2]University of Florida, Balm, FL, [3]University of Florida, Everglades, FL (63)

Carolina Redroot: Long Term Management with Sequential Residual Herbicide Applications in Cranberry. Thierry E. Besancon*[1], Baylee L. Carr[2], Wesley M. Bouchelle[1]; [1]Rutgers University, Chatsworth, NJ, [2]Rutgers University - Weed Science, Chatsworth, NJ (64)

Carolina redroot is classified as a high priority weed in cranberry beds where its development is associated with open areas where vine has been killed by fairy ring disease or by any other stand opening conditions of natural and anthropomorphic origin. Carolina redroot accounts for significant cranberry yield reduction and can also hinder the industrial processing of cranberry fruits. A study was conducted between 2018 and 2021 in two different cranberry beds to evaluate Carolina redroot control efficacy and crop tolerance in response to overlapping applications of napropamide PRE and mesotrione POST. Herbicide treatments included napropamide applied at 6.7 or 10.1 kg ha^{-1} either as a single application at the spring dormant (SD) stage within one wk of the winter flood removal, or as a split application between the SD stage and 30 d later. Single napropamide treatments were applied either alone or followed by a single application of mesotrione at 280 g ha^{-1} in mid-June or repeated in mid-July. Equally split napropamide treatments were followed by a single or a double mesotrione application. Dichlobenil at 4.5 kg ha^{-1} split-applied and followed by a single mesotrione application was used as a standard. Each individual plot received the same herbicide treatment for three consecutive years. Single napropamide application at 10.1 kg ha^{-1} followed by two applications of mesotrione POST provided 64% Carolina redroot control 16 wk after treatment (WAT) at the SD stage whereas similar treatment with napropamide only provided 53% control. In the absence of POST mesotrione, Carolina redroot control was = 20%. Splitting of napropamide application did not improve Caroline redroot control. By the end of the study, Carolina redroot density and biomass decreased by 93% and 97%, respectively, with single application of napropamide at 10.1 kg ha^{-1} followed by two applications of mesotrione POST as compared to the weedy control. Similar napropamide application followed by a single application of mesotrione POST still provided = 88% reduction of Carolina redroot density and biomass. Cumulated commercial yield was 34% and 61% in high and low productivity beds, respectively, following single napropamide treatment at 10.1 kg ha^{-1} followed by one or two mesotrione application as compared to the weedy control. In the high productivity bed, splitting napropamide application at 6.7 or 10.1 kg ha^{-1} did not significantly improve cranberry yield compared to the weedy control despite lower Carolina redroot density and biomass, suggesting potential phytotoxicity of napropamide applied during the elongation phase of cranberry uprights.

Evaluation of Weed Removal Timing for Three Sweetpotato (*Ipomoea batatas* L.) Cultivars. Emmanuel G. Cooper*[1], Stephen L. Meyers[1], Jeanine Arana[2]; [1]Purdue University, West Lafayette, IN, [2]Purdue University/Horticulture, West Lafayette, IN (65)

Weed interference is a serious challenge for many sweetpotatoes producers globally and can greatly reduce yield and production efficiency. Field experiments were conducted at the Samuel G. Meigs Horticulture Research Farm, Lafayette, IN and at the Southwest Purdue Agricultural Center, Vincennes, IN, in 2022 to evaluate the effects of different weed removal timings on the growth and yield of three sweetpotato cultivars. The experiment design was a split plot with weed establishment treatments as the main plot factor. Weeds were removed by hand and allowed to establish and compete with the crop beginning at 0 (weedy check), 14, 21, 28, 35, or 42 days after transplanting (DAP). Additionally, a weed-free check was included for comparison. The subplot factor was cultivar (Covington, Monaco, and Murasaki). Sweetpotatoes were hand-transplanted into raised bed plots 6 m long with a within and between-row spacings of 0.3 m and 1.98 m on center, respectively; resulting in 20 plants $plot^{-1}$. Weed control, canopy cover, weed count, and weed height data were collected during the growing season using a 1 ft^2 quadrat. Sweetpotatoes storage roots were graded and weighed as jumbo, US number one, and canners at harvest (112 DAP). Data were pooled across locations as there was no significant treatment-by-location interaction. Total yield reduction and yield reduction for US number one grade roots best fit a three-parameter log-logistic curve for the effect of weed establishment timing. As weed establishment was delayed from 0 to 42 DAP, predicted total yield reduction as percentage of the weed-free check decreased from 76 to 0.4 %, 64 to 0.09 %, and 89 to 6 % for Covington, Murasaki, and Monaco respectively. US number one yield for removal timing of 28 DAP was reduced by 5%, 1%, and 26% for Covington, Murasaki, and Monaco respectively. High reduction of yield for the Monaco cultivar can be attributed to reduced stand count (average of 14 plants $plot^{-1}$). No differences were observed between cultivars for sweetpotato canopy cover at 7 weeks after transplanting (WAP). However, at 15 WAP, as weed establishment was delayed from 0 to 42 DAP, predicted sweetpotato canopy increased from 0 to

87%, 0 to 98%, and 0 to 89% for Covington, Murasaki, and Monaco respectively. These results demonstrate that canopy cover of sweetpotato can be a factor to determine yield reduction as Murasaki with the highest canopy had the least yield reduction. Findings from this study suggest that the threshold of 10% total yield reduction will be achieved maintaining sweetpotato fields weed-free for 24, 20, and 36 DAP for Covington, Murasaki, and Monaco respectively. This study also confirmed previous research that suggest maintaining less interference of weeds in sweetpotato production from 2 to 6 WAP.

Assessing the Risk of Injury Induced by Chloroacetamide and Diphenylether Mixing in Cole Crops Production . Wesley M. Bouchelle*[1], Elizabeth C. Maloney[2], Lynn M. Sosnoskie[2], Thierry E. Besancon[1]; [1]Rutgers University, Chatsworth, NJ, [2]Cornell University, Geneva, NY (66)

While herbicides are significant components of cole crop production programs, the limited number of registered products and their narrow spectrums of control can result in significant in-season escapes and require the need for costly hand-weeding. Because of crop injury potential, tank-mixes of *S*-metolachlor, a WSSA Group 15 herbicide, and oxyfluorfen are not advised. There are also recommendations to avoid applying oxyfluorfen in cabbage in the same season as Group 15 products. Field studies were conducted in spring of 2022 in New York and fall of 2022 in New Jersey to evaluate weed control and crop tolerance response to herbicide programs by chloroacetamide and diphenylether mixing in cole-crops. Treatments included an untreated weedy control, oxyfluorfen (18 g ha^{-1}) pre-transplant alone or followed by S-metolachlor or acetochlor (105 g ha^{-1}) one-day post-transplant, S-metolachlor or acetochlor (105 g ha^{-1}) alone or mixed with oxyfluorfen (18 g ha^{-1}) one-day post-transplant, S-metolachlor or acetochlor (105 g ha^{-1}) one-day post-transplant followed by oxyfluorfen (18 g ha^{-1}) fourteen days after transplanting. Treatments were arranged as a four replications split-plot design with herbicide treatments as the main plot and cole-crops (cabbage and broccoli) as the subplot. Weed coverage averaged for untreated cabbage and broccoli showed complete coverage (100%). When chloroacetamide and diphenylether herbicides mixes were applied weed coverage averaged only (4%) 28 days after transplanting. Injury averaged (16%) for cabbage and (20%) for broccoli when oxyfluorfen was applied post-transplant, regardless of tank mixing herbicides. Conversely, injury was less than 4% for cabbage and 7% for broccoli when S-metolachlor and acetochlor were applied alone post- transplant. Injury following herbicide applied alone post-transplant was greater with S-metolachlor (3%) than acetochlor (0.3%) for cabbage, but was = 0.7% for broccoli, regardless of herbicide. Average individual weight of cabbage (1.4 kg) and broccoli (275 g) heads was significantly greater than that for the untreated control (0.6 kg, and 135 g for cabbage and broccoli respectively) regardless of the herbicides applied.

Evaluation of Weed Control Efficacy and Crop Safety of the PPO-inhibiting Herbicide, Tiafenacil in California Orchard Cropping Systems. Guelta Laguerre*; Brad Hanson, UC Davis Department of Plant Sciences, Davis, CA (67)

Evaluation of Weed Control Efficacy and Crop Safety of the PPO-Inhibiting Herbicide Tiafenacil in California Orchard Cropping Systems Guelta Laguerre and Brad Hanson, Department of Plant Sciences, University of California, Davis Tiafenacil is a protoporphyrinogen oxidase (PPO) inhibitor that blocks chlorophyll biosynthesis and causes protoporphyrin accumulation, a highly photodynamic intermediate. Glufosinate inhibits glutamine synthetase (GS), a key enzyme for amino acid metabolism and photorespiration. Both herbicides ultimately lead to plant death through accumulation of reactive oxygen species (ROS). Field studies were conducted to determine the crop safety of tiafenacil on young almond, walnut, prune, and pistachio trees, as well as the weed control efficacy of broadleaf weeds and grass weeds relevant to California orchards. Tiafenacil was applied at 74 and 148 g ha^{-1} three times a year as a tree row strip treatment in prune, walnut, and pistachio that were less than one-year-old at the time of the first application. Tiafenacil was applied at 74, 148, and 222 g ha^{-1} to the tree row of one-year-old almond trees. Treatments were imposed in the spring of 2020, then reapplied in 2021 and 2022 such that all plots received a total of 7 treatments over a two-year period. Although no yield measurement was taken, fruits appear to be normal. There was no visual injury on any of the young trees between 30 and 700 days after initial treatment. Similarly, there was no treatment effects on tree trunk diameter among treatments even at the highest rate of tiafenacil. In a separate efficacy study on weed control efficacy, tiafenacil at 12 g ha^{-1} performed similarly with tiafenacil plus glufosinate in most instances, but control of both broadleaf and grass weeds numerically improved when tiafenacil was in mixture with glufosinate. In a greenhouse efficacy study, tiafenacil at 12 g ha^{-1} alone provided 98-100% control of barnyardgrass and 95-98% control of junglerice. There was no significant difference among tiafenacil alone or tiafenacil plus glufosinate. Results indicate that tiafenacil is an effective tool to help manage broadleaf and grass weeds relevant to California orchards. Orchard crop safety appears adequate with rates equivalent to 3-fold commercial use rates in almond and 2-fold in prune, walnut, and pistachio not affecting plant growth.

Evaluating Vision-Guided Spraying for Use in Vineyards. Bethany M. Schulteis*[1], Elizabeth C. Maloney[2], Emily J. Sausser[3], Terry Bates[1], Lynn M. Sosnoskie[2]; [1]Cornell University, Ithaca, NY, [2]Cornell University, Geneva, NY, [3]Lebanon Valley College, Annville, PA (68)

Grape suckers are undesirable because they are excess vegetation that diverts crop nutrients away from desirable tissues. Sucker removal can be achieved 1) by hand, which is time-consuming and expensive, 2) mechanically, which may be physically damaging to the vines, and 3) chemically, as a banded spray, using post- emergence contact herbicides to eliminate unwanted growth. While chemical sprays are efficient and effective tools for managing suckers, many growers want to limit herbicide use because of environmental impact concerns and changing public perception about pesticides. Indiscriminate sprays are also wasteful when suckers are not present and/or when weeds are not emerged between the vines. One possible strategy for reducing total herbicide applications and minimizing damage potential is the use of precision spray technology to target unwanted tissue. In 2022, we conducted a research trial with the specific goals of 1) evaluating the efficacy of herbicide applications using a vision-guided sprayer compared to a continuous banded treatment for grape sucker control and 2) determining if there is a difference in the amount of herbicide applied between the two application systems. The trial was initiated on May 25, at the Cornell Lake Erie Research Extension Laboratory (CLEREL) in weed-free Concord grapes. A CO2 pressurized backpack sprayer was used to make the continuous banded applications; the boom consisted of two 11002 nozzles set 48 cm apart and held at a height of 48 cm. The backpack application was made at a travel speed of 4 km hr-1. Vision-guided sprayer treatments were made using a customized Weed-It Quadro system with 4 sensor-nozzle units conveyed on a Polaris Sportsman ATV driven at 8 km hr-1. The Weed-It Quadro sensors detect the chlorophyll fluorescence and turn on individual, associated nozzles to deliver a targeted spray to living plant tissue. Sensors and spray nozzles on the Weed-It Quadro were also positioned at a height of 48 cm. Rely 280 (4.21 l ha-1), Aim EC (0.15 l ha-1), and DCC3825 (tiafenacil) (0.11 l ha-1) were applied using the backpack sprayer and the vision-guided sprayer at a spray volume of 187 l ha-1 to suckers no larger than 12.5 cm in length. Plots were 24 to 29 vines in length; vines were spaced 2.4 m apart within rows with between row separation of 2.7 m. Each herbicide-by-application system combination were replicated three times; an untreated check was also included for comparison. Water-sensitive spray cards were placed on select vines, and on the ground between select vines, at herbicide application, to evaluate spray deposition between precision and banded treatments. Sucker number and biomass per vine were quantified on June 23, 2022, approximately 1 month after herbicide applications. Data was analyzed with mixed models ANOVA using R statistical software. All herbicide by application system treatment combinations reduced ($P < 0.05$) sucker numbers and biomass per vine relative to the untreated check (2.2 suckers and 55.6 g vine-1). Results showed significant differences ($P < 0.05$) in sucker number and biomass vine-1 among herbicides. Averaged over spray systems, Rely 280 (1.8 suckers and 22.8 g vine-1) was less effective at sucker burndown compared to the Aim EC (1.2 suckers and 8.5 g vine-1) and DCC3825 (1.0 suckers and 8.3 g vine-1) treatments. Averaged over herbicides, there were no significant differences ($P > 0.05$) between the banded (1.3 suckers and 12.9 g vine-1) and Weed-It Quadro (1.4 suckers and 13.5 g vine-1) applications. Spray deposition between vines was reduced 35 to 50% in plots treated with the Weed-It Quadro relative to the broadcast application of herbicide. These results indicate that using a vision-guided sprayer, such as a Weed-It Quadro, could be useful for managing grape suckers while reducing the amount of applied herbicide in a vineyard.

Interference of Palmer Amaranth (*Amaranthus palmeri*) in Stevia (*Stevia rebaudiana*). Stephen Ippolito*[1], Katherine M. Jennings[1], David W. Monks[1], Sushila Chaudhari[2], David L. Jordan[1], Levi D. Moore[1], Colton D. Blankenship[1], Kira C. Sims[1], Fnu Chitra[1]; [1]North Carolina State University, Raleigh, NC, [2]Michigan State University, East Lansing, MI (69)

Stevia (*Stevia rebaudiana* Bertoni) is prized as a zero-calorie sweetener. After its relatively recent legalization as a food additive in the US, demand for stevia has increased. Prior research has shown that stevia yield and quality is reduced by competition for light under shade cloth. Therefore, field studies were conducted in Clinton, NC in 2021 and 2022 to determine the effect of Palmer amaranth (*Amaranthus palmeri* S. Wats) density on stevia yield. Stevia was planted at a density of 6.5 plants m^{-1} of row on raised beds covered in white polyethylene mulch. Palmer amaranth was established at 0, 0.3, 0.7, 1, 1.3, 2, 3.3, and 6.5 plants m^{-1} of row. Yield loss was fit to a rectangular hyperbola model. Predicted yield loss to Palmer amaranth competition was 16 and 66% at .3 and 6.5 plants m^{-1}, respectively.

Influence of Growth Stage and Dicamba Rate on Eggplant, Cucumber, and Snap Bean Tolerance and Yield Response. Maggie H. Wasacz[1], Lynn M. Sosnoskie[2], Mark VanGessel[3], Thierry E. Besancon*[4]; [1]Rutgers University, Wall Township, NJ, [2]Cornell University, Geneva, NY, [3]University of Delaware, Georgetown, DE, [4]Rutgers University, Chatsworth, NJ (70)

Following the introduction of dicamba-resistant (DR) soybean in 2017, concerns regarding dicamba off-target movement (OTM) onto sensitive crops, including vegetables, has increased. Field trials were conducted in New Jersey, New York, and Delaware to evaluate cucumber ('Python'), eggplant ('Santana'), and snap bean ('Caprice' and 'Huntington') injury and yield response to simulated dicamba drift rates. Crops were exposed to dicamba applied at 0, 0.056, 0.11, 0.56, 1.12, 2.24 g ae ha^{-1}, representing 0, 1/10,000, 1/5,000, 1/1,000, 1/500, and 1/250 of the maximum soybean recommended label rate (560 g ae ha^{-1}), respectively. Dicamba was applied either at the early vegetative (V2) or early reproductive (R1) stages. Minimal to no injury, vine growth reduction or yield loss was noted for cucumber. Dicamba was more injurious to eggplant with up to 22% to 35% injury 2 WAT at rate = 1.12 g ae ha^{-1}; however, only the highest dicamba rate caused 27% reduction of the commercial yield compared to the nontreated control. Eggplant also showed greater sensitivity when dicamba exposure occurred at the R1 than at theV2 stage. Snap bean was the most sensitive crop investigated in this study. Injury 2 WAT was greater for Caprice with dicamba = 0.56 g ae ha^{-1} applied at V2 compared to R1 stage, whereas similar difference occurred as low as 0.056 g ae ha^{-1} for Huntington. Compared to the nontreated control, reduction in plant height and biomass accumulation occurred for both cultivars at dicamba rate = 0.56 g ae ha^{-1}. Dicamba applied at 1.12 g ae ha^{-1} or greater resulted in 30% yield loss for Caprice whereas Huntington yield dropped 52% to 93% with dicamba = 0.56 g ae ha^{-1}. Caprice bean yield was not influenced by dicamba timing of application. Conversely, Huntington bean yield decreased by 8% following application at R1 compared to V2 stage.

Effect of S-metolachlor and Flumioxazin Herbicides on Sweet potato Treated with and without Activated Charcoal Applied Through Transplant Water. Colton D. Blankenship*[1], Katherine M. Jennings[1], David W. Monks[1], David L. Jordan[1], Stephen L. Meyers[2], Jonathan R. Schultheis[1], David H. Suchoff[1]; [1]North Carolina State University, Raleigh, NC, [2]Purdue University, West Lafayette, IN (71)

Flumioxazin and S-metolachlor are widely used on conventional sweet potato hectarage in North Carolina; however, some growers have recently expressed concerns about potential effects of these herbicides on sweet potato yield and quality. Previous research indicates that activated charcoal can improve crop safety and reduce herbicide injury in some conditions. Field studies were conducted in 2021 and 2022 to determine whether flumioxazin applied pre-planting and S-metolachlor applied before and after transplanting negatively affect sweet potato yield and quality when activated charcoal is applied with transplant water. The studies consisted of five herbicide treatments by two activated charcoal treatments. Herbicide treatments included two rates of flumioxazin, one rate of S-metolachlor applied immediately before and immediately after transplanting, and no herbicide. Charcoal treatments consisted of activated charcoal applied at 9 kg ha-1 and no charcoal. No visual injury was observed. There was no effect of herbicide or charcoal treatment on No. 1, marketable (sum of No. 1 and jumbo grades), or total yield (sum of canner, No. 1, and jumbo grades). Additionally, shape analysis conductedon calculated length-to-width ratio (LWR) for No. 1 sweet potato roots found no effect from flumioxazin at either rate on sweet potato root shape. However, both S-metolachlor treatments resulted in lower LWR of No. 1 sweet potato roots in 2021. Results are consistent with those of prior research and indicate that flumioxazin and S-metolachlor are safe for continued use in sweet potato at registered rates.

Tiafenacil Alone and in Combination with Preemergence Herbicides for Burndown and Residual Weed Control in Grapes. Elizabeth C. Maloney*, Lynn M. Sosnoskie; Cornell University, Geneva, NY (72)

Weeds are problematic in vineyards because they compete with the crop for water and nutrients. Additionally, weeds can serve as alternate hosts for pests and pathogens, contaminate mechanically harvested fruit, and interfere with vineyard operations. Weed growth can also impact the microclimates around vines, facilitating disease development and increasing the risk of spring frost. The critical period of weed competition is from bloom until veraison, although effective weed management begins in early spring (before crop budbreak and weed emergence) with the application of residual herbicides. The goal of this project was to describe the efficacy of spring-applied herbicide tank mixes to (1) eliminate emerged weeds and (2) extend in-season residual weed control following spring hill down. The study was initiated on 25 May 2022 at Cornell's Lake Erie Research and Extension Laboratory (CLEREL) in Seyval Blanc (3309 rootstocks) and Concord grapes. Soil at the site is a Chenango gravel loam (3.0-3.5% OM and 6.0-6.4 soil pH). Vine rows were spaced 2.7 m apart; individual vines were spaced 2.4 m apart within each row. Individual treatment plots were three vines in length and the under-row area was 0.9 m across. Treatments included applications of carfentrazone (as Aim EC, 0.15 l/ha) or tiafenacil (as Gamma, 0.11 l/ha) alone or in combination with flumioxazin (as Chateau, 0.88 kg/ha) or rimsulfuron (as Matrix, 0.29 kg/ha). All herbicide applications were directed to the base of the vines (on both sides of the vine row) using a twin nozzle (11002) shielded boom and a CO2-pressurized backpack sprayer calibrated to deliver at a rate of 187 l/ha. Standing weed cover was 15% or less at application. Per plot weed cover by class (e.g., total, annual broadleaf, annual grass, and perennial) was assessed using a scale ranging from 0% (no weeds present) to 100% (complete cover of the entire plot) at 14 and 28 days after treatment (DAT). On 28 June 2022, weed biomass was harvested from two 0.25 m^2 quadrats per plot and weighed. Averaged across grape varieties and reps, total weed cover was 30% at 14 DAT in the untreated check (UTC); annual broadleaf (e.g., AMAXX, AMBEL, CHEAL) weed cover was 17%, annual grass (e.g., DIGXX, SETXX, ECHCG) weed cover was 6%, and perennial (e.g., SOLCA) weed cover was 6%. Solo cafentrazone (17% cover) and tiafenacil (15% cover) treatments were not statistically different (P >0.05) from the untreated check with respect to total weed cover. Mean total weed cover was significantly reduced (P <0.05), relative to the untreated check, when rimsulfuron and flumioxazin were partnered with the burndown products (1% to 7% cover). Mean annual broadleaf weed cover was significantly reduced (P <0.05), relative to the check by all herbicide treatments (0% to 4% cover). Mean annual grass weed cover was not impacted by herbicide treatment, except in the tiafenacil plus rimsulfuron and flumioxazin plots (0%), which were significantly different (P <0.05) from the control. There were no statistical differences (P >0.05) among treatments with respect to mean perennial weed cover. Averaged across grape varieties and reps, total weed cover was 60% at 28 DAT in the untreated check (UTC); annual broadleaf weed cover was 34%, annual grass weed cover was 14% and perennial weed cover was 10%. Like the 14 DAT observation, mean total weed cover was significantly reduced (P <0.05), relative to the untreated check, when rimsulfuron and flumioxazin were partnered with carfentrazone and tiafenacil (11% to 28% cover). Mean annual broadleaf weed cover was significantly reduced (P <0.05), relative to the check, by all herbicide treatments (1% to 13% cover). Greater reductions in annual broadleaf weed cover were observed when carfentrazone and tiafenacil were partnered with flumioxazin and rimsulfuron (1% to 6% cover), as compared to carfentrazone and tiafenacil applied alone (13% cover). Mean grass weed cover was significantly reduced (P <0.05), relative to the untreated check, only when tiafenacil was partnered with flumioxazin and rimsulfuron (1% to 3% cover). There were no statistical differences (P >0.05) among treatments with respect to perennial weed cover. Mean weed biomass at 28 DAT was 206 g for the untreated check. Although carfentrazone or tiafenacil plus flumioxazin or rimsulfuron reduced biomass accumulation 42% to 85%, relative to the check, the differences were not significant (P >0.05). 2022 results agree with 2021 observations - tank mixes of carfentrazone or tiafenacil with either flumioxazin or rimsulfuron can reduce weedy vegetation, particularly annual broadleaves, under vines and provide residual weed suppression during the early part of the grape growing season. Perennial species will be largely unaffected and must be managed using alternative strategies.

In Search of New Weed Management Tools in Spinach, Cilantro, and Parsley for Georgia Farmers. A Stanley Culpepper*, Jenna C. Vance, Taylor M. Randell-Singleton; University of Georgia, Tifton, GA (73)

Vegetable production in Georgia currently has a farm gate value exceeding $1.2 billion. Weed control is a major production challenge in many of the more than 35 specialty crops that make up this farm gate value. Research efforts to identify and register additional herbicides that could be used safely by farmers is warranted. Studies were conducted to determine spinach, cilantro, and parsley response to various herbicides during 2021 and 2022. For each study, soils contained at least 87% sand and less than 0.7% organic matter, with production including overhead irrigation as needed. During the first 42 days after seeding cilantro (Santo), crop injury was at most 5% during any weekly evaluation from preemergence (PRE) applications of Linex at 1 pt/A, Caparol at 1 or 2 pt/A, and Warrant at 1 pt/A. Maximum injury ranged from 10 to 20% with Linex at 2 pt/A, Warrant at 2 pt/A, Prowl H20 at 24 oz/A, Treflan at 1 pt/A, or a three-way mixture of Caparol at 1 pt/A + Dual Magnum at 8 oz/A + Linex at 2 pt/A when applied PRE. Severe injury ranging from 30 to 68% was observed with Dual Magnum PRE at 12 or 24 oz/A. Evaluating treatments identical to those in the cilantro experiment, parsley (Jade) was more sensitive to several herbicides when compared to cilantro. For example, Dual Magnum at 12 or 24 oz/A killed the crop, Warrant at 2 pt/A caused 65% injury, and the combination of Caparol + Dual Magnum + Linex injured the crop up to 71%. For Treflan, Prowl, Linex at both rates, and Warrant at 1 pt/A, parsley injury was similar to that observed in cilantro. The spinach (Rushmore) experiment consisted of a different set of treatments. Ro-Neet, which is labeled for spinach use in some states but not in Georgia, caused at most 3% injury when applied PRE at 1, 2, 3, or 4 qt/A. Treflan at 1 pt/A, Dual Magnum at 12 oz/A, and Warrant at 24 oz/A PRE injured the crop less than 5% throughout the first 38 days after planting while injury from Prowl H20 at 1.5 pt/A was severe reaching 68%. Similar to the response with Prowl, severe injury ranging from 66 to 100% was observed with either Linex or Caparol when applied PRE at 1 or 2 pt/A. This study also included four postemergence treatments applied to spinach at 1.5 inches in height having 2 leaves; Dual Magnum at 12 oz/A injured the crop less than 5%, Prowl at 24 oz/A caused 31% injury, and Linex at 1 or 2 pt/A killed the crop.

Current Status of Weed Management in Washington State Vegetable Production. Rui Liu*[1], Tim Waters[2]; [1]Washington State University, Prosser, WA, [2]Washington State University, Pasco, WA (74)

Washington State is one of the leading states for vegetable production in the United States. Majority of the production comes from the irrigated agricultural regions in the Columbia Basin. Vegetable varieties include sweet corn, potatoes, onions, carrots, green peas, etc.Weeds pose a serious challenge to vegetable production. They compete for nutrients, sunlight, water, and other resources, which results in vegetable yield reduction and quality degradation. Herbicides are often used as effective tools for weed management, however, options are limited for vegetable crops. The process of labeling new herbicides in vegetable crops has been difficult and time-consuming. There is an increasing need for Integrated Weed Management (IWM) programs in Washington State vegetable production. The aim of this presentation is to provide an overview of the current status of weed management approaches and to explore future weed science research directions for vegetable production in Washington State.

Defoliation Timing and Growth Responses of Seedling Field Bindweed and Honeyvine Milkweed.
Melissa McClements*[1], Lynn M. Sosnoskie[2]; [1]Cornell, Geneva, NY, [2]Cornell University, Geneva, NY (75)

Field bindweed (Convolvulus arvensis L.) is a perennial broadleaf vine in the Convolvulaceae, or morningglory family. bindweed produces vertical roots that can reach depths of 6 to 9 m, depending on soil type. It also has an extensive, adventitious, lateral root system that usually occupies the top 30 cm of soil. Lateral roots are reported to grow 35 to 100 cm away from the parent plant before secondary vertical roots are formed. Field bindweed also produces long-lived seeds that can lie dormant in the soil for dozens of years. While the root system of field bindweed is primarily responsible for facilitating the species' survival and spread, the species can also produce significant quantities of long-lived seed. Many management recommendations suggest controlling field bindweed at the seedling stage when the species is most susceptible to control measures. However, this developmental phase may be short-lived because seedlings can rapidly develop root buds that can form new crowns and tap roots. Studies were conducted at Cornell AgriTech to describe field bindweed regrowth following seedling defoliation to describe the regenerative potential of the species and determine when the species assumes characteristics of a perennial plant. Bindweed seed collected from Wenatchee, WA (2018), Fresno, CA (2019), and Geneva, NY (2020), were scarified for 30 seconds with boiling water. Bindweed seed were planted in soilless media in 19 l plastic pots and grown in a greenhouse set to 25 C, 12:12 light:dark. Aboveground biomass was removed from emerged plants at either 2, 4, 6 or 8 weeks after seedling emergence (WAE) by clipping at the soil line. Samples were then destructively harvested 2 weeks after the defoliation treatments and the above and belowground biomass weighed. Each population by defoliation timing treatment was replicated four times. Across the three populations, 50% to 100% of bindweed seedlings emerged for two weeks survived defoliation treatments and were able to establish new crowns. All bindweed emerged for four or more weeks regrew successfully after top growth was removed. Bindweed management recommendations state that seedlings are easy to control relative to mature vines. However, results from this study suggest that bindweed seedlings that are emerged less than four weeks can survive events that totally remove aboveground biomass.

Hemp and Weed Competition in New York. Lynn M. Sosnoskie*, Elizabeth C. Maloney; Cornell University, Geneva, NY (76)

Despite a history of production dating back to European colonization, hemp can, essentially, be considered a new crop for the US because of a previous legal moratorium on crop production. This suspension also impacted research opportunities; consequently, there is limited knowledge available regarding the impact of weeds and weed management on hemp growth, development, and yield. This includes evaluations of the performance and safety of herbicides for the control of unwanted vegetation. On 12 and 13 June 2022, we established trials to 1) describe the critical weed free period (CWFP) in transplanted hemp The trial was conducted at Cornell's AgriTech campus in Geneva (42.88°N, 77.01°W). Soil at the site is a Honeoye loam with 6.3 soil pH and 2.7% organic matter content. Dominant weeds at the site were common ragweed (Ambrosia artemisiifolia, AMBEL), yellow foxtail (Setaria pumila, SETPU), and barnyardgrass (Echinochloa crus-galli, ECHCG); minor, infrequently occurring species included white clover (Trifolium repens, TRFRE), Canada thistle (Cirsium arvense, CIRAR), and volunteer sorghum (Sorghum bicolor). Hemp varieties included Cherry Wine (High Grade Hemp Seed, Longmont, CO), Sour Space Candy (Jack Hempicine LLC, Independence, OR), PhotoCBD (QuickSpectrum) (Phylos Bioscience INC, Portland, OR), and Bubbatonic (Kayagene LLC, Holister, CA). CWFP treatments (hand weeded for 0, 1, 2, 4 and 6 weeks, weed free) were replicated five times. Individual plots were 12 m wide by 15 m with 9 plants per variety planted in four rows (one row for each variety) spaced 3 m apart. In-row plant spacing was 1.2 m. The trial was harvested 15 to 21 September 2022, approximately 16 weeks after it was established. Averaged across all varieties, hemp biomass (kg/plant) was significantly ($P<0.05$) affected by the duration of weed competition. Per plant yields were reduced >80% when weeds were allowed to compete with hemp season-long/almost season-long (0.53 kg/plant); mean weed cover exceeded 90%. Weed competition that was allowed to start at 4 weeks after transplanting resulted in approximately a 30% yield loss (2.80 kg/plant); mean weed cover ranged from 10 to 20%. Biomass production (3.91 kg/plant) was maximized when weeds were suppressed for at least 6 weeks after transplanting; weed cover in these plots was less than 5% at harvest. Ultimately, individual hemp plants, even in weed free plots, did not grow as large as expected due to environmental factors; the site received 12.5 cm of rainfall of which 50% occurred during the first two weeks after transplanting. The field was not irrigated, and drought stress was significant mid to late season. During the time the trial was being conducted, 13.31% to 47.92% of the Northeast US was classified as D0 (abnormally dry, potential slowing of crop growth) and 2.38 % to 23.76% of the region was classified as D1 (moderate drought, some damage to crops). However, results indicate that transplanted hemp is sensitive to competition and preventing weed establishment for six weeks was necessary to reduce competitive interactions. The trial will be repeated in 2023 to better describe the critical weed free period under NY production conditions.

Dicamba Residue Persistence in Processing Tomato. Stephen L. Meyers*[1], Jeanine Arana[2], Brandi C. Woolam[3], Nathaly Belen Vargas Arroyo[1], Laura D. Rodriguez Baquero[4], Luz Aide Cardona Giraldo[1]; [1]Purdue University, West Lafayette, IN, [2]Purdue University/Horticulture, West Lafayette, IN, [3]LSU AgCenter, Alexandria, LA, [4]University of Wisconsin-Madison, Madison, WI (77)

There is zero tolerance for dicamba and dicamba metabolite residue in tomato (*Solanum lycopersicum* L.) fruit following exposure to dicamba. Field trials were conducted in 2020 and 2021 to determine the persistence of dicamba and metabolite (5-hydroxy dicamba and 3,6-dichlorosalicylic acid [DCSA]) residue in processing tomato shoots and fruits. Dicamba was applied 49 d after transplanting at 0, 0.53, 5.3, and 53 g ae ha^{-1} . Tomato plants were harvested 5, 10, 20, 40, and 61 d after treatment (DAT). No 5-hydroxy dicamba was recovered from any sample. In 2020, the DCSA metabolite was detected from tomato shoot tissue when dicamba was applied at the 53 g ha^{-1} rate at 0 (14 µg kg^{-1}), 5 (3 µg kg^{-1}), and 20 DAT (5 µg kg^{-1}) and from tomato fruit tissue at 53 g ha^{-1} at 20 (2 µg kg^{-1}) and 61 DAT (2 µg kg^{-1}). In 2021, DCSA was not detected from tomato shoot or fruit tissues at any harvest date. By 5 DAT, dicamba was only detected from tomato shoot tissues treated with 53 g ha^{-1}. At 0 DAT, dicamba residue was detectable only from tomato fruit on plants treated with 53 g ha^{-1}. Tomato fruit dicamba residue from plants treated with 5.3 g ha^{-1} had a predicted peak of 19 µg kg^{-1} at 11.3 DAT. Tomato fruit dicamba residue from plants treated with 53 g ha^{-1} decreased from 164 to 8 µg kg^{-1} from 5 to 61 DAT. Furthermore, this study confirms that dicamba is detectable from tomato fruits at 61 DAT following exposure to 5.3 or 53 g ha^{-1} dicamba. Growers who suspect dicamba exposure should include tomato fruit tissue with their collected sample or sample tomato fruits separately.

Weed Surveys of Lima Bean (*Phaseolus lunatus)* Fields in United States Production Systems. Yudai K. Takenaka*[1], Jed Colquhoun[2], Scott Lukas[3], Pavle Pavlovic[4], Barbara Scott[5], Mark VanGessel[5], Martin Williams[6]; [1]University of Illinois at Urbana-Champaign, Urbana, IL, [2]University of Wisconsin-Madison, Madison, WI, [3]Oregon State University, Hermiston, OR, [4]University of Illinois, Champaign, IL, [5]University of Delaware, Georgetown, DE, [6]USDA-ARS, Urbana, IL (78)

In order to understand the scope of weed problems in lima bean (Phaseolus lunatus), weed communities were surveyed near the time of crop harvest from 2019 to 2022 in two major U.S. production regions. A total of 67 fields were surveyed: 33 fields in the Midwest region and 34 fields in the Atlantic region. Crop management practices used in the fields were obtained from collaborating farmers and vegetable processors. Approximately 52 weed species were observed, with significant differences in weed communities between the two regions. In the Midwest, dominant species included amaranth species (Amaranthus spp.), carpetweed (Mollugo verticillata), common chickweed (Stellaria media), common lambsquarters (Chenopodium album), common purslane (Portulaca oleracea), fall panicum (Panicum dichotomiflorum), and velvetleaf (Abutilon theophrasti). In the Atlantic region, dominant species included annual bluegrass (Poa annua), carpetweed, common chickweed, fall panicum, ivyleaf morningglory (Ipomoea hederacea), nightshade species (Solanum spp.), and smooth pigweed (Amaranthus hybridus). Widely adopted mechanical weed control methods included preplant tillage and interrow cultivation. The most widely used preemergence herbicides were halosulfuron-methyl, pendimethalin, and *S*-metolachlor. The most widely used postemergence herbicides were bentazon, clethodim, imazamox, and sethoxydim. Despite adoption of several chemical and mechanical control methods, this survey revealed extensive weed problems in many production fields. Greater diversification of integrated weed management systems is needed, particularly targeting Amaranthus species and fall panicum.

Michigan's Experience with Sugarbeet Tolerance to Acifluorfen. Brian J. Stiles II*, Gary Edward Powell, Christy L. Sprague; Michigan State University, East Lansing, MI (79)

The lack of postemergence herbicide options to control the glyphosate-resistant *Amaranthus spp.*, Palmer amaranth and waterhemp, is a concern for sugarbeet growers. In the past, sugarbeet has shown some tolerance to the Group 14 herbicide, acifluorfen. In soybean, acifluorfen's strengths include control of *Amaranthus spp.*, eastern black nightshade, and common ragweed. Field research (2018-2022) on sugarbeet tolerance to acifluorfen was conducted to justify and support Section 18 labeling of Ultra Blazer in Michigan sugarbeet. Early studies examined acifluorfen combinations with glyphosate (0.4 kg ae ha^{-1}) at rates ranging from 0.14 to 0.43 kg ai ha^{-1} to sugarbeet at the 2-, 6-, and 12-leaf stages. This research showed that acifluorfen should only be applied after sugarbeet was at the 6-leaf stage. We have continued to examine the effects from multiple applications of acifluorfen to 6- and 12-leaf sugarbeet and combinations with the tank-mixture partners of clopyralid, ethofumesate, s-metolachlor, dimethenamid-P, or acetochlor at 6-leaf sugarbeet. In many cases, the addition of tank-mix partners has increased sugarbeet injury and in the case of acifluorfen combination with ethofumesate sugarbeet yield loss was also observed. When acetochlor was tank-mixed with acifluorfen sugarbeet injury was less. While sugarbeet injury is a concern with acifluorfen, applications to sugarbeet at the 6-leaf stage or greater could be an option for postemergence glyphosate-resistant waterhemp control.

Efficacy of Organic Turf and Landscape Herbicides in Montgomery County, Maryland. Kelly M. Nichols*; University of Maryland Extension, Derwood, MD (80)

In 2019, a law went into effect in Montgomery County in central Maryland which prohibited the use of synthetic pesticides on home lawns. In the fall of 2022, a preliminary study was conducted in order to compare the efficacy of a few organic herbicides. Five organic herbicide treatments were included: ammoniated soap of fatty acids (30 L ai ha^{-1}), chelated iron [iron hydroxyethylenediaminetriacetic acid (FeHEDTA)] at three different rates (8.6, 10.6, and 21.3 L ai ha^{-1}), and caprylic acid + capric acid (10.8 + 8.8 L ai ha^{-1}). One synthetic treatment of 2,4-D ester + mecoprop-p (MCPP) + dicamba + carfentrazone-ethyl (400 g ae ha^{-1} + 150 g ae ha^{-1} + 40 g ae ha^{-1} + 30 g ai ha^{-1}) was also included. Treatments were placed in a randomized complete block design with four replications. The first application was made on September 16, 2022, and a second application of the organic herbicides was made four weeks later on October 14, 2022. Dandelion (*Taraxacum officinale* F.H. Wigg) and buckhorn plantain (*Plantago lanceolata* L.) were the two most prevalent weeds. Weed counts were taken three days before the first application and two weeks after the second application. Visual ratings were taken three days after each application as well as weekly up to one week after the second application. The organic herbicides provided 20% or less control of buckhorn plantain throughout the study. The middle and high rates of FeHEDTA provided 71 and 86% control, respectively, of dandelion at three days after the first application. However, this quickly dropped to 3 and 35% at one week after the first application. At three days after the second application, all rates of FeHEDTA provided significantly similar control (61-99%). At the end of the study (one week after the second application), all rates of FeHEDTA, along with the synthetic treatment, provided significantly similar control (65-99%). Weed counts indicated a similar or increased number of dandelion and plantain across most plots. The high rate of FeHEDTA and the synthetic treatment were the only two treatments that decreased the number of buckhorn plantain; however, none of the weed counts were significantly different across treatments. While fall applications are not ideal for contact herbicide applications on perennial plants, FeHEDTA may have a role as a part of an organic weed control program for lawns. Future research will be expanded to include additional herbicides, rates, and spring and summer application times. kellyn@umd.edu

Ethofumesate Resistant Annual Bluegrass in Oregon Grass Seed Production Systems. Vera Vukovic*, Aaron J. Patton; Purdue University, West Lafayette, IN (81)

Prolific seed production of annual bluegrass allows for rapid development of herbicide resistance. Ethofumesate resistant annual bluegrass plants have been known to exist in grass seed production in

Oregon, but their prevalence and distribution are not well documented. Therefore, a dose-response experiment was initiated to determine the potential level of ethofumesate resistance in *Poa annua* L. in seed production systems. Seed from 55 annual bluegrass populations was obtained from three sources: seed production fields (31 populations), seed cleaning process (6 populations), and seed testing prior to retail distribution (18 populations). Additionally, two populations, one with known resistance and one with known susceptibility, were identified in preliminary testing and used as controls in this experiment. Individual seedlings from each population were transplanted into separate cone-tainers, grown to a size of 2–3 tillers in the greenhouse, and then sprayed with ethofumesate. Ten doses of ethofumesate were applied using a compressed air track spray chamber: 0, 0.6, 1.1, 2.8, 5.6, 8.4, 11.2, 16.8, 22.4, and 44.8 (with 1.1 to 2.2 kg ai ha^{-1} as the label application rates for perennial ryegrass). Biomass of plants was collected 28 days after application. The resistance to susceptible ratio of populations across all sources was ranging from 0.5 to 5.5, with the most resistant population with an effective dose to kill 50% of the plants (ED50) at 13.1 kg ai ha^{-1}. The highest level of resistance from annual bluegrass populations found in production fields, removed during the seed cleaning process, or found in seed testing lots had ED50 values of 12.1, 13.1, and 9.4 kg ai ha^{-1}. These results indicate that annual bluegrass resistance is found in seed production fields and some seed lots. Strategies for managing herbicide resistance should be implemented by growers.

Glufosinate and Clethodim Efficacy on Bermudagrass. Joe C. Neal*, Christopher D. Harlow; North Carolina State University, Raleigh, NC (82)

Demand has increased for landscape weed control recommendations that do not include glyphosate. Glufosinate is non-selective but generally regarded to have lower efficacy on perennial grasses like bermudagrass. Clethodim is labeled for postemergence grass control in landscape plantings but requires multiple applications. Tank mixes of glufosinate and clethodim have been reported to be antagonistic, but the antagonism was overcome with higher doses. Glufosinate and clethodim efficacies on bermudagrass were evaluated, both individually and in combination. The experiment was conducted at the North Carolina State University Horticultural Field Laboratory in Raleigh. Two doses of each herbicide were included. The doses selected were the lowest and highest labeled doses for weed control in landscape settings. Nonionic surfactant was included in all treatments. Herbicides were applied with a CO_2-pressurized sprayer using TeeJet 8008 flat fan nozzles and calibrated to deliver 50 GPA. A single application was made on August 10, 2021. The test area was an established low-maintenance lawn area dominated by common bermudagrass, *Cynodon dactylon*. Grass control was evaluated for two months after application, and percent green vegetation was estimated using the Canopeo app. Both doses of glufosinate visibly reduced green cover within 3 days and reduced green cover to less than 10% by 7 days. Regrowth following glufosinate applications was observed starting 13 days after application. Clethodim symptom development was slower, with over 80% green cover remaining at 7 days and about

50% green cover by 13 days. By 40 days after application, bermudagrass had fully recovered from all treatments. The addition of clethodim did not improve bermudagrass control with glufosinate. However, the addition of clethodim to the low dose of glufosinate did improve crabgrass control. No antagonism between glufosinate and clethodim was observed in this study. These results confirm prior observations that control of perennial grasses like bermudagrass without glyphosate likely will require multiple herbicide applications.

Site-Specific Treatment of Kyllinga in Golf Fairways Using a Remotely Piloted Aerial Application System. Ubaldo Torres*[1], Daniel E. Martin[2], Bholuram Gurjar[1], Bishwa B. Sapkota[1], Andrew W. Osburn[1], Chase M. Straw[1], Muthukumar V. Bagavathiannan[1]; [1]Texas A&M University, College Station, TX, [2]United States Department of Agriculture, College Station, TX (83)

Green kyllinga (*Kyllinga brevifolia* Rottb.) is a common and problematic weed in managed turfgrass systems. The growth habit and morphology of this species create ununiform appearance and rough surface characteristics in golf courses. Once established, green kyllinga may spread rapidly, making it difficult to control. Site-specific treatment using a remotely piloted aerial application system (RPAAS) may be an effective tool for controlling green kylllinga. Additionally, aerial RGB images captured using an Unmanned Aerial System (UAS) can be utilized for weed detection and subsequent treatment using the RPAAS. An experiment was conducted in 2021 at Briarcrest Golf Course in Bryan, TX to investigate two objectives: 1) detect green kyllinga from UAS-based aerial RGB imagery, and 2) evaluate the effectiveness of using an RTK-GPS guided RPAAS for site-specific herbicide treatment of green kyllinga in golf course fairways. Treatments included 1) an RPAAS-based herbicide application with weeds located using manually obtained geo-coordinates, 2) an RPAAS-based herbicide application with weeds located based on image analysis derived geo-coordinates, 3) herbicide application using a backpack sprayer, and 4) an untreated check. Sulfosulfuron (Certainty®) herbicide was used in the experiment for the aerial and ground applications. First, the accuracy of RPAAS targeting green kyllinga patches based on exact coordinates obtained through manual recording was evaluated. Additionally, visible plant injury was recorded at 4 weeks after treatments, and efficacy between the RPAAS treatments and conventional backpack application were compared. The overall accuracy of green kyllinga detection using image analysis was 78%. Compared to backpack applications, the efficacy of aerial applications based on weeds detected using image analysis was 12% lower. No differences were observed between the backpack and the RPAAS-based application with weeds located manually, suggesting that further improvements are necessary with the weed detection model used here. This experiment provided necessary directions for further improving the detection accuracy for green kyllinga, and demonstrates the promise of using RPAAS for site-specific weed management in turfgrass systems.

Non-Target Site Resistance Mechanisms in Annual Bluegrass (*Poa annua*). Ronald R. Rogers*, Travis Gannon, Khalied Ahmed, Mathieu LeCompte; North Carolina State University, Raleigh, NC (84)

Annual bluegrass (*Poa annua* L.) is a cool season annual grass that is problematic in turfgrass systems. Annual bluegrass ranks third globally in terms of the number of herbicide sites of action for which a weed species has developed resistance. The presence of herbicide-resistant annual bluegrass increases agronomic, economic, and social pressure on turfgrass managers. Mechanisms of herbicide resistance in weeds can be classified in two different manners, target site resistance (TSR) and non-target site resistance (NTSR). NTSR is characterized by reduced herbicide absorption and/or translocation or increased metabolism as well as sequestration of herbicide in a particular plant part. TSR and NTSR mechanisms can combine at the individual level to produce populations with higher resistance levels. Four unique annual bluegrass populations, three of which were suspected resistant, were collected from the North Carolina, South Carolina, and Tennessee. These populations underwent a comprehensive dose response study to confirm resistance and susceptibility to glyphosate. To determine if these populations exhibited NTSR mechanisms, an absorption and translocation study utilizing ^{14}C-glyphosate was conducted. Absorption studies showed rapid absorption of ^{14}C-glyphosate 4 hours after treatment for the Tennessee population when comparted to the susceptible population (58.6% > 30.1%). Additionally, greater recovery of ^{14}C-glyphosate in the treated leaf was observed for the Tennessee population when compared to the susceptible population (70.7% > 50.5%).

Characterization of Potential Quinclorac-Resistant Smooth Crabgrass (*Digitaria ischaemum*) Population. Claudia A. Rutland*[1], Bridgette C. Johnson[2], James D. McCurdy[3], Joseph S. McElroy[2]; [1]Auburn University Department of Crop, Soil, and Environmental Sciences, Auburn, AL, [2]Auburn University, Auburn, AL, [3]Mississippi State University, Starkville, MS (85)

Quinclorac resistance in turfgrass systems presents a significant issue for weed management practices, as it removes a beneficial grass-in-grass herbicide from use. Smooth crabgrass (Digitaria ischaemum) is among the species evolving resistant to auxin mimic herbicides, quinclorac included. A potential resistant smooth crabgrass population, dubbed Tier17, was identified at the Auburn University Sports Surface Field Laboratory and a dose-response experiment was conducted in order to confirm its resistance status. An increasing rate of Drive XLR8 was applied to a putative resistant population and a susceptible Digitaria ciliaris population in a randomized complete block design with four replications. Digitaria ciliaris was selected for the susceptible population as no known susceptible Digitaria ischaemum population was available. Percent injury ratings were taken at 7, 14, 21, and 28 days after treatment (DAT), and fresh weight was taken at 28 DAT. Based on the data collected, the Tier17 population can be considered resistant and can be used for further molecular studies.

Halosulfuron and Sulfentrazone Combinations for False-green Kyllinga (Kyllinga Gracillima) Control.

Daniel P. Tuck*[1], Matthew T. Elmore[1], Charles A. Silcox[2]; [1]Rutgers University, New Brunswick, NJ, [2]Amvac Chemical, Lincoln University, PA (86)

Halosulfuron and Sulfentrazone Combinations for False-green Kyllinga (*Kyllinga gracillima*) Control. D.P. Tuck*[1], M.T. Elmore[1], C.A. Silcox[2]; [1]Rutgers University, New Brunswick, NJ, [2]AMVAC Chemical, Lincoln University, PA Field experiments were conducted in 2019 and 2020 to evaluate combinations of halosulfuron and sulfentrazone for postemergence false-green kyllinga (*Kyllinga gracillima*) control in cool-season turfgrass. Treatments consisted of halosulfuron-only, sulfentrazone-only, and sulfentrazone + halosulfuron mixtures at various rates, application numbers, and application intervals. Imazosulfuron and sulfentrazone + carfentrazone were included for comparison. The 2019 experiment was initiated on 17 June at the Rutgers Adelphia Research and Extension Farm in Freehold, NJ, on perennial ryegrass (*Lolium perenne*) and false-green kyllinga of three local biotypes planted in 2015. The 2020 experiment was initiated on 9 June at the Four Seasons Golf Course in Lakewood, NJ, on a mixed stand of cool-season turfgrass naturally infested with false-green kyllinga. Both sites were maintained at 6.4 cm mowing height. Treatments were arranged in a randomized complete block design with four replications and applied using a CO_2-powered single AI9504EVS nozzle sprayer with a water carrier of 420 L ha^{-1} and NIS at 0.25% v/v. False-green kyllinga control was evaluated visually on a 0 (i.e., no control) to 100 (i.e., complete control) percent scale relative to a nontreated control. Data were analyzed using the GLIMMIX procedure in SAS ($P = 0.05$) and Fisher's protected LSD test was used to separate means. In 2019, when applied singly, sulfentrazone + halosulfuron at 210 g ha^{-1} + 35 g ha^{-1} and sulfentrazone-only at 210 g ha^{-1} controlled false-green kyllinga more than halosulfuron-only (35 to 47 g ha^{-1}). A single application of sulfentrazone-only at 280 g ha^{-1} controlled false-green kyllinga more than a single application of sulfentrazone + halosulfuron at 280 g ha^{-1} + 47 g ha^{-1}. When applied in six biweekly applications, sulfentrazone + halosulfuron (35 to 70 g ha^{-1} + 6 to 12 g ha^{-1}) controlled false-green kyllinga more than sulfentrazone-only (35 to 70 g ha^{-1}). At 16 weeks after initial treatment (WAIT), six biweekly applications of sulfentrazone-only at 35 and 70 g ha^{-1} resulted in 36% and 82% false-green kyllinga control, respectively. All sequential sulfentrazone + halosulfuron programs provided >90% false-green kyllinga control. Six biweekly applications of sulfentrazone + halosulfuron at 35 + 6 g ha^{-1}, respectively, applied the least amount of yearly active ingredient while providing excellent false-green kyllinga control. In 2020, single applications of all treatments except for imazosulfuron resulted in poor false-green kyllinga control at 17 WAIT. Three sequential applications of sulfentrazone + halosulfuron at 140 + 23 g ha^{-1}, respectively, controlled false-green kyllinga more than two sequential applications. At 17 WAIT, six biweekly applications of sulfentrazone + halosulfuron resulted in 97% false-green kyllinga control compared to 75% control for six biweekly applications of sulfentrazone-only. Six biweekly applications of sulfentrazone provided similar control to three sequential applications of sulfentrazone + halosulfuron. Among sulfentrazone + halosulfuron treatments with a yearly active ingredient total of 420 + 70 g ha^{-1}, respectively, six biweekly applications provided greater false-green kyllinga control than two monthly applications and three biweekly applications.

Phenotypic and Reproductive Trait Diversity of *Poa annua* L. (Annual Bluegrass) in Common Garden Experiments Across Multiple Ecoregions in the United States. Andrew W. Osburn*, Muthukumar V. Bagavathiannan; Texas A&M University, College Station, TX (87)

Annual bluegrass (*Poa annua* L.) is commonly considered one of the most problematic weeds in various turfgrass systems. This weed drastically increases management costs and greatly reduces the aesthetic value of turf. Annual bluegrass is a high-risk species for developing resistance to multiple herbicide modes of action (MOA), with documented resistance to as many as 12 MOAs. This species exhibits diverse growth habits and has been found infesting all major landmasses, including Antarctica, demonstrating a high adaptive potential. However, not much is known about the phenotypic diversity in annual bluegrass across various environmental gradients in the U.S. turf systems. In the Spring of 2021, ten unique populations of annual bluegrass (originating from across the U.S.) grown in seven common gardens across a geographical gradient in the U.S. were studied to determine, among others, the reproductive and seed trait diversity in the populations. In each garden, each population had 4 replications, and 10 individuals within each replication (thus 40 plants total per population). Once plants began to senesce, seed from each plant was harvested and total seed production per plant was evaluated. Up to 100 seed from each plant was placed in an incubation chamber for 28 days and then treated with 1.0% tetrazolium hydrochloride solution to evaluate germination potential and dormancy. Additionally, seedlings from each plant were placed on growth papers to determine seedling vigor. Results suggest that the growing environment had little influence on the germination or dormancy of annual bluegrass, nor did it influence the seedling vigor. Rather, there were very high levels of intra-population diversity, which might be fundamental to their adaptive potential in any given environment. Results provide unique insights into the diversity characteristics of this troublesome species, suggesting that robust management programs are vital to thwart the rapid adaptation of this species to management practices.

Mowing Height Influences False-Green Kyllinga (*Kyllinga gracillima*) Establishment in Perennial Ryegrass Turf. Matthew T. Elmore*, Daniel P. Tuck; Rutgers University, New Brunswick, NJ (88)

The objective of this experiment was to evaluate the effect of mowing height and nitrogen (N) fertilizer on false-green kyllinga (*Kyllinga gracillima*) encroachment into perennial ryegrass (*Lolium perenne*) turfgrass. False-green kyllinga was established vegetatively into kyllinga-free perennial ryegrass in 2015. From spring 2017 to 2019, three different mowing height treatments (2.8, 6.4, and 10.0 cm) were evaluated in a complete factorial with two different N fertilizer programs (50 or 100 to 150 kg N ha^{-1} yr^{-1}). Mowing height treatments comprised whole plots (3 by 4 m) and fertilizer treatments were applied to sub plots (3 by 2 m). Kyllinga was established in each sub plot with six 5-cm diameter plugs of three biotypes (two plugs of each biotype). Plots were maintained by weekly mowing during the growing season with a push-behind rotary mower bench-set to the mowing heights. Fertilizer for the low N treatment (50 kg N ha^{-1} yr^{-1}) was applied in two annual applications at 25 kg ha^{-1} in spring (May) and autumn (late September to mid-November). Fertilizer for the high N (100 kg N ha^{-1} yr^{-1} in 2017 and 150 kg N ha^{-1} yr^{-1} in 2018 and 2019) treatment was applied in two to three annual applications at 50 kg ha^{-1} in spring, early summer, and autumn. The area covered by each kyllinga plug was evaluated by measuring the shortest and longest diameter of each plug 1 August, 6 November 2017, and 31 May 2018. On these dates, the percent increase of decrease in area of the kyllinga plug was calculated compared to initial measurement from May 2017. After May 2018, presumably due to false-green kyllinga establishment from seed, individual kyllinga plugs were no longer discernable and grid intersect counts (six grid counts of 100 intersects each per sub plot) were used to determine percent false-green kyllinga cover. Plug area increased from May to August 2017 more at the 6.4 cm mowing height (220%) than at 2.8 or 10.0 cm mowing height (115 and 163%, respectively). One kyllinga biotype increased in area 224%, more than the two other biotypes (127 and 134%) from May to August 2017. From August 2018 to August 2019, kyllinga cover according to grid intersect counts was greatest at the 2.8 cm mowing height and decreased as mowing height increased. False-green kyllinga cover was 61% at the 2.8 cm mowing height and 35% at the high mowing height in August 2018. Kyllinga cover increased slightly with nitrogen rate in August 2018 and May 2019 but not August 2019. A mowing-by-nitrogen-rate interaction was detected only in August 2019 but was of little biological significance. This research indicates kyllinga cover increases as perennial ryegrass mowing height decreases. This is presumably due to greater kyllinga establishment from seed when turfgrass is less competitive.

Injury to Nursery Crops from Herbicide Residues in a Contaminated Irrigation Water. Ada C. Montoya*, Joe C. Neal, Wesley Everman, Christopher D. Harlow, Travis Gannon; North Carolina State University, Raleigh, NC (89)

Container production of nursery crops is an intensive system requiring numerous management inputs including frequent irrigation. Most container nurseries use surface ponds as source water for overhead irrigation and each day apply about 20,000 gallons of water per acre. Recently, surface ponds at several nurseries in North Carolina were contaminated with herbicides applied to nearby rights of way sites resulting in extensive damage and crop losses. Diagnosis of causes was hampered by limited information on the bioactivity of low concentrations of herbicides in irrigation water. Therefore, this investigation was conducted to compare crop injury on four common crop species from log-dilutions of broad-spectrum herbicides in an irrigation water. *Thuja* x 'Green Giant', *Hypericum inodorum* 'Miracle Grandeur', and *Rosa x* 'Knockout Double Red' were the selected for the test because they exhibited significant injury in nursery sites and to represent a diversity of plant types. Green beans (*Phaseolus vulgaris*) were also included as an easily available bioassay species. The crop plants were potted to 4-L plastic pots using a hammer milled pine bark substrate and top dressed with slow-release fertilizer. Plants were grown in containers for about 8 weeks to ensure they had established root systems before treatment. Beans were seeded and were about 4 inches tall at the time of treatment. Herbicides included in the study were aminocyclopyrachlor, imazapyr and metsulfuron methyl. Detection limits in water for these herbicides are: 1 ppb, 0.2 ppb, and 0.008 ppb, respectively. Setting the detection limit as the 1X value, pots were drench-irrigated with 0.1X, 1X, 10X, or 100X dilutions, 250 ml of treatment solution per pot. Treatments were applied three times over a 10-day period; on other days pots received about 1.5 cm of overhead irrigation. Injury to plants was visually evaluated and above-ground fresh weights were measured. Hypericum, roses, arborvitae and green beans were injured by aminocyclopyrachlor at all concentrations, including 0.1X. Species differed in response to imazapyr. Hypericum and rose were injured at the 1X dilution, but arborvitae was not injured at concentrations <100X. No injury was observed from any metsulfuron-methyl treatments. These data suggest that the detection limits for metsulfuron-methyl in water are sufficiently sensitive to be useful in crop injury diagnostic testing. However, the data also suggest that the detection limits for aminocyclopyrachlor or imazapyr in water are not low enough to determine if injurious concentrations are present in irrigation water.

Guardrail Bareground Applications in KY: Spring or Fall Timing? Control Next Season? Joseph Omielan*; University of Kentucky, Lexington, KY (90)

For highway safety guardrails need to be kept clear of visual obstructions. Usually that means maintaining a vegetation free zone underneath them. Applications of broad spectrum residual herbicides in the spring have become the mainstay for bareground maintenance operations in combination with a broad spectrum post-emergent herbicide like glyphosate. Fall applications using newer less soil mobile chemistries may have some advantages. The trial was established beside cable barrier near Morehead, KY in 2021. In the spring 24 treatments with 3 replications were arranged in a randomized complete block design and in the fall 8 treatments in common were established as well. Treatments were applied at 233 L/ha onto 1.2 m by 3.2 m plots on May 27 and Sept 14, 2021. All common treatments included

Roundup ProMax or Rodeo (glyphosate) for post-emergence control. Our long-term standard treatment has been the older, high use rate herbicide Sahara (diuron + imazapyr). Newer low use rate products and combinations tested included Escort (metsulfuron), Milestone (aminopyralid), Method (aminocyclopyrachlor), Perspective (aminocyclopyrachlor + chlorsulfuron), Viewpoint (aminocyclopyrachlor + metsulfuron + imazapyr), and Esplanade (indaziflam). Visual % bareground ratings were taken 63 (7/29), 110 (9/14), 158 (11/1), and 357 (5/19/2022) days after treatment (DAT) for the spring application and 48 (11/1), 247 (5/19/2022), 309 (7/20), and 377 (9/26) DAT for the fall application. Data were analyzed using ARM software and treatment means were compared using Fisher's LSD at p = 0.05. All common treatments with residual herbicides had more bareground than the control at the first assessment date at both times of application (63 and 48 DAT for spring and fall, respectively). By the last assessment for the spring application (357 DAT) only two treatments had more bareground (53 to 47%) than the control (22%). These were combinations with the high rate of Esplanade and either Milestone or Viewpoint. By the last assessment for the fall application (377 DAT) the three treatments with the most bareground (48 to 63%) were those combinations with the high rate of Esplanade and Method or Milestone or Viewpoint. Similar levels of bareground control can be achieved a year after either spring or fall application..

Evaluation of Hemp Dogbane (*Apocynum cannabinum*) Response to Auxinic Herbicides. Kayla L. Broster*, John D. Byrd, Jr.; Mississippi State University, Starkville, MS (91)

Hemp dogbane (*Apocynum cannabinum*) is a native perennial of North America, commonly found in pastures, roadsides, and non-cropland areas. It has been noted that dogbanes can be toxic to livestock, which makes finding effective control options necessary for producers to prevent harm or death of livestock, and to encourage growth of desired forages. Previous studies have focused on control with auxin herbicides. A field study was conducted at three locations in Clay Co., Mississippi, with dense well-established populations of hemp dogbane. A hayfield and fallow field were used for the years 2020, 2021, and 2022. An additional site was added in 2022 on a population along the roadside. Population density was measured by taking individual stem counts within a meter square, with 2 subsamples per plot, prior to treatment application. Eighteen treatments arranged in a randomized complete block design with four replications were applied to hemp dogbane, that was juvenile to mid-flower, with a CO_2 backpack sprayer calibrated to deliver 281 L ha^{-1}. Treatments included: Vastlan (triclopyr choline salt) at 4.67 and 2.34 L ha^{-1}, Remedy Ultra (triclopyr butoxy ester) at 4.67 and 2.34 L ha^{-1}, Garlon 3A (triclopyr triethylamine) at 6.23 and 3.12 L ha^{-1}, Trycera (triclopyr acid) at 6.52 and 3.26 L ha^{-1}, MezaVue (aminopyralid + picloram + fluroxypyr) at 2.34 and 1.17 L ha^{-1}, DuraCor (aminopyralid + florpyrauxifen-benzyl) at 1.46 and 0.73 L ha^{-1}, Method (aminocyclopyrachlor) at 1.32 and 0.66 L ha^{-1}, Surmount (picloram + fluroxypyr) at 7.01 and 3.5 L ha^{-1}, and Tordon K (picloram) at 4.67 and 2.34 L ha^{-1}. Visual injury ratings were made at two and four weeks after application (WAT). Data were analyzed using R

Studio (Version RStudio 2022.07.1), subjected to ANOVA, to compare treatments across location and year, then means separated using Fisher's LSD (a = 0.05). Differences in responses across populations and years, lead to locations and year being analyzed separately. There were no differences between treatments at the roadside location, due to the variability in response. At the fallow field location, years were analyzed separately, but 4WAT DuraCor and Method were less effective than all other treatments, except for Tordon K at 2.34 L ha^{-1} in 2021, for control of hemp dogbane. The hayfield location had similar results in 2020 4WAT with DuraCor and Method the least injurious. DuraCor performed similarly in 2021 at 4WAT. In 2022 4WAT, DuraCor at 0.73 L ha^{-1} and Garlon 3A at 3.12 L ha^{-1} resulted in less control in comparison to all treatments except for DuraCor at 1.46 L ha^{-1}. In terms of differences in stand counts, no differences were observed two years out at either the hayfield or fallow field locations. However, at the fallow location, DuraCor at 0.73 L ha^{-1} had less reduction in stand from all treatments one year after application. These results show that the response of hemp dogbane to herbicides may be population dependent and vary from year to year. However, visual observation would suggest the use of Surmount, MezaVue, Remedy, or Trycera for the control of hemp dogbane.

Environmental Variables Influence Native Plant Composition of Virginia Roadside. Shawn Askew*, Clebson G. Goncalves; Virginia Tech, Blacksburg, VA (92)

A presidential memorandum in 1994 set policy for all federal lands to incorporate native plants in landscaping and pursue avenues to promote native plant use and suppress exotic invasive species across aligned state government agencies. Since this time, the Federal Highway Administration has issued several policy statements to encourage native plant use on roadsides. Virginia has the third largest roadway system in the U.S. spanning seven different EPA Level III Ecoregions. Pursuant to federal policy and public pressure, the Virginia Department of Transportation funded a project to assess roadside vegetation with respect to plant origin status and environmental factors that may foster some plant communities over others. At 490 sites, we identified 67,330 plants comprising 616 unique species. The Shannon Diversity Index was calculated for 2,450 10-m transects, and indicated that plant biodiversity was higher on low-maintenance distal backslopes compared to high-maintenance road edges, shoulders, and ditches. Plant biodiversity was also higher on secondary roads compared to primary roads. Aside from management intensity, another contributor to lower plant biodiversity in road edges and ditches is likely that unique habitats favor a smaller selection of species. The road edge and ditch were characterized by extreme soil compaction and moisture. In addition, the roadway is a source of deicing salt, automobile exhaust, and other possible pollutants. When averaged across 490 sites statewide, soil in the road edge had an average pH of 6.9 and salinity of 0.43 ds m^{-1}, and these parameters decreased 0.21 and 0.04, respectively, with each meter distance away from the pavement. The road edges were dominated by a select group of plants, such as *Eragrostis pectinacean, Hordeum pusillum, Polygonum aviculare, Eleusine indica,* and *Poa annua,* which are known to thrive in compacted sites. Road edges also had lower vegetative cover than all other transects, with a statewide average of 56% cover on all roads and only 46% cover on primary roads. Many ditches were predominately aquatic plants, such as a variety of sedges and rushes, *Diodia virginiana, Persicaria punctata,* and *Polygonum caespitosum var. longisetum,* which were typically not found as frequently in other transects. Barring the ditch, which had dramatically less compaction than other transects, soil compaction, as measured by soil penetration pressure, decreased as the distance from the pavement increased. Soil moisture, as measured by volumetric water percentage, increased from the road edge to the ditch and then decreased from the ditch to the distal backslope. Soil temperature was not significantly influenced by the topographical transect. As the primary roadside species sown by VDOT, *Schedonorus arundinaceus* (tall fescue) was found most frequently on all topographical transects and ranged from 158 sites on distal backslopes to 379 sites (77%) of the state on shoulders. It was noted that two recently introduced invaders from Asia, *Microstegium vimineum* and *Arthraxon hispidus*, were among the top 10 most frequently encountered introduced grasses on all transects, except the road edge. Another recent invasive grass species from Africa, *Anthoxanthum odoratum,* was among the top five most commonly introduced grasses in all transects and the second most common in ditches, backslopes, and distal backslopes.

Does PRE Herbicide Timing Affect the Control of Foxtail Species (*Setaria* Spp.) in Southern Forages.

Nicholas T. Basinger*[1], Chris Bocz[1], Sambit Shome[2], Hannah C. Lindell[3]; [1]The University of Georgia, Athens, GA, [2]University of Georgia, Athens, GA, [3]The University of Georgia Crop and Soil Science, Athens, GA (93)

Foxtail species (*Setaria* spp.) are becoming increasingly problematic in southeastern forages. Often forage producers recognize they have a foxtail problem when the plant produces a seed head resembling its namesake. PRE herbicides often are an effective option to limit the impact of foxtail in pastures but due to weather and other issues, the timing of these applications is in most cases too late to control early emerging cohorts. This study was conducted at the J. Phil Campbell Research and Education Center (JPCREC) in 2022 in a foxtail-infested bermudagrass pasture. Applications of indaziflam (70 g ai ha^{-1}), pendimethalin (4480 g ai ha^{-1}), and hexazinone (840 g ai ha^{-1}) were applied bi-weekly beginning February 1, with the last application on April 1. Cumulative emergence counts were conducted biweekly beginning April 1 through September 1 within ¼ m^2. Three harvests were conducted in June, July, and September and botanical separations (foxtail, bermudagrass, other) occurred from 2 ¼ m^2. Foxtail cumulative emergence was reduced with pendimethalin when compared to other herbicide treatments and the non-treated check regardless of timing. February 1 applications of pendimethalin delayed foxtail emergence until later in the season. The composition of bermudagrass was greatest for the first two harvests for indaziflam and pendimethalin treatments, but foxtail composition across all treatments was similar for the third harvest. Results suggest that producers may need to apply pendimethalin earlier than their current application timings. But may need to implement other methods of control (mowing, mid-season herbicide application) to further reduce seed production of later emerging foxtail cohorts.

Do Forage Grasses Tolerate Graminicides? Dwight Lingenfelter*, John M. Wallace; Penn State University, University Park, PA (94)

For many years, weed control in grass pastures and hayfields in Pennsylvania and the Northeast region was primarily focused on broadleaf weeds. However, within the past decade or so, more farmers are struggling with weedy annual grasses such as large crabgrass (*Digitaria sanguinalis*), yellow foxtail (*Setaria lutescens*), Japanese stiltgrass (*Microstegium vimineum*), roughstalk bluegrass (*Poa trivialis*), and others. Currently, there are no effective POST herbicides that provide control of these and other grass species in a grass forage setting. Pendimethalin, quinclorac, and aminopyralid have some activity on weedy grasses however, not all species are controlled, and there can be issues with application timing, product cost, and other negative factors. Therefore, field studies with an emphasis on herbicides that could potentially provide foliar control of weedy grasses were conducted in 2021 and 2022. Two studies were conducted in an established orchardgrass (*Dactylis glomerata*) field to evaluate crop tolerance – one was an initial herbicide screen of numerous products and the other was a rate titration study of selected herbicides. Other studies were established in a non-crop area infested with Japanese stiltgrass to study efficacy. Each study was arranged in a randomized complete block design with three replications. Treatments were applied to orchardgrass in the spring (avg. 8 inches tall) and to the stiltgrass in early summer (1 to 4 inches tall). Herbicides were applied with a small-plot, CO2-backpack sprayer system that delivered 15 GPA thru TeeJet AIXR110015 nozzles. Treatments included in one or both studies: metsulfuron (0.028 lb ai/A), imazapic (0.03, 0.06, 0.09, and 0.13 lb), imazethapyr (0.03, 0.05, 0.06, and 0.09 lb), imazamox (0.047 lb), chlorsulfuron (0.047 lb), primisulfuron (0.036), mesosulfuron (0.013 lb), pyroxsulam (0.016 lb), pinoxaden + fenoxaprop (0.08 lb), clodinafop (0.06 lb), quizalofop (0.014 and 0.028 lb), florpyrauxifen (0.026 lb), and topramezone (0.02 lb). Appropriate adjuvants were included where necessary. Visual crop injury and weed control ratings were collected. Preliminary results from the herbicide screen/orchardgrass tolerance study indicate that early season ratings (i.e., before June 8 cutting) for all treatments caused 20 to 95% visual crop injury, except for chlorsulfuron, florpyrauxifen, and topramezone (<15%). However, ratings of regrowth after harvest, indicated that all herbicides caused less than 15% crop injury by late July, except mesosulfuron and quizalofop (37-90%). Late season results from the rate titration study of imazapic and imazethapyr revealed that imazapic caused 9 to 35% orchardgrass injury, while imazethapyr caused 0 to 9% injury; each was respective to the increasing treatment rates. Results from the 2021 stiltgrass control study indicate that only imazapic and imazethapyr provided greater than 80% control by late season, all other treatments only provided 20 to 70% control. A rate titration study of imazapic and imazethapyr in 2022 revealed that imazapic provided 90 to 98% control of Japanese stiltgrass across all rates, while imazethapyr provided 76 to 94% control. In summary, additional studies of herbicide tolerance on several grass forage species and applications timings will be required. Furthermore, fine tuning of herbicide rates and application timings to control problem grassy weeds will be necessary before determining which herbicides should be recommended for this utility. Currently, selected grass forages are listed on the imazapic (Plateau) and imazethapyr (Pursuit) labels, so using these to provide control of weedy grasses in grass forages might be the most feasible option. Many of the other herbicides evaluated would require additional tolerances and approvals before supplemental labeling could occur.

Invasive Aquatic and Native Marine Vegetation on the Mississippi Barrier Islands. Gray Turnage*, Andrew Sampe; Mississippi State University, Starkville, MS (95)

Seagrass beds serve as feeding and spawning habitat for many species of marine fauna. Barrier island lagoons may act as refugia for seagrass species that are vital components of marine ecosystems in the Gulf of Mexico and the Mississippi Sound. Lagoons may also be invaded by aquatic invasive plant species (AIS) capable of surviving brackish environments (10-20 ppt salinity). Loss of ecosystem function due to AIS has been documented in many brackish and saline habitats, suggesting that AIS invasion could further reduce seagrass abundance in island lagoons such that these sites no longer function as nursery populations that replenish seagrass beds in the Mississippi Sound. The goal of this project was to survey the aquatic and marine vegetation in select lagoons on the four major barrier islands (Petit Bois, Horn, Ship, and Cat) of Mississippi to 1) determine the aquatic/marine plant community of each lagoon and 2) determine if AIS were present in these systems. Twelve lagoons and six reference sites in the MS Sound were identified and surveyed in 2020 using the point intercept method to generate baseline density of seagrasses. All lagoons had vegetation present with Shoal grass (native) and torpedograss (AIS) being the most prevalent species. Seagrass was recorded in 58% of lagoons with 70% of those populations potentially serving as nursery populations for offshore sites adjacent to each island and approximately 40% serving as nursery sites across the Mississippi Sound. However, AIS were present in 66% of lagoons and co-occurred with seagrass populations in 25% of lagoons. To our knowledge, this is the first work to document AIS co-occurring with seagrasses on the Mississippi Barrier islands. Future reduction of AIS in lagoons may allow for subsequent colonization by seagrasses which could further increase the abundance of nursery seagrass populations in the Mississippi Sound.

Vegetation Damage by Ecosystem Disturbance Plants '*Sicyos angulatus* and *Humulus japonicus*' in Korea. Soo-Hyun Lim*; Korea National Open University, Seoul, South Korea (96)

Sicyos angulatus (Bur-cucumber) and *Humulus japonicus* (Japanese hop) are an invasive, noxious, and problematic vine weeds and mainly distributed in non-cropping areas including field margin, roadside, parkside in Korea. Due to its early seedling emergence starting from February and vigorous vining growth, it has caused serious vegetation damage and soil erosion. However, little information is available on how much they affect vegetation. Therefore, vegetation analysis was conducted to investigate the impacts of *S. angulatus* and *H. japonicus* on vegetation diversity. Two neighboring sites, one with *S. angulatus* and *H. japonicus* infested and the other without *S. angulatus* and *H. japonicus* were selected in each location and their vegetation parameters including plant density and plant cover of all plant species in 1 m x 2 m quadrat were assessed. The total number of locations was 50. Using the vegetation parameters assessed, Simpson's diversity index and other indices associated with vegetation diversity were calculated. As a results, the sites with *S. angulatus* and *H. japonicus* invested showed much lower vegetation diversity than the sites without *S. angulatus* and *H. japonicus*. Interestingly, the sites with *S. angulatus* and *H. japonicus* infested showed much lower proportion of perennial plants, while the sites without *S. angulatus* and *H. japonicus* maintained relatively higher proportion of perennial plants. Our finding thus suggests that the interference of *S. angulatus* and *H. japonicus* reduce vegetation diversity with a particular decrease in perennial plants. The decreased perennial plants in the site may eventually cause soil erosion. Once perennial plants were eliminated from a site by *S. angulatus* and *H. japonicus* interference, it takes much longer to restore perennial plants in the site, so *S. angulatus* and *H. japonicus* management should also consider not only how to control them but also how to restore vegetation with perennial plants.

P23: Join the WSSA Plant Names Committee. Theresa Piskackova*; Czech University of Life Sciences Prague, Prague 6 - Suchdol, Czechia (Czech Republic) (97)

The plant names committee is responsible for updating nomenclature of common and scientific names used within the WSSA and maintaining the composite list of names. Over the past two years the committee undertook an extensive review of the composite list of names, especially to compare discrepancies between the USDA plants database and the WSSA composite list of names. Our next task will be to make a full review of the scientific names and synonyms that are referenced in the list. If this is work that you would be interested to help with, consider joining the P23 committee.

WeedR: An R Package and Shiny Application for Weed Scientists. Lucas Kopecky Bobadilla*[1], Patrick Tranel[2]; [1]University of Illinois, Champaign, IL, [2]University of Illinois, Urbana, IL (98)

Experimental design and data analysis in weed science are critical steps toward reliable and reproducible research allowing the delivery of crucial findings to society. However, both steps can take a long time to conduct, especially when the use of programming languages like R is necessary. R is a powerful and open-source programming language that contains essential tools for weed scientists to analyze data. The learning curve for mastering R can discourage future users, shifting them towards paid/subscription software solutions. Our goal was to start the development of a free R package for weed scientists that do not intend to master R but still desire to use its power to conduct their analyses. To achieve this goal, we built a Shiny app (designated WeedR), which provides an interactive way for users to explore data and results in real-time through their web browsers. WeedR consists of a dashboard containing options for users to develop their experimental design, run a linear regression with posthoc analysis, and build dose-response analysis via the DRC package with a one-function step. WeedR will also provide support for genetic segregation analysis for structured populations to define inheritance patterns of weedy traits. For advance R users, WeedR functions were built fully within the Tidyverse, allowing users to customize data frames and plots. By using WeedR, weed scientists will have access to the power of R without any coding experience. A future upgrade will focus on adding more functionalities, improving functions, and hosting the app on a server to allow more users.

Weeds of the Northeast - 2nd Edition. Joe C. Neal*[1], Richard H. Uva[2], Joe M. DiTomaso[3], Antonio DiTommaso[4]; [1]North Carolina State University, Raleigh, NC, [2]Seaberry Farms, LLC, Federalsburg, MD, [3]UC Davis, retired, Davis, CA, [4]Cornell University, Ithaca, NY (99)

Weeds of the Northeast, 2nd Edition is a practical guide to the identification of over 500 common, ecologically, and economically important weeds of the northeastern United States, Mid-Atlantic States, upper Mid-West, and southern Canada. It is also a reference book for those aspects of weed biology and ecology important to weed management. Since first publication of *Weeds of the Northeast* in 1997, many weeds have expanded their range or prevalence in the northeastern United States. For example, waterhemp, Japanese stiltgrass, and garlic mustard were rare in the region in 1996 but are now widespread. These and many other species of importance have been added to this second edition. Furthermore, species of importance in the mid-Atlantic states and upper Midwest have been added to improve the utility of the book in those regions. Since the publication of the first edition there have been many changes in plant nomenclature. In this second edition we followed the naming conventions in the *Flora of North America* and the WSSA composite list of weeds. For genera that are not yet published in the *Flora of North America*, the most recently published authoritative source was used. The book contains several tools for identifying weeds including a dichotomous key based upon vegetative characters and comparative tables for related species. Each species account includes descriptions of seedling and mature traits, modes of reproduction and spread, common habitats and distributions, photographs of habit and identifying characteristics, and a brief description of similar and related species. The 1st edition has been an important tool in weed management education and decision-making. It is the authors' hope that this 2nd edition will be even more useful to weed science educators and weed management practitioners.

Comparing the Accuracy of Available Weather Resources forPesticide Applicators in Texas. Rohith Vulchi*[1], Zachary S. Howard[1], Don Conlee[1], Scott A. Nolte[2]; [1]Texas A&M University, College Station, TX, [2]Texas A&M AgriLife Extension, College Station, TX (100)

Favorable weather conditions are necessary for successful synthetic auxin herbicide application. Accurate weather-measuring instruments and web/mobile applications are available to applicators which provide reliable information to determine if weather conditions are favorable for chemical applications at the field level. A study conducted across the US by Nolte et.al, 2021 concluded handheld devices showed equal/greater accuracy compared to digital apps in predicting weather at closer distances to state-maintained weather stations. In this study, we compared different weather tools against state-maintained weather stations in Texas up to 15 miles for air temperature, wind speed, and wind direction. Weather tools behaved similarly with regard to air temperature up to 10 miles away from the state-maintained weather station. Digital weather apps and websites showed greater accuracy with regard to wind direction and also for wind speeds up to 15 miles compared to handheld devices. Hand-held weather tools were less accurate with wind direction at farther distances. Increased variation with distance could be because of the inability of handheld devices to capture data from wide areas, where digital tools can be helpful. These results indicate digital tools are reliable to gather weather information for fields farther from state-maintained weather stations compared to hand-held devices.

Things That Work: Integrated Weed Management Teaching Modules That Stood the Test of Time. David A. Mortensen*; University of New Hampshire, Durham, NH (101)

The Importance of Communicating Agricultural Science and Technology – the Role of CAST and You!
Anthony L. Witcher*[1], Jill Schroeder[2]; [1]Tennessee State University, Mcminnville, TN, [2]New Mexico State University, Las Cruces, NM (102)

The Council for Agricultural Science and Technology (CAST) was established in 1972 as a nonprofit group and is composed of many scientific/professional societies, universities, nonprofit organizations, companies, and individuals. CAST convenes and coordinates networks of experts to assemble, interpret, and communicate credible, unbiased, science-based information to policymakers, the media, the private sector, and the public. The wide distribution of CAST publications to non-scientists enhances the education and understanding of topical agricultural issues to the public. At a time when the internet has made it more difficult to separate accurate and inaccurate information, it has never been more critical for reputable sources to deliver trusted, non-partisan information from authorities in their respective fields. Our primary work is the publication of task force reports, commentaries, special publications, and issue papers written by volunteer experts from many disciplines covering topics related to Food Sciences, Agricultural Technology, Animal Sciences, and Plant and Soil Sciences. Weed science is well represented on the CAST Board of Representatives Plant Agriculture and Environmental Issues Work Group with members from WSSA, NEWSS, NCWSS, WSWS, and APMS. Board representatives welcome input from the members of our organizations on how CAST can enhance communication of science to more audiences and on topics of concern that need to be addressed through CAST publications. Becoming a CAST individual member is easy, free for students, and one of the best ways to stay informed on the science and technology of food and modern agriculture.

Undergraduate Student Research Experience -- from Proposal to Poster Presentation in *AGRN 479 Weed Science*. Mark L. Bernards*; Western Illinois University, Macomb, IL (103)

Weed Science (AGRN 479) is a capstone course in the crop science curriculum at Western Illinois University (WIU). Students who enroll in this class have completed *Integrated Pest Management* (AGRN 373). Major projects in AGRN 373 include making a pictorial collection of weeds, insects and diseases, and completing multiple lab activities/assignments focused on sprayer calibration and pesticide-related calculations. After teaching AGRN 479 several years, I wanted to make it distinct from AGRN 373 (e.g., less emphasis on a weed collection and sprayer calibration and math). In addition, a new greenhouse facility had been built on the WIU farm 7 minutes from campus. My goal was to create an experience in AGRN 479 where each student could apply the scientific method and participate in a unique research project. Over the next few years the following framework was developed, primarily using the weekly 2-hour lab period. Week 1 – As a class we read and discuss an article from a WSSA journal to demonstrate the IMRAD format and how to effectively read a research paper. Students brainstorm questions they have related to weed biology and management, and are paired into teams of two. Week 2 – Research teams come with a specific research question they would like to study. This lab period is held in a

Malpass Library classroom. A university librarian teaches how to conduct effective searches using article databases (such as AGRICOLA), and how to find journal articles through the library's resources. Students apply what has been taught, and leave the library with at least 5 articles related to their research question. Week 3 – Team meetings are scheduled with the instructor. Teams come with a brief written research proposal, including background information, a specific research question and hypothesis, and a potential treatment list. The proposal is reviewed and critiqued, and treatment and materials lists are finalized. Week 4 – Teams submit final written proposals, and studies are implemented. Weeks 5-10 – Postemergence herbicide treatments (where part of the protocol) are applied using a single-tip chamber sprayer in the greenhouse, and treatment effects are measured at predetermined intervals. Week 11 – Team meetings are held with the instructor to analyze data, interpret the statistical analysis, and discuss ways of presenting the data in the poster. Week 12 – Team meetings are held with the instructor to review and revise the poster. Finalized posters are sent for printing. Week 13 – Students practice presenting their research to other class members, and then present their posters at the WIU Undergraduate Research Day poster session. Week 14 – Research posters are hung in the hallway outside the classroom. Some general observations are: 1) not all projects show treatment differences, but are still valuable for understanding that the scientific method; 2) some projects produce interesting and useful results that have been incorporated into classes and outreach talks; 3) it is time-intensive for the instructor, especially during weeks 3-4 and 11-12; and 4) nearly all the students are proud of what they have accomplished and demonstrate greater appreciation for the scientific method and its importance in advancing weed science knowledge and practice.

Weed Management Using Unmanned Aerial Systems. Dhiraj Srivastava*[1], Vijay Singh[2], Vipin Kumar[1]; [1]Virginia Polytechnic Institute and State University, Painter, VA, [2]Virginia Tech, Painter, VA (104)

Timely monitoring and identification of weeds helps growers select the best tools for weed management tactics. However, there is a need to integrate conventional and new-age weed management practices to establish more efficient and sustainable weed management strategies than the technologies that have been used hitherto. Recent advances in Unmanned Aerial Systems (UAS) and machine learning (ML)/ artificial intelligence (AI)-based technologies have opened new opportunities for real-time weed monitoring and precision management of crop fields to address some of the challenges faced by growers these days. This research emphasizes developing a digital library of aerial images of weeds and crops representative of diverse weather conditions to build reliable weed detection models. The current research study was conducted between 2021 to 2022 in the Eastern Shore region of Virginia where UAS-based red, green, and blue (RGB) images were acquired at different growth stages of Soybean and Common ragweed (Ambrosia artemisiifolia) at research sites and farmers' fields. The objective was to implement the different classifications and object detection method architectures for evaluating the best model for real-time deployment. For classification, our custom architecture based on depth-wise separable convolutions and Bayesian optimization outperformed state-of-the-art transfer learning Big Transfer (BiT) model. However, the object detection approach would be the best model for real-time deployment in weed detection considering speed, accuracy, and memory consumption. Therefore, 'You Only Look Once'(YOLOv6) was tested and found to be the best-performing object detection model to detect common ragweed in soybean with precision and recall greater than 81% at an average inference speed of 7.56 milliseconds on NVIDIA GeForce RTX 2080 GPU. Further, weed pixels were correlated with crop yield to understand if the yield of soybean can be predicted ahead of time by estimating the number of common ragweed pixels in the image. This research will enable efficient tracking of weed populations at the farm-scale and time for effective weed management.

Designing and Testing an Ultra-Low Volume Spray System for Site-Specific Herbicide Application Using Ground Robots. Bholuram Gurjar*, Sunil Kumar, Joe Johnson, Robert Hardin, Muthukumar V. Bagavathiannan; Texas A&M university, College Station, TX (105)

A new micro-volume spray prototype was developed for the precise application of herbicides using ground robots. The concept combines an array module of micro-volume nozzles, an electronic circuit for controlling individual nozzles, and the morphology of target plants. The spray droplets are formed by a series of injectors controlled by the electric circuit. When the system passes over the target plant, it opens only those valves directly over the plant, while the rest will remain closed. The specific objectives of this study were to: 1) develop an array of nozzles and electronic control systems; 2) develop a computational fluid dynamics (CFD) model to simulate the complex flow conditions, droplet stability, and leaf retention properties; and 3) evaluate the weed control efficacy of the developed system. Palmer amaranth (*Amaranthus palmeri*) and johnsongrass (*Sorghum halepense*) were the target weed species in this study, representing a troublesome broadleaf and a grass species, respectively, and glufosinate herbicide was used to control them at the 3- to 4-leaf growth stage using the micro-volume spray system. The developed system can deliver a minimum volume of 6.28 microliters per 2 milliseconds (ms) at 2.5105 Pa, with a sub-centimeter accuracy on the target plant. The developed control circuit has a minimum response time of 7.2 ms, which makes the entire system much faster, i.e., nozzles can be opened and closed in less than 1 ms. The droplets change their shape and tend to spread after falling on the leaf surface. Due to gravity and inclined leaf surface, the sprayed liquid may move downwards. In this regard, CFD allows for the prediction of droplet stability and leaf retention properties. CFD also helps to predict the influence of droplet diameter, height of injection, and the effect of leaf surface roughness on coverage and spray efficiency. Results showed that the micro-volume spray system developed here had 90 to 95 percent weed control efficacy for Palmer amaranth and johnsongrass. This micro-volume spray system represents great progress toward addressing the environmental impacts and health risks associated with synthetic herbicides in conventional agriculture, while providing a valuable tool for effectively applying non-synthetic herbicides in organic systems. This prototype will be field-tested and optimized for site-specific herbicide applications using an unmanned ground vehicle.

Clethodim Antagonism from Dicamba on Volunteer Corn Control in Soybean Was Resolved by Using a Dual Tank/Boom Sprayer. Marcelo Zimmer*[1], Bryan G. Young[2], William G. Johnson[1], Tristen H. Avent[3], Jason K. Norsworthy[3], Diego J. Contreras[4], Wesley Everman[4], Michael M. Houston[5], Lauren M. Lazaro[6], William L. Patzoldt[6]; [1]Purdue University, West Lafayette, IN, [2]Purdue University, Brookston, IN, [3]University of Arkansas, Fayetteville, AR, [4]North Carolina State University, Raleigh, NC, [5]Blue River Technology, Greenville, MS, [6]Blue River Technology, Sunnyvale, CA (106)

Volunteer corn (*Zea mays* L.) is one of the most important weeds in soybean production in the Midwest. High infestations of volunteer corn in rotational crops can occur due to corn harvest conditions associated with high wind speeds, hail damage, and defective or improper adjustment of harvest equipment. Herbicide options for control of volunteer corn in soybean fields depend on the herbicide-tolerance traits of corn hybrids planted in the previous growing season. Many farmers utilize corn hybrids with stacked resistance to both glyphosate and glufosinate, limiting herbicide options for postemergence volunteer corn control. ACCase-inhibiting herbicides, such as clethodim or quizalofop, are often used to control volunteer corn that is tolerant to both glyphosate and glufosinate. Previous research indicated that the addition of synthetic auxin herbicides, such as 2,4-D or dicamba, antagonizes clethodim efficacy on grass weeds. Furthermore, research has shown that the addition of glyphosate to postemergence clethodim applications can further antagonize the control of glyphosate-resistant volunteer corn. The widespread adoption of dicamba- and 2,4-D-resistant soybeans across the United States has resulted in increased reports of failed ACCase herbicide applications to control volunteer corn and other grasses. One approach for reducing the antagonism of postemergence herbicides is by splitting herbicide applications in time and/or space. The effectiveness of splitting herbicide applications in space depends on whether the antagonistic response is caused by tank incompatibility or by a physiological response within the target weed. A field experiment was conducted at four locations (Arkansas, Indiana, Mississippi, and North Carolina) to evaluate the control of volunteer corn with clethodim-based postemergence treatments when applied using a dual tank/boom sprayer developed by Blue River Technology in comparison to tank mixtures using a single tank/boom. Three clethodim rates (25.5, 51, and 102 g ai ha^{-1}) plus *S*-metolachlor (1390 g ai ha^{-1}) were applied alone as broadcast treatments, or in combination with dicamba (560 g ae ha^{-1}) plus glyphosate (1260 g ae ha^{-1}) in either a single tank/boom or as split applications using the dual tank/boom sprayer. The addition of dicamba + glyphosate to clethodim + *S*-metolachlor in the same spray solution reduced volunteer corn control at 21 days after treatment by 6 to 71% compared to clethodim + *S*-metolachlor applied alone. Splitting the application of clethodim + *S*-metolachlor from dicamba + glyphosate by using the dual tank/boom system completely resolved the antagonism. Data for volunteer corn height and density reduction at 28 to 35 days after application showed similar trends. These results suggest that the antagonistic response caused by dicamba plus glyphosate on the control of volunteer corn with clethodim is caused by tank incompatibility rather than a physiological plant response in the target weed. In addition to providing site-specific herbicide application, the dual tank/boom system may alleviate herbicide compatibility problems associated with herbicide mixtures.

The Critical Period for Weed 'Seed' Control: the Concept and its Application. Charles M. Geddes*[1], Adam S. Davis[2]; [1]Agriculture and Agri-Food Canada, Lethbridge, AB, Canada, [2]University of Illinois, Urbana, IL (107)

Increasing herbicide-resistant weed issues globally warrants reevaluation of fundamental weed control objectives. While the mitigation of crop yield loss remains critically important, reducing the proliferation of herbicide-resistant weeds warrants limiting their seed from returning to the soil seedbank. The critical period for weed 'seed' control (CPWSC) builds upon the well-accepted concept of the critical period for weed control, or the period of crop growth and development during which the crop must remain weed-free to avoid unacceptable yield losses. Instead, the CPWSC shifts the focus away from the crop and on to problematic weed species. The CPWSC is the period of the growing season during which weed management can mitigate unacceptable weed seed production and return to the soil seedbank. The CPWSC provides a framework to target problematic weeds during phenological stages critical for seed development. Like the critical period for weed control, the CPWSC is derived from overlapping two component curves: (i) the critical period for weed emergence, and (ii) the critical period for seed production. Preliminary research at Lethbridge, AB in 2021 identified the CPWSC for seven problematic weed species: green foxtail (*Setaria viridis*), kochia (*Bassia scoparia*), lambsquarters (*Chenopodium album*), redroot pigweed (*Amaranthus retroflexus*), wild buckwheat (*Fallopia convolvulus*), wild mustard (*Sinapis arvensis*), and wild oat (*Avena fatua*). The CPWSC at a 20% seed production threshold ranged from June 15 to August 06 (985–2089 GDD, T_{base} = 0°C) for wild buckwheat, June 27 to July 01 (12121306 GDD) for wild mustard, July 12 to August 06 (15432089 GDD) for kochia, July 16 to August 05 (1632–2061 GDD) for lambsquarters, July 20 to July 24 (17121803 GDD) for redroot pigweed, July 31 to August 12 (19612210 GDD) for green foxtail, and was nonexistent for wild oat [minimum seed threshold of 21.5% by targeting July 03 (1357 GDD)].

Light Quality Manipulation as a Seedbank Management Tool for Palmer Amaranth (*Amaranthus palmeri*). Sarah E. Kezar*, Shuyang Zhen, Muthukumar V. Bagavathiannan; Texas A&M University, College Station, TX (108)

Seed dormancy is a critical determinant of weed species persistence in the soil seedbank, and dormancy regulation governs the proportion of germinable seedbank. Among the various environmental factors, the ratio between red (R) and far-red (FR) levels in the sunlight is known to alter seed dormancy in plants. In this research, we investigated seed dormancy dynamics of Palmer amaranth (*Amaranthus palmeri*) in response to light quality differences across three separate studies. The first experiment explored if maturing seeds on female Palmer amaranth inflorescence can be manipulated by light treatments: R, FR, R followed by FR, FR followed by R, high R:FR ratio, low R:FR ratio, and natural light. These light quality treatments were implemented at different durations of: 5 seconds, 30 seconds, 1 minute, 10 minutes, and 30 minutes. An aluminum-frame chamber was designed with mounted R and FR light bars and controls to implement the above treatments at nighttime on Palmer amaranth inflorescences at the seed development and seed maturity stages. After harvesting, a germination test was conducted with germination counts recorded every 4 days for 24 days. Findings from this experiment showed that seed on the mother plants can perceive and retain light quality until germination, regardless of the duration. Treatments containing R light had a significant effect on promoting germination by up to 57%, whereas the ones with FR light promoted dormancy up to 51%. Seed development stage or the duration of light treatment, however, did not influence seed germination or dormancy. The second experiment entailed mounting the R and FR light bars onto a cultivator (operated at noon), whereas in the third experiment, a rototiller was used and tested at 4 different times of the day (9 am, 12 pm, 5 pm, and 9 pm). Each field experiment had two experimental runs. Overall, the field experiments revealed that Palmer amaranth emergence is promoted by R light by up to 38% and suppressed by FR light up to 82%, consistent with the findings of the first experiment. The treatment time of day created a differential response in emergence patters from spectral variation in the light environment. Findings show promise for manipulating seed dormancy and impacting seedbank dynamics of Palmer amaranth through alteration of light quality, which can serve as an additional tool in the weed management toolbox for this troublesome species.

Quantifying Long-term Competitiveness of Weed Species in a Cereal-legume Rotation. Jose L. Gonzalez-Andujar*[1], Maria J. Aguilera[2], Rene Van Acker[3]; [1]Instituto de Agricultura Sostenible (CSIC), Cordoba, Spain, [2]Departmento de Física Aplicada, Universidad de Córdoba, Cordoba, Spain, [3]Ontario Agricultural College, University of Guelph, Guelph, ON, Canada (109)

Weeds are a permanent constraint to crop productivity in agriculture being a major biotic constraint in the agricultural systems. Weeds compete with crop species, causing severe losses, potentially they are responsible for 34% of world production lossesDue to the importance of the effect of weeds on the crop, there has been great interest in establishing the competitive capacity of each species to optimize their control. Weed scientists have developed numerous ways to measure competition and rank its effects. This work presents a new methodology based on the population dynamics theory to establish the competitiveness of weed species. A population model was used to infer the intensity and importance of the interactions ofPapaver rhoeas L., Veronica hederifolia L., Descurainia sophia L. and Fumaria spp. in a wheat-legume rotation. Data to fit the model were obtained from a long-term experiment (1985-2017) in a semi-arid Mediterranean climate. The results showed the competitiveness ranking varied with the rotation. In cereal was FUM>PAP>VER>DES being interspecific competition greater in Fumaria and Papaver than intraspecific competition. For instance, Fumaria ssp. strongly affected all species and D. sophia was affected for the other species. The ranking in legume rotation was VER>FUM>DES>PAP being in all the cases intraspecific competition greater than interspecific competition.This study highlight the power of simple demographic approaches for analysing the competitive dynamics of multispecies weed communities. The knowledge of the competitive capacity of each species allows a better management of the weed community focused on a context of sustainable management.

Characteristics of Self-fertilization in *Lilium formosanum* Var. *pricei* . Sawa Seoka*, Hidehiro Inagaki; Shizuoka university, Shizuoka, Japan (110)

Formosa lily (*Lilium formosanum* Wallace.) is the only weed species of *Lilium* species. It has the characteristics of weediness, such as fast growth and small size at flower. We developed small potted novel horticultural lily varieties using these weediness characteristics. However, obtaining plants stably with a short plant height from the common Formosa lily is difficult because of its high plasticity. Therefore, this study focused on the dwarf Formosa lily (*L. formosanum* var. *pricei*), a variant of Formosa lily. Dwarf Formosa lily is characterized by its smaller plant size compared to the common Formosa lily, and we proposed that it would be useful for developing small potted lilies. In this presentation, we introduce the self-fertility of dwarf Formosa lily, which is a characteristic of weediness that is also an important factor for horticultural uses, such as seed multiplication and breeding. We investigated pollination types of dwarf Formosa lily flowers treated (bagged, stamen removed, bagged and stamen removed, and no treatment) 2–3 d before flowering. Capsule enlargement was 100% in the no-treatment plots, while it was 66.7% in the stamen-removed plot and the bagged plot. This suggested that the dwarf Formosa lily was not self-incompatible and had characteristics of both self-pollination and cross-pollination. The pollen/ovule ratio of the Formosa lily was classified as partially selfing, while that of the dwarf Formosa lily was classified as fully selfing. In the germination tests of the dwarf Formosa lily, the germination rate of the cross-fertilized seeds was higher than that of the self-fertilized seeds. The optimum temperature for germination of the dwarf Formosa lily was 15–20 ?. Furthermore, light conditions did not affect the germination rate of the dwarf Formosa lily, indicating that the dwarf Formosa lily seeds do not exhibit photogermination. In the common Formosa lily, the final germination rate between cross-fertilized seeds and self-fertilized seeds did not differ. However, the germination speed of the cross-fertilized seeds tended to be faster than that of the self-fertilized seeds. Moreover, the final number of germinations in the common Formosa lily was higher than that in the dwarf Formosa lily. Additionally, the speed of germination was slower in the dwarf Formosa lily than in the common Formosa lily. These results suggested that the common Formosa lily could germinate at a higher temperature than the dwarf Formosa lily.

The Role of Cytochrome P450 in Herbicide Metabolism. Fatemeh Abdollahi*; Postdoctoral researcher, Fort Collins,, CO (111)

Target-site and non-target-site herbicide tolerance are caused by the prevention of herbicide binding to the target enzyme and the reduction to a nonlethal dose of herbicide reaching the target enzyme, respectively. There is little information on the molecular mechanisms involved in non-target-site herbicide tolerance, although it poses the greater threat in the evolution of herbicide-resistant weeds and could potentially be useful for the production of herbicide-tolerant crops because it is often involved in tolerance to multiherbicides. An, *arabidopsis thaliana*, cytochrome P450 (CYP706A3) heterologously expressed in yeast. The microsomal fractions could catalyze the metabolism of pendimethalin herbicide but not oryzalin. Analysis of the metabolites demonstrated that the CYP706A3 hydroxylated the pendimethalin to non-phytotoxic compounds. Kinetic parameters K_m and K_{cat} for pendimethalin was 436.3 (±140.6) and 2.34 (±0.44) in vitro, respectively. Genetically engineered Arabidopsis line overexpressing CYP706A3 (ectopic expression under control of CaMV 35S promoter showed a significant gain of herbicide resistance.

In the Ruins: the Neglected Link Between Archaeology and Weed Science. Alice C. Wolff*[1], Sophie Westbrook[2], Antonio DiTommaso[2]; [1]Cornell University/Medieval Studies, Ithaca, NY, [2]Cornell University, Ithaca, NY (112)

This poster articulates the ways archaeology has and can contribute to the study of weed ecology and science in the past and present, particularly in cereal cropping systems. A small part of this poster is devoted to defining "archaeobotany" (the archaeological study of human-plant relationships) and giving a brief history of the field. The majority of this poster covers recent research areas at the intersection of weed ecology and archaeobotany. Topics include the use of functional weed ecology to study the impact of factors like irrigation, crop sowing time, and cultivation intensity on weed populations; the use of archaeological weed remains to study long-term histories of soil fertility and fertility management practices; historical weed management practices; and the use of archaeobotany to study landscape change and management in history.

Rational Design of an Artificial Weed for High-Throughput Soil Bin Physical Weed Control Tool Testing. Sonora Ortiz*[1], Eric R. Gallandt[1], Bashir Khoda[1], Margaret R. McCollough[1], Daniel C. Brainard[2]; [1]University of Maine, Orono, ME, [2]Michigan State University, East Lansing, MI (113)

Physical weed control tool testing can be performed faster and more effectively in a controlled environment such as a soil bin system. To support high-throughput testing, soil bin experiments often rely on surrogate or artificial weeds instead of real weed species. Generally, artificial weeds range in sophistication from wooden dowels or golf tees to more complex branched structures that simulate basic root systems. Unfortunately, the simplest designs do not sufficiently mimic real weeds, while the more innovative designs have not been well characterised. The objective of this research was to investigate the relationship between anchorage force and two basic characteristics of a silicone artificial weed (SiAW): root length and number of branches. Anchorage force was studied due to the likely correlation between uprooting resistance and cultivation efficacy. Three different root lengths (2.5, 4.0, and 5.5 cm) and numbers of branches (1, 2, and 3) were tested and both were positively correlated with anchorage force. These data provide a menu of sorts for the rational design of artificial weeds that closely mimic the uprooting resistance of various real weed species. The next steps in this research are to examine how SiAW root surface area and anchorage force are related, then compare cultivation efficacy among SiAWs, surrogate weeds, and real weeds in soil bin and field experiments.

Cereal Rye Residue Effects on the Germination of Troublesome Southeastern Weeds. Annu Kumari*[1], Steve Li[1], Andrew J. Price[2]; [1]Auburn University, Auburn, AL, [2]USDA-ARS, Auburn, AL (114)

Weed seed germination and early growth stage are critical parts of the weed life cycle controlled by environmental and genetic factors. Therefore, weed control strategy should focus on the most susceptible parts of the weed cycle to maintain sustainability and reduce chemical herbicide use. Cover crops have been increasingly adopted to suppress weed germination and vigorous vegetative growth. A greenhouse experiment was conducted to evaluate the germination and growth response of several key weeds in the Southeast to various levels of cereal rye residue. Seeds of palmer amaranth, sicklepod, morning glory, and crabgrass were mixed with organic garden soil and placed over the top of the tray.

The soil flats were covered uniformly by four different biomass of rye straw. Plant growth was quantified through weed counting and recording of dry weight. The results illustrated that morning glory was least responsive to increasing biomass, and palmer was the most responsive due to small seed sizes. While germination and growth rate of crabgrass and sicklepod have fluctuated with different levels of biomass residue during this greenhouse study.

Elevated CO_2 and Temperature Effects on Growth and Photosynthetic Responses in Purple and Yellow Nutsedge (*Cyperus rotundus* L. and *C. esculentus* L.). Srinivasa Rao Pinnamaneni*, Krishna N. Reddy; Crop Production Systems Research Unit USDA-ARS, Stoneville, MS (115)

Global atmospheric carbon dioxide (CO_2) and air temperatures are steadily rising and likely to have profound effects on biology of weed species. Growth chamber experiments were conductedin 2022 to determine effects of elevated CO_2 and temperatures on growth and photosynthetic responses in two invasive weeds, purple nutsedge (*Cyperus rotundus* L.) and yellow nutsedge (*C. esculentus* L.). Plants were grown in pots, irrigated, fertilized as needed. In one experiment, plants were subjected to five day/night temperatures of 20/12, 25/17, 30/22, 35/27, and 40/32 °C from 3-day old propagules to until harvest. In another experiment, plants were subjected to three CO_2 levels of 420, 570, and 720 ppm at two day/night temperatures of 30/22 and 35/27 °C to assess CO_2 and temperature interactions. A randomized complete block design with 5 to 6 replications was used and the experiments were repeated. Purple and yellow nutsedge shoot biomass was harvested after 5 and 7 weeks of growth, respectively. After clipping one set of both species were allowed to regrow for two weeks and shoot biomass recorded. Data on shoot number, plant height, shoot biomass, root biomass, regrowth shoot number, regrowth shoot height, regrowth shoot biomass, regrowth root biomass, chlorophyll index, stomatal conductance, net photosynthesis rate, transpiration rate, and instantaneous water use efficiencywere collected. A temperature regime of 30/22 °C appears to be optimum for both species and purple nutsedge was more impacted by the temperature deviation on either side of the 30/22 °C for shoot production, shoot biomass, and root biomass in the first and regrowth harvests. The net photosynthetic rate, transpiration rate, and stomatal conductance declined with either increase or decrease in temperatures. Both species recorded significantly higher shoot number, shoot height, shoot and root biomass at elevated CO_2 levels. The species differences were conspicuous as yellow nutsedge recorded higher shoot number, shoot biomass, root biomass, and regrowth root biomass than purple nutsedge by 59%, 103%, 59% and 23%, respectively. The speciesdifferences were insignificant for shoot height, regrowth shoot number and regrowth shoot biomass. The temperature and CO_2 interactions were significant for most of the traits. Theflowering appeared to be compromised at elevated CO_2 levels in both species and yellow nutsedge did not flower even after 10 weeks of growth. The net photosynthetic rate andchlorophyll index did not differ among CO_2 treatments in purple or yellow nutsedge. Whereas, stomatal conductance, transpiration rate decreased with increased CO_2 levels and internal CO_2 concentration and intrinsic water use efficiency increased with increased CO_2 levels. This study demonstrated that elevated CO_2 levels had a profound positive effect on growth, while temperatures higher or lower than 30/22 °C had a negative effect on growth of both species.

Analysis of the Recently Completed Genome of *Salsola tragus* (Russian Thistle), a Tenacious Tetraploid Tumbleweed. John M. Lemas*[1], Eric L. Patterson[2], Sarah Morran[1], Nickolas A. Johnson[2], Jacob S. Montgomery[1], Victor Llaca[3], Kevin Fengler[3], Philip Westra[1], Todd A. Gaines[1]; [1]Colorado State University, Fort Collins, CO, [2]Michigan State University, East Lansing, MI, [3]Corteva Agriscience, Johnston, IA (116)

Salsola tragus is a tetraploid (2n+2n=36, total size 1.3 Gbp) tumbleweed that threatens agriculture, rangelands, and open spaces mainly in the western the United States, especially during this era of climactic flux. The goal of this research is to provide the scientific community with a genomic understanding of *Salsola tragus* (Russian thistle). In 2019, *Salsola tragus* seeds were collected in Weld County, Colorado. Collected seeds were then grown in a greenhouse setting until mature before transferring into a dark environment for 48 hours. Tissue from this population was collected and sent to Corteva©. High molecular weight DNA was extracted from leaf tissue and sequenced using PacBio HiFi to a coverage of 120x. PacBio HiFi data was initially assembled using miniasm. One scaffold per chromosome was assembled through the combination of Bionano optical mapping and a Hi-C chromatin contact mapping. Next, RNA was extracted from four tissue types for PacBio ISO-seq which was used to call gene models using a novel annotation pipeline developed by the International Weeds Genomics Consortium. This transcript library was compared to other species for gene function prediction. Each haplotype is over 1.2 Gbp (95% of known genome size) and organized into 19 scaffolds. Of this 1.2 Gbp sequence, 59% represents interspersed repeats, including retroelements (28.14%) and transposons (3.87%). Genome statistics were generated using Assemblathon, and both sequences were analyzed against a BUSCO database (95.4% complete BUSCOs of the 2,326 groups searched). The reference sequence was then analyzed through intragenomic alignment using minimap2 (v2.24-r1122). Genome and alignment visualization was preformed using Genome Ribbon (https://genomeribbon.com/) and RIdeogram (R version 4.2.1, RIdeogram v. 0.2.2). Overall, this *Salsola tragus* genome provides the scientific community with an excellent resource for future referencing.

Using Weed Extracts to Protect Sweet Potato from Deer Browsing. Ziming Yue[1], Karina Beneton[1], Presley Lee[2], Mark Hall[3], Callie J. Morris[1], Mark W. Shankle[3], Te-Ming (Paul) Tseng*[1]; [1]Mississippi State

University, Starkville, MS, [2]North Pontotoc High School, Pontotoc, MS, [3]Mississippi State University, Pontotoc, MS (117)

Competitive Interactions Between Sorghum and Johnsongrass (*Sorghum halepense*) is Influenced by Multiple Abiotic Stress Factors. Vikas C. Tyagi*, Aniruddha Maity, Nithya K. Subramanian, Muthukumar V. Bagavathiannan; Texas A&M University, College Station, TX (118)

Sorghum is an important crop grown on significant acreage around the world, and weeds present a major production constraint. In particular, johnsongrass is a troublesome weed in sorghum; due to genetic similarities between the two species, the options for selective control of johnsongrass in sorghum is very limited. The projected climate change scenarios may alter the competitive interactions between the two species, but limited information is available on the impact of various abiotic stress factors on sorghum-johnsongrass competition. Experiments were conducted under growth chamber conditions at the Norman E. Borlaug Southern Crop Improvement Controlled Plant Growth Facility, Texas A&M University, College Station, TX, at two levels each of temperature (ambient and ambient + 5°C), CO_2 (400 and 700 µmol mol-1), and soil moisture (100% and 50% field capacities), to quantify the competitive outcomes between sorghum and johnsongrass. A single plant of each species was grown either as a monoculture (1:0) or in an equal-mix combination (1:1 ratio) in pots filled with potting soil mix. Plant height, number of leaves per plant, stem thickness, shoot dry weight, and root dry weight were used to compute competitive indices such as relative yield, relative yield total, competition ratio, relative crowing coefficient, and aggressiveness index. Preliminary results indicated that abiotic stress conditions can alter the competitive interactions between the two species, but there were significant interactions among the factors for the different variables measured. The findings will be useful for predicting sorghum x johnsongrass competitive interactions under future climatic conditions and designing suitable management strategies.

Integrating Ecological Tactics to Manage Multiple Herbicide-Resistant Pigweeds in U.S. Soybean Production. Prashant Jha*[1], Jason K. Norsworthy[2], Vipan Kumar[3], Karla L. Gage[4]; [1]Iowa State University, Ames, IA, [2]University of Arkansas, Fayetteville, AR, [3]Kansas State University, Hays, KS, [4]Southern Illinois University Carbondale, Carbondale, IL (119)

Stakeholders from across Midwest, South, and Central Great Plains of the U.S. have identified pigweeds (Palmer amaranth and waterhemp) as one of the most problematic and economically damaging weed species. Both pigweeds currently pose a serious threat to sustainable crop production due to a limited effective herbicide options. Widespread multiple resistance to 6-7 different herbicide sites-of-action has increased the need and the desire for IPM-based solutions for managing pigweeds in soybean-based cropping systems of the U.S. For this stakeholder-identified research, we propose to: 1) evaluate the effectiveness of two ecologically-based tactics, namely cover crop and harvest weed seed control (chaff lining and weed seed destructor), within an integrated weed management (IWM) framework, for managing pigweed seed banks in soybean, and 2) quantify the economic benefits and risks of adopting a diversified IWM program to mitigate herbicide resistance. We will conduct multi-location (IA, IL, KS, AR) field and greenhouse/laboratory studies to develop and integrate these ecological tactics that have a high likelihood of reducing pigweed seed banks and exposure of this species to herbicides, thereby reducing selection for herbicide resistance evolution across the three major soybean production regions. Implementation and adoption of these ecologically-based IWM strategies will reduce potential environmental impacts associated with increased pesticide use. Results from this IWM project will be disseminated across geographic boundaries to meet the National IPM Roadmap Goals.

CO_2 And Temperature Effects on Herbicide-resistant and Susceptible Palmer Amaranth Biotypes.
Juliana de Souza Rodrigues*[1], Timothy L. Grey[2], Donn Shilling[3], Todd A. Gaines[4], Nicholas T. Basinger[5], John Snider[2], Viktor Tishchenko[6]; [1]University of Georgia/Department of Crop and Soil Science, Tifton, GA, [2]University of Georgia, Tifton, GA, [3]University of Georgia, Athens, GA, [4]Colorado State University, Fort Collins, CO, [5]The University of Georgia, Athens, GA, [6]University of Georgia, Griffin, GA (120)

In a climate change scenario, C3 plants are likely to benefit more from the rise in [CO_2]. C4 plants are well-known for their limited response to increases in [CO_2] (at low, high, and normal levels) regarding photosynthetic rates, which leads to growth and biomass production. However, previous research mentioned that C4 plants' responses to elevated [CO_2] are specie specific. It is necessary to consider that temperature increases also play a significant role in plant phenology, affecting seedling emergence, flowering, and plant senescence. In addition to the environmental variations previously mentioned, herbicide-resistant weeds are a considerable challenge for weed management programs. Among the cases of herbicide-resistant weeds, Palmer amaranth, a C4 plant, is well-known for its high plasticity level and quick ambiance adaptation. Due to the lack of knowledge about weed behavior under [CO_2] and temperatures, it is necessary to observe which environmental variables impact Palmer amaranth biotypes (glyphosate-resistant and susceptible) growth the most. In addition, assess the extent of these impacts on plant development. Therefore, the objective of this study was to evaluate if glyphosate resistance interferes with plant growth, development, and phenotypic adaptation to changes in [CO_2] and temperatures. Seeds of glyphosate-resistant 1 (2017; ED_{50}= 1180 a.e. g/ha) and glyphosate-resistant 2 (2020: ED_{50}=3603 a.e. g/ha) and susceptible (early 2000's; ED_{50}= 272 a.e. g/ha) biotypes, all from Georgia, US, were sowed in trays and grown in a greenhouse until they reached 8 to 10 cm in height.

Seedlings were placed individually in 5L containers with potting media and moved to four growth chambers: (1. 23/33°C, 410ppm ±20 ppm; 2. 23/33°C, 750ppm ±20 ppm; 3. 26/36°C, 410ppm±20 ppm, and 4. 26/36°C, 750ppm±20 ppm), in an RCBD (randomized complete block design). Three harvest dates (HD) were scheduled to start after ten days of the transplant, with an interval of 7 days between harvests. The data collected on each HD included height, stem diameter, number of leaves, and above-ground biomass (leaves and stem). Data were analyzed as a four-way factorial treatment structure (Temperature X [CO_2] X Biotype X Harvests (HD) performed), with the year (2021 and 2022) as a blocking factor. First, multivariate analysis identified [CO_2] as the factor that primarily affects Palmer amaranth growth. Then, a series of univariate analyses were done for each parameter, with a p-value adjusted to 0.00625 to avoid multiplicity. For height and number of leaves, plants grown under high [CO_2] were taller and among biotypes, the susceptible is taller compared to the two resistants' tested. Stem diameter did not show differences under [CO_2], temperature, and biotype tested. For the above-ground biomass, no differences were found for leaf biomass, while stem biomass was greater under high [CO_2]. In this case, when comparing biotypes, the difference found was between susceptible and resistant 1. Overall, the parameters appraised responded differently to the [CO_2] levels, being unresponsive to the temperature. In addition, differences regarding HD were expected due to plant growth. It was also noted that there are notable distinctions between glyphosate-resistant biotypes, in terms of height, number of leaves, and stem biomass. Further investigation is essential to address whether those discrepancies are related to glyphosate resistance or what could cause the variations observed. In conclusion, Palmer amaranth is an example of a C4 plant that can respond to high levels of CO_2 in the atmosphere.

Influence of Multiple Abiotic Stress Factors on Pollen Viability and Stigma Receptivity of Johnsongrass (*Sorghum halepense*). Vikas C. Tyagi*, Aniruddha Maity, Nithya K. Subramanian, Muthukumar V. Bagavathiannan; Texas A&M University, College Station, TX (121)

Johnsongrass (*Sorghum halepense*) is an aggressive, invasive, and perennial weed species infesting several crops worldwide. Though a predominantly selfing species, substantial levels of outcrossing has been reported within and between populations. Johnsongrass is a high-risk species for the development of herbicide resistance and pollen movement can transfer herbicide resistance across populations. Further, gene flow between johnsongrass and its crop relative cultivated sorghum (*S. bicolor*) can transfer traits such as herbicide resistance, which can complicate management. The abiotic stress conditions associated with future climatic scenarios may affect the pollination biology of johnsongrass, but the nature of such effects is unknown. Experiments were conducted in controlled environment growth chambers at the Norman E. Borlaug Southern Crop Improvement Plant Growth Facility, College Station, TX, to investigate the effects of temperature, CO_2, soil moisture, and herbicide stress on pollen viability, stigma receptivity, floret counts, days to flowering, and seed set percentage. Plants were grown under two levels each of the stress factors temperature (ambient and ambient + 5°C), CO_2 (400 and 700 µmol mol-1), soil moisture (100% and 50% field capacities), and herbicide (with and without quizalofop @ 0.2x the label rate). Pollen viability was determined using the MTT solution, which consisted of 1% concentration of the substrate 2,5-diphenyl tetrazolium bromide in 5% sucrose. The stigma receptivity was assessed using the H_2O_2 (conc. 25-35%) method. Preliminary results showed that the plants under high temperature stress conditions exhibited delayed flowering, low pollen count, and short pollen longevity than the plants under normal temperatures. Likewise, stigma receptivity also impacted significantly under high-temperature conditions. However, the effect of CO_2 on days to flowering, seed set percent, floret count, and stigma receptivity were found to be non-significant. Results of this study show that altered climatic variables can impact pollination biology and fecundity in johnsongrass.

Differentiation Between Yellow Foxtail(*Setaria pumilla*) and Knotroot Foxtail (*Setaria parviflora*) Via Morphological Characteristics and Genetics Markers. Mikerly M. Joseph*[1], Joseph S. McElroy[1], Claudia A. Rutland[2]; [1]Auburn University, Auburn, AL, [2]Auburn University Department of Crop, Soil, and Environmental Sciences, Auburn, AL (122)

Setaria pumila (Poir.) Roem. & Schult. (1817) and *Setaria parviflora* (Poir.) Kerguelen, respectively known as yellow and knotroot foxtail, are two problematic weed species that share morphological similarities and are often misidentified. The primary morphological differences between yellow and knotroot foxtail are found in the seed head and the lower floret, along with differences in root systems, where yellow foxtail has fibrous roots while knotroot foxtail has rhizomes. However, those criteria offer little practical help for differentiation, such as the seed head and rhizomes, which appear late in the growth stage, long after identification is required for herbicide management. Research was conducted in the herbicide resistance diagnostics laboratory at Auburn University in Alabama to differentiate yellow and knotroot foxtail beyond morphological characteristics using DNA barcoding. Thirty (30) biotypes for each species were identified in Alabama using seedhead size, ligule size, and the presence of rhizomes. DNA for each biotype was isolated from the leaves and amplified using polymerase chain reaction (PCR) with primers targeting trnH-psbA and ITS regions. Sanger sequencing results demonstrate the presence of different single nucleotide polymorphisms in the sequenced regions that can differentiate yellow and knotroot foxtail. Some biotypes that were initially identified as knotroot foxtail, were later correctly identified as yellow foxtail by DNA barcoding. This study demonstrated that applying these DNA barcodes is relevant in accurately identifying foxtail species at an early stage and thus help effectively manage them using correct herbicide combination.

Insights into Weed Aboveground and Seedbank Dynamics from a Long-Term Cropping Systems Experiment. Natasha Djuric*[1], Michel A. Cavigelli[2], Steven Brian Mirsky[2], Eugene P. Law[3], Antonio DiTommaso[1], Matthew R. Ryan[1]; [1]Cornell University, Ithaca, NY, [2]USDAARS, Beltsville, MD, [3]University of Delaware, Beltsville, MD (123)

Many changes in agricultural systems occur slowly and outcomes are influenced by weather, soil conditions, and management. Long term agricultural research (LTAR) projects are crucial for understanding the dynamics of these systems at the temporal and spatial scales needed to inform sustainable practices and policy. The Farming Systems Project (FSP), established in 1996 in Beltsville, Maryland, is a collaborative LTAR that has provided decades of research on management of corn, soybean, and wheat. The FSP compares two conventional cropping systems that vary in tillage intensity and three organic cropping systems that vary in rotational diversity. A suite of sustainability metrics have been quantified in the FSP including crop yield and profitability, soil health indicators, pest population and community dynamics, and environmental impacts. Here, twenty years of insights about weed aboveground and seedbank dynamics are synthesized. In its early stages, the FSP provided evidence for the importance of longer crop rotations in organic systems to increase phenological and growth form diversity, reducing abundance of annual dicot weeds. Later research demonstrated the benefits of assessing crop rotation sequence through a full entry design, revealing that organic rotations beginning with hay maintain lower seedbanks of troublesome weeds like smooth pigweed (Amaranthus hybridus) and common lambsquarters (Chenopodium album). FSP research subsequently focused on yield differences between conventional and organic systems through interacting factors like weed competition and nitrogen availability, and situations in which these differences may be reduced. Most recently, FSP research has highlighted how management and meteorological factors drive weed abundance through complex interactions such as planting date and precipitation in early summer. New research is now underway that aims to identify weed functional group selection across cropping systems treatments using a trait-based approach to analyzing weed community data. Results have potential to increase predictive capacity and assist in the development of robust weed management strategies that contribute to long-term sustainability goals.

Evapotranspiration of Volunteer Corn in Corn, Soybean, and Sorghum Cropping Systems. Mandeep Singh*[1], John Lindquist[2], Stevan Knezevic[2], Suat Irmak[3], Vipan Kumar[4], Amit J. Jhala[1]; [1]University of Nebraska-Lincoln, Lincoln, NE, [2]University of Nebraska, Lincoln, NE, [3]Penn State University, University Park, PA, [4]Kansas State University, Hays, KS (124)

Volunteer corn is a competitive weed in row cropping systems in the Midwestern U.S. Numerous studies have evaluated yield loss due to volunteer corn in row crops. However, literature is lacking on water loss due to volunteer corn. Therefore, a field experiment was conducted to evaluate the evapotranspiration of volunteer corn in corn, soybean, and sorghum in center-pivot and subsurface drip irrigation systems at

Clay Center, NE. Treatments were replicated four times and arranged in a strip plot design. Eight crop strips included corn, soybean, sorghum, fallow, corn + volunteer corn (VC), soybean + VC, sorghum + VC, and fallow + VC. Fallow, and fallow + VC strips were included to compare the evapotranspiration (ET) of volunteer corn alone. Volunteer corn plants were grown in a greenhouse using seeds harvested from the last year. These plants were transplanted after crop emergence at four plants per meter row length in the middle of two central rows of the plots. After VC transplanting, soil moisture sensors were installed at 0-0.30, 0.30-0.60, and 0.60-0.90 m depths to measure changes in the volumetric water content during the growing season. ET was calculated based on the soil-water balance approach. Under the center pivot irrigation system, ET of VC was the highest in sorghum (641 mm), followed by soybean (637 mm), corn (614 mm), and fallow (533 mm). Under the subsurface drip irrigation system, ET of VC was the highest in soybean (645 mm) followed by sorghum (629 mm), corn (594 mm), and fallow (437 mm). Averaged across irrigation systems, soil water depletion was 0.5 mm, 5.7 mm, and 15.2 mm more in corn+VC, soybean+VC, and sorghum+VC as compared to their respective crop only plots; although differences were statistically non-significant. Crop yields were similar across crops with and without VC, with non-significant yield losses of 3, 10, and 21% due to VC in corn, sorghum, and soybean, respectively. Overall, it is concluded that volunteer corn is capable of causing additional water and yield loss in row crops, therefore, should be managed effectively early in the growing season.

Predicting Yield Loss Using the Kropff-Spitters Model and Early Season Leaf Area Indices. Avi S. Goldsmith*, Sandra R. Ethridge, Ramon G. Leon, Robert Austin; North Carolina State University, Raleigh, NC (125)

Weed management is among the most important and time-consuming jobs in the agricultural industry. Having the ability to accurately predict the impact of weeds on yield loss is essential for a grower to plan not only control actions, but also time and price to sell grain and purchase inputs for future seasons. Current models rely on weed density as the predictor of yield loss, which does not properly account for the level of interference dependent on weed size. This can result in the overestimation of yield loss, especially later in the season when the crop has developed a competitive edge over smaller, late emerging weeds. The Kropff & Spitters model (1991) utilizes the relative leaf area of weeds as the predictor of yield loss in order to overcome the shortcomings of density. To measure the predictability of the Kropff & Spitters model an existing dataset with yield and leaf area index of corn, soybean, & cotton under two densities in weedy/weed free conditions was used. A strong correlation ($R^2 = 0.83$) between observed and predicted yield loss as well as a moderate correlation ($R^2 = 0.58$) between weed relative leaf area and observed yield loss was found. The utilization of this model will allow for accurate yield loss predictions, but automation of crop and leaf area estimation is still needed. UAS based sensors are promising tools to generate the information needed to use the Kropff & Spitters model in real time.

Fate of *Amaranthus palmeri* (Palmer Amaranth) and *A. tuberculatus* (Waterhemp) Seeds Under Cereal Rye Cover Crop Across the North Central Region. Lily A. Woitaszewski*[1], Sarah Lancaster[1], Anita Dille[1], Kevin W. Bradley[2], Mandy Bish[2], William G. Johnson[3], Bryan G. Young[4], Claudia R. Bland[3], Joseph T. Ikley[5], Rodrigo Werle[6], Nicholas J. Arneson[6]; [1]Kansas State University, Manhattan, KS, [2]University of Missouri, Columbia, MO, [3]Purdue University, West Lafayette, IN, [4]Purdue University, Brookston, IN, [5]North Dakota State University, Fargo, ND, [6]University of Wisconsin-Madison, Madison, WI (126)

Palmer amaranth and waterhemp are aggressive summer annual species that are ranked as some of the most difficult to control weeds, rightfully so, as research has confirmed many herbicide-resistant populations. Significant economic damage results if left uncontrolled in agricultural fields. Using cereal rye as a cover crop is popular in the north central region and has been found to boost soil microbial activity. Soil microbial activity, in turn, could increase weed seedbank decay; however, this is dependent on what weed species are present. A study was conducted to determine if a cereal rye cover crop can enhance seed bank decay of Palmer amaranth and waterhemp in soybean production systems. Palmer amaranth and waterhemp seeds were collected in Kansas, Indiana, Missouri, North Dakota, and Wisconsin. Seeds were then counted and placed into fine wire mesh packets and sent back to respective locations. Seeds were buried around the time of cereal rye planting in the fall of 2021 in four replications. Treatments included burial in cereal rye cover crop and in no cover crop. Kansas populations were buried alongside locally-collected seeds at six sites – one each in Indiana, Missouri, North Dakota, and two sites in Kansas and Wisconsin. All populations were buried at one site in Kansas. Seeds were removed at the time of soybean planting in spring and at time of harvest in fall of 2022 and sent to Kansas for viability testing. At seven months after burial, cover crop had no effect on the seedbank viability of seven amaranth populations across seven states. At twelve months after burial, when analyzing each population individually, there was an effect of cover crop on seed viability of Missouri and Indiana waterhemp. These two populations exhibited greater viability in cover crop than with no cover crop treatments. Moreover, seed dormancy was also impacted by burial location and removal timing. These findings are contrary to our hypothesis, that an expected increase in microbial activity would result in seedbank decay. These data will provide a basis for further research involving seedbank decay in presence of cover crops. (Lwoitaszewski2@ksu.edu)

Building Reference Genome of Canada Thistle (Cirsium Arevense) and Identifying Genomic Regions Behind Sex Determination. Sofia Marques-hill*[1], Todd A. Gaines[2]; [1]Colorado State University - Agricultural Biology, Fort Collins, CO, [2]Colorado State University, Fort Collins, CO (127)

Identifying Genes Associated with Conferring Clopyralid Resistance in Common Ragweed (*Ambrosia artemisiifolia*) Using RNA Sequence Transcriptome Analysis. Nash D. Hart*[1], Erin E. Burns[2], Eric L. Patterson[2]; [1]Michigan State University, Durand, MI, [2]Michigan State University, East Lansing, MI (128)

Ambrosia artemisiifolia (common ragweed) is a globally distributed, difficult to control weed species that can cause extensive crop yield reductions unless appropriately managed. Clopyralid is a synthetic auxin herbicide commonly used to control *A. artemisiifolia* and other weeds in the Asteraceae family. In 2018, a population of *A. artemisiifolia* was discovered in a Michigan Christmas tree farm that is highly resistant to clopyralid, surviving at clopyralid doses thirty-two times the recommended field use rate. Chemical weed control is a mainstay in most agricultural systems in the United States and herbicide resistance threatens its effectiveness; therefore, it is essential to understand the mechanism of resistance that allows weed species to become resistant to herbicides. To this end, we have begun investigating potential resistance mechanisms in this clopyralid resistant *A. artemisiifolia* population using RNA-seq. As no well-curated genome or transcriptome currently exists for this species, we first built a reference transcriptome de novo using Trinity and reads from the susceptible population. Next, we aligned all reads from one resistant and one susceptible population using the program HISAT2. We identified all assembled Aux/IAA annotated transcripts using BLAST and manually screened the read alignments for polymorphisms (SNPs, InDels, etc.) that distinguished resistant from susceptible individuals. We specifically evaluated the sequence motif known as "the degron" which has previously been shown to be involved in target site resistance (TSR) to other auxinic herbicides including 2,4-D and dicamba in other dicot species. We also performed a whole transcriptome differential expression analysis to identify resistance mechanisms that involve changing gene expression (i.e. non-TSR and target site over-expression). Lastly, we performed a Gene Ontology (GO) enrichment analysis on the RNA-seq differential expression data to identify global differences in the transcriptomes of these two biotypes. After searching the entire genome there were Twenty-three transcripts containing the Aux/IAA degron sequence. Nine out of Twenty-three transcripts containing the sequence contained a polymorphism. However, zero of the transcripts had a significant polymorphism that correlated perfectly with the resistant phenotype. Differential expressional analysis highlighted 270 genes that were significantly differentially expressed, with 129 being overexpressed in the resistant population and 141 being overexpressed in the susceptible. Out of the 270 genes, there were 109 genes that had a log fold change of greater than two or less than negative two, interestingly of these 57 genes were overexpressed in the susceptible population and only 52 genes overexpressed in the resistant population. Within the 109 genes, there are many genes of interest that directly relate to the auxin pathway, such as auxin response and ABC transporter genes, as well as genes associated with non-target site resistance, such as cytochrome P450s. Ultimately, understanding clopyralid resistance in *A. artemisiifolia* and the potential for cross resistance to other auxinic herbicides is of critical importance for continued agricultural productivity in the North Central region as technologies like Enlist, Xtend, and XtendFlex soybean become more common.

Chlorophyll Content and Fluorescence Fail to Detect Crop Injury from Physical Weed Control Tools.
Anthony J. Codega*[1], Ruthie Clements[1], Jordan W. Parks[1], Rebecca Champagne[2], Eric R. Gallandt[1]; [1]University of Maine, Orono, ME, [2]The University of Maine, Orono, ME (130)

Farmers rely on physical weed control tools to selectively kill weeds while avoiding or minimizing crop damage. Crop damage is often determined by comparing end of season crop biomass or yield, but such measurements are far from the time of cultivation and poor statistical power may obscure damage to crops. Chlorophyll content and chlorophyll fluorescence (Fv/Fm) are commonly used as measurements of plant stress, and could be a viable means of detecting low levels of crop injury from physical weed control. Field experiments were performed to determine the utility of chlorophyll content and fluorescence measurements as indicators of crop damage. The test crop was bush bean (*Phaseolus vulgaris*) 'Provider' from Johnny's Selected Seeds. Physical weed control treatments included finger weeders and torsion weeders at three adjustments each to represent suboptimal, optimal, and aggressive tool settings. Root pruning using a rice knife was used as a positive control, and careful hand weeding was used as a negative control. Measurements of chlorophyll content and Fv/Fm were taken on the day of cultivation and 2, 7, and 14 days after cultivation. Crop biomass was also collected 14 days after cultivation, split into above- and below-ground fractions, dried, and weighed. There were no significant differences found among weeding treatments and between control. Unfortunately, measurements chlorophyll content and of light adapted Fv/Fm are not useful due to their high variability under field conditions.

Effective Anthraquinone Derivative Dose; A Weed-based Deer Repellent for Soybean. Ncomiwe A. Maphalala*[1], Ziming Yue[2], Te-Ming (Paul) Tseng[2]; [1]Mississippi State University-Department of Plant and Soil Sciences, Starkville, MS, [2]Mississippi State University, Starkville, MS (131)

A Novel Allele of *IAA16* is Associated with Dicamba Resistance in *Bassia scoparia*. Jacob S. Montgomery*[1], Neeta Soni[2], Sarah Morran[1], Franck E. Dayan[1], Philip Westra[1], Eric L. Patterson[3], Todd A. Gaines[1]; [1]Colorado State University, Fort Collins, CO, [2]Corteva Agriscience, Indianapolis, IN, [3]Michigan State University, East Lansing, MI (132)

The recent introduction of genetically modified crops resistant to the herbicide dicamba has increased the usage of dicamba in agricultural settings. This increase has caused accelerated evolution of dicamba resistance in weedy species due to increased selection pressure. A dicamba-resistant population of *Bassia scoparia* (M32) was identified in an herbicide resistance survey conducted in 2012. Dose response studies confirmed dicamba resistance within M32 (R:S 11) of approximate equal magnitude conveyed by the previously reported G127N allele of *IAA16* (R:S 15). Sanger sequencing of dicamba-resistant M32 plants revealed the absence of the G127N substitution in the *IAA16* gene. Experiments using radiolabeled dicamba to quantify herbicide uptake and translocation showed that M32 plants behaved similarly to plants homozygous for the G127N allele of *IAA16*. Populations segregating for dicamba resistance were produced via a bi-parental cross between individuals of M32 and a dicamba-sensitive reference population (7710). Approximately 300 plants each from two segregating F3 families were screened with a delimiting rate of dicamba (560 g ae ha^{-1}). Ratios of alive and dead plants suggest dicamba resistance was simply inherited. A newly developed genotype-by-sequencing protocol was used on individuals from these segregating populations to identify one Quantitative Trait Locus (QTL) underlying this trait. Transgenic Arabidopsis lines heterologously expressing candidate genes within the identified QTL are being developed to understand the role of each gene in dicamba resistance. Understanding the genetic/physiological basis of resistance will inform future herbicide design and allow for rapid diagnosis of dicamba resistance in *Bassia scoparia*. Senior author: Todd.Gaines@colostate.edu

A Fungal Pathogen (*Colletotrichum shisoi*) Reduces Biomass, Flowering, and Competitive Ability of Invasive Beefsteak Plant (*Perilla frutescens*). Michael R. Fulcher*; United States Department of Agriculture, Agricultural Research Service, Fort Detrick, MD (133)

Beefsteak plant is an invasive weed in North America where it excludes native vegetation in natural areas and can cause respiratory illness in grazing livestock. An adventitiously introduced fungal plant pathogen (*Colletotrichum shisoi*) is being evaluated for biological control activity. The effects of pathogen infection on perilla biomass, flower production, and competition with native plants were recorded under controlled conditions. A dose-response experiment determined a concentration of $1x10^7$ fungal spores/mL aqueous suspension applied to perilla seedlings reduced biomass by 88% relative to plants treated with a sterile water control. Application of the same spore concentration to adult perilla plants at the onset of floral development reduced aboveground plant biomass by 34% and the number of flowers produced at plant maturity by 45%. A modified partial additive competition experiment was performed using perilla, bee balm (*Monarda didyma*), and mountain-mint (*Pycnanthemum virginianum*) grown with and without application of the pathogen. The native plants produced 80% less biomass in the presence of healthy perilla compared to growth alone, and application of the pathogen restored native plant biomass to levels observed in the absence of perilla. These results suggest severe pathogen infections might decrease the invasiveness of perilla by reducing fecundity and competitive ability. Contact: Michael.Fulcher@usda.gov

Invasive Weeds Serving as Microbial Reservoirs and Their Role in Pathogen Evolution and Emergence. Matthew A. Tancos*, Michael R. Fulcher; United States Department of Agriculture, Agricultural Research Service, Fort Detrick, MD (135)

Invasive weeds encroach on fields, forests, and pastures, damaging agricultural productivity and agroecosystem health. Moreover, non-native invasive weeds contain highly diverse microbial assemblages that potentially influence pathogen evolution and disease epidemics by their widespread distribution and proximity to agricultural systems. Genetically diverse weedy plants have been demonstrated to harbor more virulent and diverse pathogens relative to genetically uniform crops, but the role of invasive weeds in pathogen evolution and disease dynamics remains unexplored. Increased microbial genomic plasticity, diverse microbial interactions, and widespread distribution of invasive weeds may foster conducive avenues for microbial evolution and pathogen emergence in agroecosystems. Understanding the genomic plasticity of weed pathogens and their emergence in invasive weed systems is vital to predicting host jump events and spillover to agriculture from invasive weed reservoirs. Monitoring weed-microbial interactions through genomics and host-range studies will illuminate the role invasive weeds play as ancillary hosts and how they may provide conducive microenvironments for pathogen evolution and emergence. As a case study, comparative genomics of *Xanthomonas campestris* strains isolated from diverse cruciferous weeds and crops highlight increased rates of recombination, genomic plasticity, and host range potential of weed-collected strains. Managing invasive weed infestations and identifying weedy plant species capable of supporting high-impact, agronomically-damaging phytopathogens could reduce landscape level disease pressures and assist with improved weed management approaches.

Mycorrhizae Mediate the Control of Broomrapes (*Orobanche* Spp.) in Lentils and Sunflowers. Cumali Ozaslan*[1], Antonio DiTommaso[2]; [1]Dicle University, Diyarbakir, Turkey, [2]Cornell University, Ithaca, NY (137)

This study was carried out to investigate the effects of arbuscular mycorrhizal fungi (AMF) on parasitic broomrapes (*Orobanche crenata* and *Phelipanche aegyptiaca*) in red lentil and sunflower crops in Diyarbakir province, Turkey in 2012. Five different red lentil cultivars (i.e., 'Firat 87', 'Sakar', 'Seyran 96', 'Çagil', and 'Yerli Kirmizi') were included in the lentil experiment and one sunflower cultivar was used in the sunflower experiment. The sunflower or lentil plants were treated with AMF (15 g mycorrhizal soil per plant) or no AMF. Control treatments without AMF and broomrapes were included in the study for comparison. The results revealed that AMF application reduced broomrape germination and infestation, compared to the control without AMF. Broomrape infestation was reduced by 60% in lentil and 55% in sunflower. AMF application increased root length and yield-related traits of lentil and sunflower. Overall, this study demonstrated that AMF can limit damage from broomrapes and enhance yield in lentil and sunflower. Further studies involving different rates of AMF application and AMF species are warranted.

Root Architecture Study for Allelopathic Traits in Cotton for Weed Suppression. Worlanyo Segbefia*[1], Varsha Singh[1], Parth Sharma[1], Ncomiwe A. Maphalala[2], Te-Ming (Paul) Tseng[1]; [1]Mississippi State University, Starkville, MS, [2]Mississippi State University-Department of Plant and Soil Sciences, Starkville, MS (138)

Root System Architectural Traits and Genes Linked to Allelopathic Effects in Weedy Rice (Oryza Sativa f. Spontanea). Ncomiwe A. Maphalala*[1], Auriana Tucker[2], Brooklyn C. Schumaker[2], Swati Shrestha[2], Shandrea D. Stallworth[2], Nilda Roma-Burgos[3], Te-Ming (Paul) Tseng[2]; [1]Mississippi State University-Department of Plant and Soil Sciences, Starkville, MS, [2]Mississippi State University, Starkville, MS, [3]University of Arkansas, Fayetteville, Fayetteville, AR (139)

A Tale of 20 Years: What Can be Learned About Glufosinate Tolerance When Comparing Palmer Amaranth Populations from 2001 Vs. Now? Pamela Carvalho-Moore*[1], Jason K. Norsworthy[1], Fidel Gonzalez Torralva[1], Leonard B. Piveta[1], Maria Carolina C R Souza[2], Tom Barber[3], Thomas R. Butts[3]; [1]University of Arkansas, Fayetteville, AR, [2]University of Arkanas, Fayetteville, AR, [3]University of Arkansas System Division of Agriculture, Lonoke, AR (140)

Glufosinate resistance was confirmed in Palmer amaranth [*Amaranthus palmeri* (S.) Wats.] populations from Arkansas. Later, the amplification and overexpression of chloroplastic glutamine synthetase (*GS2*) was confirmed to be the resistance mechanism in one highly resistant population. It is uncertain when this variation in the gene occurred, and the objective of this study was to verify if *GS2* copy number has been increasing over the years across Palmer amaranth populations with reduced susceptibility to glufosinate in comparison to accessions collected more than twenty years ago. Ten accessions from 2001 (A01 accessions) and thirteen accessions collected in 2020 or 2021 (A20 accessions) were selected for this experiment. A total of 75 plants per accession were sprayed with glufosinate at 575 g ai ha^{-1} to assess mortality. Gene copy number assay was conducted with DNA extracted from nontreated plants from A01 accessions and glufosinate survivors from A20 accessions. Three biological replications were extracted for each accession, and each DNA sample was assessed twice in each primer pair. Gene copy number was calculated relative to reference gene. Glufosinate mortality considerably decreased when comparing Palmer amaranth populations from 20+ years ago to those recently collected. All A20 accessions had at least three survivors, while A01 accessions had 100% mortality. Three clusters were formed with the mortality data. Cluster 1 comprised accessions with mortality ranging from 96 to 100% and included two A20 accessions and all A01 accessions. Cluster 2 and 3 only included A20 accessions. Mortality ranged from 82 to 93% in cluster 2. Cluster 3 included two accessions with low mortality (52% and 68%). Relative to the reference gene, no significant difference was detected for gene copy number among the accessions when grouped by collection year (2001 vs. 2020/21). Also, there was no correlation between mortality and *GS2* copy number. This result indicates that increased glufosinate tolerance is due to unidentified mechanism(s), and investigations are ongoing.

A Mutation in a Transcription Factor Imparts Non-target Site Herbicide Resistance in Rice. Srishti Gupta*; Colorado State University, Fort Collins, CO (141)

Chemical mutagenesis using sodium azide/methylnitrosurea was performed to develop quizalofop-p ethyl (QPE) resistant lines. The TSR1 line had a mutation in acetyl-CoA carboxylase, the target site of QPE. On the other hand, the NTSR1 line had a novel mutation imparting non-target site resistance to group 1 herbicide. Genetic mapping identified a single nucleotide polymorphism in an uncharacterized putative zinc finger transcription factor. Transcription factors (TF) play important role in many biological processes like cell growth, division, and response to abiotic factors, but no mutation in TFs has been associated with herbicide resistance to date. The objective of this research is to understand the molecular mechanism through which a mutation in a transcription factor imparts herbicide resistance. The degree of resistance of TSR1 and NTSR1 was assessed with dose-response curve experiments using QPE, relative to a wild-type rice line. This experiment was done in the greenhouse and QPE was applied at doses ranging from 2.4 to 1,000 g/ha with 0.1% NIS to 3 weeks old plants. Injury rating and biomass were measured 21 days after treatment. The rate of QPE metabolism was measured to determine the basis of resistance in wild-type rice, NTSR1 rice and TSR1 rice collected 1,2,4 and 8 days after treatment using LC-MSMS analysis. This was followed by identifying putative metabolites in plant extracts using LC-MSMS. Putative NTSR genes that may be regulated by the mutated transcription factor and contribute to herbicide resistance were investigated using RNA Seq analysis. In greenhouse assays, the NSTR1 and TSR1 lines were 1.9 and 2.4 more tolerant to QPE than the wildtype. On the other hand, a TSR1/NTSR1 double mutant line was 6.8 times more tolerant to QPE than wildtype.

CRISPR/Cas9 Editing of a *CYP81A* Homoeolog in *Triticum aestivum*.. Olivia A. Landau*[1], Natalya Nersesian[2], Thomas Clemente[2], Dean E. Riechers[3]; [1]University of Illinois, Urbana, IL, [2]University of Nebraska, Lincoln, NE, [3]University of Illinois-Urbana, Urbana, IL (142)

Members of the CYP81A family of cytochrome P450s (P450s) catalyze the detoxification of several herbicide classes, including synthetic auxin herbicides. While the CYP81A family has been examined in several grass crop and weed species, research is lacking for *Triticum aestivum* (allohexaploid bread wheat). Previous phenotypic and metabolic experiments in our lab indicated genes on wheat chromosome 5A significantly contribute to natural halauxifen-methyl (HM) tolerance, a synthetic auxin herbicide. Following HM de-esterification, natural tolerance in wheat is achieved through rapid detoxification of the biologically active form, halauxifen acid (HA), and detoxification is enhanced by the herbicide safener, cloquintocet-mexyl (CM). Furthermore, RNA-Seq and RT-qPCR experiments indicated expression of a *CYP81A* located on chromosome 5A (*CYP81A-5A*) was significantly induced by CM. Since P450s are commonly involved with Phase I detoxification of synthetic auxins, we hypothesized the *CYP81A-5A* encodes an HA-detoxifying enzyme. To address this hypothesis, clustered regularly interspaced short palindromic repeats (CRISPR)/CRISPR-associated nuclease 9 (Cas9) technology was employed to edit *CYP81A-5A* with the goal of creating wheat lines with altered expression or function/activity of *CYP81A-5A*. The CRISPR plasmid was designed to deliver Cas9 and five guide RNAs (sgRNAs) that targeted the coding region or promoter of *CYP81A-5A* using the spring wheat cultivar, CB037. These efforts resulted in T_0 mutants with edits in *CYP81A-5A*, which were self-fertilized to produce T_1 progeny. To create Cas9-free, *CYP81A-5A*-edited progeny, T_1 plants were crossed with unaltered CB037 to produce F_1 progeny. Among 49 F_1 plants (screened by PCR and Sanger sequencing), 14 plants possessed edits in *CYP81A-5A* but all plants were also heterozygous for Cas9. These F_1 progeny were self-fertilized to create F_2 progeny, and among 20 F_2 plants, three Cas9-free plants possessed edits in *CYP81A-5A*. All three Cas9-free plants possessed a mutant allele with a single nucleotide insertion within the promoter region but varied in their zygosity for the mutant allele with two individuals being heterozygous and one being homozygous. The 5B and 5D homoeologs of *CYP81A-5A* in these plants were also sequenced due to the possibility of off-target editing by CRISPR/Cas9, which revealed both homoeologs were unedited in all plants. Progeny that are homozygous and heterozygous for the mutant allele will be examined further for alterations to herbicide tolerance, agronomic and developmental traits, and gene expression patterns in response to CM or other stress-inducing agents.

Reactive Oxygen Species in Herbicidal Activity: What We Know. Catherine J. Traxler*[1], Todd A. Gaines[1], Peter Luemmen[2], Franck E. Dayan[1]; [1]Colorado State University, Fort Collins, CO, [2]Bayer Crop Sciences, Frankfurt, Germany (143)

A natural consequence of stress in plants is the generation of reactive oxygen species (ROS). When plants are exposed to stressful conditions, ROS accumulates, and plants initiate a cascade of defense response mechanisms. This production of ROS can be caused by many biotic and abiotic stressors, such as high light exposure, salt stress, temperature stress, and the application of phytotoxic compounds like herbicides. When ROS production is so high that it overwhelms its protective antioxidative defense mechanisms, cell component damage and lipid peroxidation can occur, leading to membrane breakdown and eventual cell death. The new HRAC classification of herbicides has already provided a section of herbicides that implicate ROS as primary drivers of damage in their modes of action. But many other herbicides have ROS accumulation and oxidative damage as secondary aspects to their primary modes of action. This review aims to outline our current understanding of ROS's

involvement in different herbicide modes of action and provide an overview of oxidative damage as a result of herbicidal application.

Activation of a Singlet Oxygen Signaling Pathway by Competition Cues in Arabidopsis Thaliana. Nicole Berardi[1], Sasan Amirsadeghi[1], Clarence Swanton*[2]; [1]University of Guelph, Guelph, ON, Canada, [2]University of Guelph, Department of Plant Agriculture, Guelph, ON, Canada (144)

Oxidative stress responses of *Arabidopsis* to low red (R) to far-red (FR) signals (R:FR ~ 0.3), generated by a biological weedy and an artificial source of FR light, were compared with a weed-free control (R:FR ~1.4). In the low R:FR treatments, induction of the shade avoidance responses coincided with increased singlet oxygen ($^1O_2$) production and decreased levels of superoxide and superoxide dismutase activity. Although the increase of $^1O_2$ was not due to protochlorophyllide accumulation and did not result in cell death, treatments with the $^1O_2$ generator 5-aminolevulinic acid increased sensitivity to cell death. Transcriptome responses minimally resembled those reported in four *Arabidopsis* $^1O_2$ generating systems such that only a few genes (6 out of 1931) were consistently up-regulated supporting the specificity of

$^1O_2$ signaling. Moreover, suppressors of jasmonate accumulation including the $^1O_2$-responsive amidohydrolase *ILL6* and the sulfotransferase *ST2a,* which is involved in the prioritization of elongation growth versus defense, were consistently up-regulated. Our data support a model in which photoreceptors link low R:FR light cues to the JA signaling pathway. Repression of bioactive JAs via the amidohydrolase *ILL6* and sulfotransferase *ST2a* may promote shade avoidance (versus defense) and $^1O_2$ acclimation (versus cell death) responses to competition cues.

Herbicidal Activity of Chalcones. Raphael M. Garrido*[1], Franck E. Dayan[1], Rosana M. Kolb[2], Luis O. Regasini[3]; [1]Colorado State University, Fort Collins, CO, [2]São Paulo State University, Assis, Brazil, [3]São Paulo State University, São José Do Rio Preto, Brazil (145)

Chalcones are multifunctional molecules with several molecular targets and a broad spectrum of biological activities. They also have good phytotoxic potential and are relatively easy to synthesize. The herbicidal activity and mode of action of chalcones was evaluated. An initial screen tested the herbicidal activity of 33 chalcones on *Digitaria insularis* (sourgrass), *Rottboellia cochinchinensis* (itchgrass), *Urochloa decumbens*(signalgrass), *Amaranthus viridis* (slender amaranth), *Bidens pilosa* (hairy beggartick), *Conyza canadensis*(horseweed), *Emilia fosbergii* (Florida tasselflower), *Mucuna pruriens* (mucuna bean) and *Raphanus raphanistrum* (wild radish). Subsequently, dose-response curves were conducted with chalcones CH1 and CH2 and P1 precursor. Agar solutions of CH1, CH2 and P1 were prepared at 0, 5, 10, 25, 50 and 100 μM in 0.1% ethanol. Square Petri dishes with *Arabidopsis thaliana* seeds were placed vertically in a growth chamber for 21 days under 110 ± 5 μmol/m/s light with a cycle of 12 h light and 12 h dark. After that time, root lengths were measured and IC_{50} were calculated. Phenylalanine ammonia-lyase (PAL) activity was determined spectrophotometrically by measuring the formation of cinnamic acid at 290 nm. The effect of the chalcones on membrane integrity and photosynthetic electron transport was assessed on cucumber cotyledon discs by measuring either change in conductivity of the bathing medium or chlorophyll fluorescence, respectively. Non-linear regression was used to analyze the dose-response curve data using the drc package in R Studio. CH1 was the most potent chalcone at inhibiting root growth, with IC_{50} at 17.3 μM. The P1 and CH2 were less active, with IC_{50} value of 72.3 μM and 31.4 μM, respectively. Compounds P1 and CH2 presented IC_{80} close to 100 μM, whereas chalcone CH1 at 56.1 μM. In the three mode of action experiments, the compounds did not differ from the control, supporting that the chalcones and the precursor P1 do not alter PAL enzyme activity, or destabilize membranes or affect photosynthetic efficiency. An artificial intelligence algorithm suggested that these compounds may inhibit cellulose biosynthesis, but additional laboratory experiments will be necessary to confirm this. While the mode of action of chalcones remains elusive, our activity results show that chalcones and the precursor have potential as herbicides.

Phytotoxicity of Sorgoleone Revealed by Expression of the Biosynthetic Pathway in *Nicotiana benthamiana*. Zhiqiang Pan*, Joanna Bajsa-Hirschel; USDA-ARS, University, MS (146)

Sorghum (*Sorghum bicolor*) is known to possess allelopathic properties and produces allelochemicals such as benzoquinone sorgoleone which accounts for much of the allelopathic activity of the genus. The biosynthesis of sorgoleone involves the polyketide synthase-mediated production of an alkylresorcinolic intermediate derived from 16:3?[9,12,15] fatty acyl-CoA starter unit. This intermediate is then methylated by SAM-dependent *O*-methyltransferases and dihydroxylated by cytochrome P450 enzymes to yield dihydrosorgoleone which then undergoes autoxidation upon rhizosecretion, yielding the bioactive sorgoleone. All the enzymes involved in the biosynthesis of this compound from palmitoleyl-CoA have been identified and functionally characterized. To examine the possibility of producing sorgoleone by expression of the pathway in heterologous systems, a multigene vector was constructed which contains all the genes (SbDES2, SbDES3, SbARS2, SbOMT3, CYP71AM1) required for sorgoleone biosynthesis from palmitoleyl-CoA. The DNA construct was deployed in *Nicotiana benthamiana* leaf tissues via Agrobacterium-mediated transient expression and successfully directed the *in planta* accumulation of sorgoleone, as detected by GC-MS. The leaf areas infiltrated with the multigene construct expressing these sorghum genes developed necrotic lesions, indicating that the compound possesses significant cytotoxicity in plant cells, one of the characteristics of sorgoleone and other allelochemicals. The results suggest that the genes for the biosynthesis of this compound could be used for developing plant-incorporated weed resistant traits.

An Omics Approach to Investigate the Resistance Mechanism to ACCase Inhibitors in Feather Fingergrass (Chloris Virgata*)*. Luiz Augusto Inojosa Ferreira*[1], Rubem Silvério de Oliveira Jr[2], Jamil Contantin[2], Caio A. Brunharo[3]; [1]Penn State University, State College, PA, [2]State University of Maringa, Maringa, Brazil, [3]Penn State University, University Park, PA (147)

Feather fingergrass (*Chloris virgata* Sw.) is an important weed species in tropical regions and is present in crops such as soybeans, corn, cotton, and sugarcane. With the evolution of glyphosate resistance, ACCase inhibitors have become an alternative tool for feather fingergrass management. In this study, we investigated target site and non-target site resistance mechanisms to ACCase inhibitors in feather fingergrass by conducting dose-response experiments with ACCase inhibitors (clethodim, haloxyfop, and pinoxaden), genome sequencing and annotation, and evaluating ACCase copy number variation, gene expression, and mutations. Dose-response experiments revealed cross-resistance to haloxyfop-methyl and pinoxaden, with resistance indices of 432 and 3, respectively. For clethodim, populations showed resistance indices higher than 1, but the field rate was able to reduce the dry biomass by 90%. The draft assembly, produced with short-read Illumina sequences, contained 370M bp, with BUSCO completeness of 97% (3.6% duplication). K-mer analysis estimated a genome size of 360M bp. The heterozygosity estimate of 0.06% suggests a self-pollinating mating system. We annotated over 36k genes with short-read RNA-seq. *ACCase* copy number variation and enhanced gene expression were ruled out as resistance mechanisms because no differences were observed between biotypes. A mutation was detected in the ACCase gene that encodes for position 2027, resulting in a tryptophan-to- cysteine substitution. We are currently investigating the role of reduced absorption, translocation, and enhanced metabolism on the resistance phenotype, in addition to developing quick molecular tests for resistance detection. Our results will shed light on the mechanisms of ACCase resistance in feather fingergrass and generate genomics resources for the scientific community.

Resistance of Sagittaria Montevidensis Biotypes to Acetolactate Synthase Inhibitor and Bentazon Herbicides. Lariza Benedetti[1], Marcus Vinícius Fipke[1], Vívian Ebeling Viana*[1], Diogo da Silva Moura[2], Edinalvo Rabaioli Camargo[1], José Alberto Noldin[3], Luis Antonio de Avila[1]; [1]Federal University of Pelotas, Pelotas, Brazil, [2]Centro de Ensino Superior Riograndense, Marau, Brazil, [3]Empresa de Pesquisa Agropecuária e Extensão Rural de Santa Catarina, Itajaí, Brazil (148)

Sagittaria montevidensis (SAGMO) is an aquatic weed often found in rice paddy fields. The extensive and repetitive use of chemical control, mainly ALS and PSII inhibitors, resistant SAGMO populations have been reported with multiple resistance these MOAs. This study aimed at investigate the mechanism of resistance of the ALS and PSII multiple resistant SAGMO 32 population. To verify the resistance, the SAGMO 32 population (resistant) and SAGMO 35 (susceptive) were analyzed and dose response curve (DLC) using ALS (penoxsulam and imazethapyr + imazapic) and PSII (bentazon) inhibitors were performed. Also, the gene sequence of *ALS* and *D1*(*psbA*) were analyzed using Sanger sequencing. The DLC showed that the SAGMO 32 display resistance to penoxsulam, imazethapyr + imazapic and bentazon herbicides. SAGMO 32 showed a nucleotide substitution in *ALS* gene, relative to the 574 position. We found a Trp574Leu and our data suggests that SAGMO 32 is heterozygous for the mutation. No changes were observed for *D1*(*psbA*) gene, suggesting a non-target resistance mechanism. Thus, new studies should be performed to verify if the resistance to PSII inhibitors is associated to metabolism of the herbicide.

Protoplast Generation from Cell Suspension Culture in Waterhemp [*Amaranthus tuberculatus* (Moq.) J. D. Sauer]. Peter E. Beerbower*, Robert Sabba, Michael J. Christoffers; North Dakota State University, Fargo, ND (149)

Waterhemp [*Amaranthus tuberculatus* (Moq.) J.D. Sauer] has developed resistance to at least seven herbicide modes of action (MoAs): acetolactate synthase (ALS) inhibitors, protoporphyrinogen oxidase (PPO) inhibitors, hydroxyphenylpyruvate dioxygenase (HPPD) inhibitors, photosystem II (PSII) inhibitors, 5-enolpyruvylshikimate-3-phosphate (EPSP) synthase inhibitors, very long-chain fatty acid (VLCFA) synthesis inhibitors, and synthetic auxin herbicides. The emergence of resistance to most common herbicide MoAs for waterhemp is increasing reliance on a dwindling number of alternatives, in turn likely contributing to selective pressure for resistance to remaining herbicide options. The spread of herbicide-resistant waterhemp and related *Amaranthus* spp. throughout the midwestern United States is a major challenge of growing importance in weed management. To expand available research tools for addressing this issue, our research has focused on development of a protocol for generating waterhemp protoplasts from cell suspension culture, facilitating future molecular and genetic research for waterhemp and other *Amaranthus* spp. Currently, this process has three species-specific aspects to be optimized: 1) generation and proliferation of waterhemp cell suspension cultures, 2) viability of waterhemp protoplasts following cell wall degradation, and 3) effective transformation with foreign DNA/RNA/protein. Presently, we have developed a method for the generation of waterhemp cell suspension cultures and are preparing to implement this across additional waterhemp biotypes. Enzymatic cell wall digestion of cultured cells has successfully generated live protoplasts, with viability confirmed using fluorescein diacetate (FDA) staining. Observations thus far suggest that waterhemp protoplasts may be sensitive to the manufacturer of cell wall degrading enzymes. Further, waterhemp protoplasts have been found to be under oxidative stress, suggesting that the inclusion of antioxidants in culture media may be beneficial. Fluorescent microscopy has confirmed successful polyethylene glycol (PEG)-mediated transformation of waterhemp protoplasts with a fluorescent Cas9/GFP fusion protein, but successful transient expression of a transgene for fluorescent protein has not been achieved. Future research will focus on generating cell suspension cultures of different waterhemp biotypes and optimization of protoplast cell culture conditions for transgene expression. Gene expression in protoplasts is necessary to both express foreign DNA transformed into the cell and for endogenous gene expression necessary for cell wall regeneration and cellular division. Successful establishment of this protocol will facilitate genetic engineering techniques, such as CRISPR-mediated gene editing, to target waterhemp genes for herbicide resistance research.

Agrematch AI Platform Leads to the Discovery of AS9057 as a Novel Natural Herbicide. Alyssa Twitty*[1], Yosef Geva[2], Eyal Simonovsky[2], Ifat Shub[2], Yael Phillip[2], Franck E. Dayan[1]; [1]Colorado State University, Fort Collins, CO, [2]Agrematch, Rehovot, Israel (150)

As global agricultural demand increases, many of the issues faced by growers increase subsequentially. Included amongst these issues is weed control, as pest plants can stunt the yield of agricultural plants by competing for resources. Thus, innovative means of weed control, new mode of action discovery, and the discovery of new compounds to manage resistant weeds are more important than ever before. A new artificial intelligence system developed by a startup company, Agrematch, predicts the herbicidal activity and the mode of action of new synthetic and natural compounds. Compound AS9057 was predicted to have new mode of action by this platform, but due to its chemical composition and behavior it was speculated to inhibit PSII. This research aims to assess this assumption. The first step in testing this was performing a dose response of AS9057 on both wild type plants sensitive to PSII inhibitors and plants with a Ser264 mutation imparting resistance to some PSII inhibitors. Photosynthetic electron transport data was then collected for AS9057 and compared with a variety of controls. The data from these two experiments suggested that AS9057 was not inhibiting PSII. More data was then collected to deduce the mode of action of AS9057. AS9057 causes loss of membrane integrity, as measured by increased electrolyte leakage which was associated with accumulation of reactive oxygen species. Additional experiments are on-going to further characterize the mode of action of AS9057.

Transcriptomic Response to Glyphosate in Rapid Response Giant Ragweed(*Ambrosia trifida*). Crystal D. Sparks*[1], Christopher Van Horn[2], Roland S. Beffa[3], Todd A. Gaines[1], Franck E. Dayan[1], Philip Westra[1]; [1]Colorado State University, Fort Collins, CO, [2]Bayer Crop Science, St. Louse, MO, [3]Senior Scientist Consultant, Frankfurt, Germany (151)

There are two biotypes of *Ambrosia trifida* (giant ragweed) that are resistant to glyphosate. This study focuses on a rapid response biotype that displays death of mature leaves within 48 hours of treatment, followed by remarkable regrowth. Previous work evaluated target site resistance by polymerase chain reaction (PCR) amplification of the glyphosate target gene, 5-enolpyruvylshikimate-3-phosphate synthase (EPSPS) and did not identify a target-site mutation, under the assumption that this species has a single EPSPS copy. Autoradiography with radiolabeled glyphosate revealed this biotype has reduced glyphosate translocation, hypothesized to be due to the rapid death of mature leaves quickly limiting source to sink translocation. In this study analysis of short-read RNA sequencing was used to evaluate transcriptomic response to glyphosate treatment. All glyphosate applications were 0.8 kg ha^{-1} acid equivalent, in Generation 4 Research Track Sprayer (DeVries Manufacturing) with a flat tip nozzle at 40 PSI. Differential gene expression analysis initially indicated that the rapid response may be driven by signaling of jasmonate, a phytohormone with roles in response to several biotic and abiotic stresses. Metabolomic analysis revealed similar phytohormone response to glyphosate in resistant and susceptible individuals, implicating requirement of additional pathways or signals for resistance. We present several cellular and metabolic pathways that are enriched in resistant individuals and attempt to place them in the context of glyphosate resistance. Analysis of transcript sequence also identified a second locus of EPSPS. This second EPSPS gene contained a mutation of proline 106 to serine, an allele known to confer glyphosate resistance in other species. Further work, such as genetic mapping, is currently underway to characterize the glyphosate resistance mechanism of the rapid response *Ambrosia trifida*.

Relative Soybean Response from 4 Auxin Herbicides Applied POST. Thomas C. Mueller*[1], Larry Steckel[2]; [1]University of Tennessee, Knoxville, TN, [2]University of Tennessee, Jackson, TN (152)

Auxin herbicides have been a key component to weed management in cropping systems and non-crop settings. Dicamba, 2,4-D, aminopyralid, and aminocyclopyrachlor (ACP) were sprayed over top of non-auxin resistant soybean at 0.001X, 0.01X, and 0.1X of a target rate to simulate tank contamination or possible particle drift. Visual evaluations, leaf samples, and high-quality images were taken at 3, 7, 14, and 21 days after application. Leaf samples were analyzed using extraction followed by liquid chromatography-mass spectrometry to determine herbicide concentration. Herbicide concentration in leaf samples decreased while symptomology increased over time. At 21 DAT, all 0.1X treatments had greater than 20% crop injury with 2,4-D indicating the least. Auxin herbicides can reduce yields at low concentrations. These herbicides have the same or a similar mode of action which makes discerning symptomology difficult even if the exact dosage and time frame is known. Analytical analysis is needed to differentiate between auxin herbicide present. Comparing the slopes for the respective herbicides at the lowest herbicide rate, dicamba had a much greater linear increase in soybean response over time compared to ACP, aminopyralid, and 2,4-D. Dicamba sensitivity to soybean is well documented and these findings are consistent with the extreme sensitivity of soybean to dicamba. This report also documents the relative activity of the other herbicides when soybeans are exposed at a range of concentrations. The observation that the herbicide concentration in the treated leaves is rapidly declining while the apparent soybean response is increasing over time shows the dilemma of trying to obtain soybean leaf samples and verify which herbicide would be responsible for a given soybean response. This is an example where herbicide chemical analysis would be more definitive in conclusively showing which herbicide caused the damage. Given that detectable auxin herbicide residue in soybean is transitory over time, Extension agents and State Department of Agriculture inspectors often rely more upon visual symptomology than herbicide chemical analysis to determine potential (dicamba) soybean exposure.

Cereal Rye Cover Crop Impacted the Fate of PRE-emergence Herbicides Flumioxazin and Pyroxasulfone in Soybean. Jose J. Nunes*[1], Eric J. Miller[2], Lily A. Woitaszewski[3], Shelby E. Lanz[4], Nicholas J. Arneson[1],

John M. Wallace[5], Karla L. Gage[6], Sarah Lancaster[3], Thomas C. Mueller[4], Rodrigo Werle[1]; [1]University of Wisconsin-Madison, Madison, WI, [2]Southern Illinois University, Carbondale, IL, [3]Kansas State University, Manhattan, KS, [4]University of Tennessee, Knoxville, TN, [5]Penn State University, University Park, PA, [6]Southern Illinois University Carbondale, Carbondale, IL (153)

The use of preemergence (PRE) herbicides associated with cereal rye (*Secale cereale* L.) cover crop (CC) has been shown to improve waterhemp (*Amaranthus tuberculatus* (Moq.) Sauer.) control in soybean production systems. Moreover, delaying CC termination until soybean planting (i.e., planting green) can optimize CC biomass production and weed suppression but also impact the fate of PRE herbicides. However, limited research is available on the fate of PREs sprayed over living CC in the planting green system. Thus, the objective of this study was to investigate the impact of CC management (no-till, CC early terminated, and CC planting green) on the fate of the PREs flumioxazin and pyroxasulfone, and waterhemp control in soybean production. Soybean was planted in May of the 2021 and 2022 growing seasons in Illinois, Kansas, Pennsylvania, and Wisconsin. A flumioxazin (70.4 g ai ha^{-1}) + pyroxasulfone (89.3 g ai ha^{-1}) herbicide premix was applied PRE at soybean planting. The cover crop was terminated with glyphosate (1,262 g ae ha^{-1}). Soil samples (three soil cores 10 cm in diameter by 0-7.6 cm of soil) were collected from no-till without PRE (check), no-till with PRE (no-till), CC terminated ~12 days before soybean planting with PRE (CC early termination), and CC terminated at soybean planting with PRE (CC plant green) at 0, 7, and 21 DAT (days after treatment). Soil samples were analyzed for flumioxazin and pyroxasulfone concentration (ng g^{-1} soil) through mass spectrometry. Waterhemp density (plants m^{-2}) was recorded at the time of postemergence (POST) herbicide application. The concentration of both PREs decreased from 0 to 21 DAT in all treatments. Higher concentrations of both PREs were observed under no-till in all sampling timings compared to the CC treatments. Spraying the PREs over a living CC (CC plant green) only reduced flumioxazin at 0 and 7 DAT and pyroxasulfone at 7 DAT compared to the CC early terminated. The increase in CC biomass accumulation at termination correlated to a decrease in herbicide concentration in the soil at 0 DAT. Nevertheless, such reductions in the concentrations of both PREs in the presence of CC biomass did not reduce waterhemp control at the time of POST application compared to no-till. Therefore, we conclude that the presence of CC biomass at the time of PRE application affected the fate of flumioxazin and pyroxasulfone but did not reduce waterhemp control compared to no-till. Lastly, a major difference was observed between no-till and the presence of CC rather than between the two CC termination timings in the concentration of both PREs in the soil.
jjnunes@wisc.edu

How Does Environment Influence VLCFA-Inhibitor Efficacy? Tyler P. Meyeres*[1], Lucas Haag[2], Peter Tomlinson[1], Sarah Lancaster[1]; [1]Kansas State University, Manhattan, KS, [2]Kansas State University, Colby, KS (154)

As weather extremes and herbicide-resistant weeds continue to challenge growers, it is important to characterize the effect of the environment on the efficacy of residual herbicides. The objective of this trial was to compare the impact of environmental conditions on the efficacy of VLCFA-inhibiting herbicides when applied at various times throughout the growing season. To fulfill this objective, a bare-ground field experiment was conducted in Manhattan, KS in 2021 and 2022. The experiment included four herbicides, acetochlor (1682 g ai ha^{-1}), *S*-metolachlor (2142 g ai ha^{-1}), dimethenamid-P (1104 g ai ha^{-1}), and pyroxasulfone (219 g ai ha^{-1}). The dominant weed species was Palmer amaranth (*Amaranthus palmeri* S. Watson). Visible weed control and weed height were recorded 1 to 8 weeks after treatment. Weed density and biomass were collected 8 WAT. Accumulated rainfall (RF) and accumulated soil growing degree days (SGDD) from application to evaluation were collected. Herbicides were applied on April 14, April 29, May 14, June 4, and June 18, 2021 and April 18, May 23, June 21, and July 12, 2022. Accumulated RF received from application to 8 WAT ranged from 11.5 to 34.9 cm and SGDD accumulated from application to 8 WAT ranged from 582.2 to 2031.0 SGDD. Prior to analysis, weed control was transformed to a binary variable, where weed control ratings =80% were considered successful (response =1) and <80% control were considered unsuccessful (response = 0). To test the effects of RF and SGDD on the probability of successful weed control, binary responses were subjected to logistic regression. Individual models were built for each residual herbicide. Accumulated RF and SGDD influenced the probability of successful weed control with applications of acetochlor. The probability of successful weed control decreased as both RF and SGDD increased. When =10 cm RF was received, probability of successful weed control was >64% regardless of SGDD. Interactions of RF and SGDD influenced the probability of successful weed control with applications of dimethenamid-P. The probability of successful weed control increased as both RF and SGDD increased. However, when =8 cm RF was combined with =2400 SGDD the probability of successful weed control was <90%. Interactions of RF and SGDD influenced the probability of successful weed control with applications of pyroxasulfone. In general, the probability of successful weed control increased as the amount of RF and SGDD increased. However, once the amount of RF reached 16 cm probability of successful weed control decreased, regardless of SGDD. Accumulated RF and SGDD influenced the probability of successful weed control with applications of *S*-metolachlor. The probability of successful weed control decreased as both RF and SGDD increased. The probability of weed control was = 83% when =10 cm RF was received regardless of the SGDD. Excessive rainfall and/or high temperatures decreased the probability of successful control for all herbicides evaluated. However, the tested VLCFA-inhibiting herbicides varied in their responses to the environment.

Impact of Blue Light on the Germination Characteristics of Grass Weed Seeds. Sarah A. Chu*, Sarah E. Kezar, Shuyang Zhen, Muthukumar V. Bagavathiannan; Texas A&M University, College Station, TX (155)

Light quality plays an important role in plant growth and physiology. Previous research has established the role of red and far-red light on various plant species, especially on seed germination. However, the influence of other wavelengths of light, such as blue light (320-490 nm), has been less investigated. The objective of this study was to determine how blue light impacts the germination rate, seed dormancy, and viability for three different weed species: barnyardgrass (*Echinochloa crus-galli*), large crabgrass (*Digitaria sanguinalis*), and johnsongrass (*Sorghum halepense*). A known number of seeds for each species was exposed to five different intensities and seven exposure times of blue light, and the germination of the treated seed were observed every 4 days for 21 days. Following this, a tetrazolium test was conducted to determine the viability of the non-germinated seed. Any seeds that did not germinate or viable were considered dead. Blue light intensity had little influence on seed germination, but high exposure time (60 sec) significantly reduced the germination of barnyardgrass (23% reduction), but not other species. With respect to seed mortality, blue light at 966 (moderate) μmol m^{-2} sec^{-1} intensity for a short exposure time (5 sec) increased by 23% the morality for johnsongrass, a response absent with other species. Preliminary results suggest that there is a significant interaction between blue light intensity, duration of exposure, and weed species on weed seed germination and viability. Further research is ongoing to gain deeper insights into these interactions.

Integrated Weed Management Using Biosolarization, Cover Cropping and Strip Tillage. Dwayne D. Joseph*[1], Alan W. Leslie[2], Kurt M. Vollmer[3], Cerruti Rr Hooks[2]; [1]University of Maryland, Chestertown, MD, [2]University of Maryland, College Park, MD, [3]University of Maryland, Queenstown, MD (156)

Biosolarization is a soil disinfection technique similar to solarization but involves the addition of organic amendments to the soil prior to the passive solar heating process. Research has shown that biosolarization can increase weed seed mortality and decrease soil pathogens. Moreover, biosolarization is compatible with organic farming and can be used in suboptimal climates where solarization isn't effective. Fruit processing by-products (pomace) are promising biosolarization soil amendments because they are rich in organic compounds, don't pose any biohazard risks, and are relatively abundant and inexpensive. In this study, the biosolarization potential of a mixture of apple and grape pomace combined with a between-row living mulch was investigated as an integrated pest management (IPM) technique. Our objectives were to demonstrate the use of biosolarization, conservation-tillage and cover cropping to: 1) reduce nematode, weed and insect pests, and 2) enhance crop yield. The study was performed at the Central Maryland Research and Education Center in Upper Marlboro, MD in organically managed eggplant. The study included four replicated treatments comprising of eggplant: 1) grown in living mulch + no-till (LM-NT), 2) interplanted with cover crops (LM), 3) grown in solarized soil (Sol), or 4) interplanted with a cover crop and grown in biosolarized soil (Biosol). The results indicated that Biosol plots had the lowest mean number of broadleaf weeds at all rating times. Broadleaf weeds were greatest in LM plots throughout the study. Sedge weeds were significantly greater in Biosol and Sol plots compared to LM and LM-NT plots at all rating times. There were significantly more grass weeds in LM plots than LM-NT, Biosol and Sol plots at 2 and 4 WAP. However, at 6 and 9 WAP, Sol plots contained significantly more grass weeds than the other treatments. Eggplant yield was greatest in LM followed by LM-NT, Biosol then Sol plots. The results tentatively show that broadleaf weed seed were inactivated by biosolarization, however sedge control was minimal. Furthermore, biosolarization may be an effective IPM technique in organic vegetable production.

Ecology and Landscape Scale Management of a Globally Invasive Weed; *Nassella trichotoma* (Serrated Tussock). Talia J. Humphries*; Federation University Australia, Ballarat, Australia (157)

Integrated Weed Management in Olive Orchards. Veronica Pedraza[1], Juan A. Lezaun[2], Irache Garnica[2], Jose L. Gonzalez-Andujar*[3]; [1]Instituto de Agricultura Sostenible (CSIC), Córdoba, Spain, [2]Instituto Navarro de Tecnologías e Infraestructuras Agrarias (INTIA), Villava, Spain, [3]Instituto de Agricultura Sostenible (CSIC), Cordoba, Spain (158)

The conventional soil management and weed control methods in olive orchards have caused, over the years, serious problems of water availability, erosion, the appearance of herbicide resistance, as well as a decrease in farmland biodiversity and the associated ecosystem services. In this context, this study highlights the importance of using integrated weed management techniques (IWM), which allows farmers to combine crop production with the protection of the agroecosystem and the conservation of the soil productive potential. A 2-year study with four IWM strategies in 3 plots and a randomized complete block design with 4 replications per strategy were established. Strategy CC included 'no tillage with chemical control' in the intra-row spacing and 'cover crops' in the inter-row spacing. The cover crops comprised spontaneous grasses (*Bromus* spp) in southern Spain and sown crucifers (*Sinapis alba*) in northern Spain. Strategy TL involved 'tillage' combined with pruning wood residues in both sampling areas of southern Spain and strategy NT 'no tillage with chemical control' in both areas of northern Spain. Effects on the weed community and olive crop were evaluated analyzing the richness, abundance, diversity (Shannon index) and equity (Pielou eveness index), as well as the olive yield and quality. Linear mixed effects models were used to test for differences between IWM strategies and were adjusted using the *lmer* function from the *lme4* library in the R environment. In all cases, the variable 'plot' was included as a random effect and the variables 'IWM strategy' and 'sampling area' were included as fixed effects. The statistical significant of the effects was obtained by the function *Anova* and *Tukey* test at a 5 % significant level. Results showed a greater richness, abundance and diversity in the CC intra-rows than

TL, as well as a greater richness and diversity in the CC inter-rows in southern Spain. In the north of Spain, a greater richness was observed in the NT inter-rows, but diversity and equity indices only showed significant differences in the CC intra-rows the second year. Moreover, yield and quality results were not affected by the IWM strategy at any location. These preliminary results seem to indicate that the introduction of IWM in olive orchards makes it possible to reconcile crop production and beneficial weed flora at a manageable threshold if a sustainable agricultural system is to be achieved in the long-term.

Atrazine Wash Off from Cereal Rye (*Secale cereale* L.) Residue Onto the Soil After Simulated Rainfall.

Lucas Oliveira Ribeiro Maia*[1], Bryan G. Young[2], William G. Johnson[1], Eileen J. Kladivko[1], Shalamar Armstrong[1]; [1]Purdue University, West Lafayette, IN, [2]Purdue University, Brookston, IN (159)

The application of soil residual herbicides at cover crop termination is often recommended as part of an integrated weed management strategy. Previous research suggests that cover crop biomass accumulation significantly reduces the amount of residual herbicide that reaches the soil at the time of application. Once intercepted by cover crops, these herbicides can only move to the soil with rainfall or irrigation. To date, limited research has been published on the effect of rainfall on residual herbicide wash off from cover crops onto the soil. A field trial was established to investigate the influence of rainfall on atrazine wash off from cereal rye residue onto the soil. The experiment followed a split-plot design with rainfall volumes (0, 12.5, and 25 mm) as the main plot and cover crop orientations (standing and roller crimped) and a fallow control as the subplot. Cereal rye was roller crimped at flag leaf growth stage, the day before herbicide application and rainfall simulation. Atrazine was applied 30 minutes before rainfall simulation. Rainfall was simulated for 20 minutes for all treatments. Plant and soil (0 to 5 cm depth) samples were taken 90 minutes after rainfall simulation and used to determine atrazine concentrations. After 25 mm of rainfall, the concentration of atrazine was reduced by 67 and 28% in standing and roller crimped cereal rye residue, respectively, compared to the cereal rye residue that did not receive rainfall. Approximately 88, 60, and 65% of the applied atrazine was recovered in the top 5 cm of soil under roller crimped, standing cereal rye, and fallow, respectively, after 25 mm of rainfall. Results from this research suggests that the physical barrier created by the roller crimped cereal rye is reducing atrazine leaching below the top 5 cm of soil while providing a slow-release of the herbicide onto the soil during rainfall. Although this slow-release effect may be beneficial from an environmental perspective, the weed control efficacy from atrazine is likely to be reduced. E-mail for contact: oliveirl@purdue.edu

Changing Labor Landscape in Northeast Weed Management for Specialty Crops. Laura Pineda-Bermudez*, Lynn M. Sosnoskie; Cornell University, Geneva, NY (160)

Specialty crops are significant contributors to the agricultural economy of the United States, with 15,656,245 acres nationwide dedicated to specialty crop production. Weeds threaten yield quantity and quality in agricultural systems through direct crop competition, by complicating and contaminating harvest, and by harboring pests and pathogens. Specialty crop growers have limited tools available to control weeds in their operations, with limited access to chemical control methods and low threshold for risk of loss or damage to their high-value crops. These operations also have the highest share of labor costs as a proportion of total expenses compared to non-specialty crop operations. Additionally, recent scarcity and rising costs for both herbicides and labor have made it increasingly difficult to solely depend on these weed control methods. Adoption of novel weed control technologies that reduce labor input will likely be the future of weed management, thus changing the labor landscape of specialty crop production.

Summer Cover Crop Suppression of Broadleaf, Grass, and Nutsedge Weeds. Moriah Williams*[1], Nathan Boyd[2]; [1]University of Florida, Saint Petersburg, FL, [2]University of Florida, Balm, FL (161)

In the past couple of years, strawberry production has been faced with the appearance of new problem weeds, including ragweed parthenium and green kyllinga. This study is one component of an integrated management approach to suppressing problem weeds in strawberry production. Cover crops have been used in sustainable agriculture to reduce erosion, runoff, and leaching, improve soil health, manage pests, manage plant pathogens, attract and conserve beneficial insects, and suppress weeds (Wang et al., 2019). The objective of this study is to assess the effectiveness of mixtures and monocultures of cowpea, sorghum-sudan grass, and sunn-hemp for suppressing target weeds on a complexity gradient. The results of this study will contribute to the development of recommendations for the management of problem weeds in Florida strawberry production. The trial was conducted in two fields at Gulf Coast Research and Education Center. Monocultures and mixtures of cowpea, sunn hemp, and sorghum-sudan grass were seeded at fifty pounds an acre. Treatments included a weedy fallow, monocultures of the selected cover crops, three bicultures of the cover crops, and a three-way mixture (polyculture). During growth, weed counts, ground cover, cover crop, weed biomass, and light penetration measurements were collected. Combinations of cowpea and sunn hemp had the highest percentage decrease of grass, sedge, and broadleaf biomass compared to other mixtures, with the next highest overall decrease being with the mixture of cowpea, sunn hemp, and sorghum-sudan grass. We conclude that the combination of sunn hemp and cowpea resulted in a synergistic effect where the cowpea filled in the understory while the sunn hemp lost leaves during growth, which helped maintain ground cover.

Field Scale Evaluation of the Harrington Seed Destructor in Alberta, Canada. Breanne D. Tidemann*[1], K. Neil Harker[1], Hugh Beckie[2]; [1]Agriculture and Agri-Food Canada, Lacombe, AB, Canada, [2]University of Western Australia, Perth, Australia (162)

Field Scale Evaluation of the Harrington Seed Destructor in Alberta, Canada. Tidemann, B.D.*[1], Harker, K.N.[2], Beckie, H.J.[2] [1]Agriculture and Agri-Food Canada, Lacombe, AB. [2]Retired. Alternative weed control strategies are necessary for management of herbicides resistant weeds. One strategy used primarily in Australia, but growing in global interest, is physical impact mills to prevent weed seedbank additions at harvest. A physical impact mill, the original Harrington Seed Destructor, was tested from 2017-2020 in 20 producer fields within an approximately 50km radius of Lacombe, AB. In each producer field a dense weed patch was identified, mapped, and divided into three replicates. Each replicate contained an untreated check treatment, and a treatment where the Harrington Seed Destructor, was operational as part of crop harvest. The mill was operated for three harvests, 2017, 2018 and 2019, with pre-harvest weed density assessments conducted, as well as pre-spraying seedling weed densities the following spring (2018, 2019 and 2020, respectively). Soil seedbank samples were also collected in spring of 2020. ANOVA statistical analysis of 2020 seedling density and total weed population was conducted for the most abundant weeds in each field using Proc. GLIMMIX in SAS 9.4. Limited significant differences between weed densities were observed between the regular harvest and the Harrington Seed Destructor treatments. Exploratory PCA analyses were conducted for seedling and seedbank weed densities, but show limited separation between treatments, although 95% confidence interval ellipses are showing some shape differences. Limitations to the study design that may have limited the ability to detect differences between treatments include high weed seedbank densities, non-optimal combine settings, limited timeframe of study, unmeasured seed losses at harvest and high weed density variability. Investigation into some of these factors could help to more accurately assess the potential of the Harrington Seed Destructor and other physical impact mills in Alberta and in Canada in general.breanne.tidemann@agr.gc.ca

Weed Suppression Provided by a Second Season Perennial Living Mulch and Self-reseeded Cover Crop. Veronica Yurchak*, Alan W. Leslie, Cerruti Rr Hooks, Bill Phillips; University of Maryland, College Park, MD (163)

Cover crop residues and living mulches have been used to suppress annual and/or perennial weed species, and thus can be incorporated as part of an integrated weed management (IWM) program. Although much research has focused on the use of conservation tillage and cover cropping for weed suppression in soybeans, surprisingly little has been conducted to determine how these practices impact weed suppression during the critical period of weed control (CPWC). In soybeans, the CPWC extends until the V3 or third trifoliate stage. Weeds emerging after this stage are unlikely to significantly impact final grain yields, and as a result, management beyond this stage is often not performed. However, continued weed management may be needed to prevent weed seeds from entering the soil seedbank. Cover crop mulches may suppress weed emergence and slow the development of weeds that successfully emerge, thus preventing them from reaching maturity and producing viable seeds. Experiments were conducted over two field seasons to investigate the use of conservation tillage in concert with an interplanted living mulch and/or cover crop residue for managing weeds during the CPWC and until canopy closure, in soybeans. Whole plot treatments included: (1) conventional till, (2) no-till with cover crop residue, (3) living mulch + cover crop residue, and (4) living mulch + winter killed residue. The split-plot factor consisted of weed management duration: (1) weed removal through canopy closure, (2) weed removal through CPWC, and (3) no weed removal. All cover crop systems suppressed weeds through the CPWC. However, reproductive stages of various species were found in some cover crop subplots where weeds were managed only through the CPWC, indicating the potential of enhancing the soil seedbank if weeds are not managed beyond the CPWC.

Can Exogenous Application of Phytohormones Involved in the Stimulus of Seed Germination Become Part of Integrated Waterhemp (*Amaranthus tuberculatus*) Management Programs? Jose J. Nunes, Nicholas J. Arneson, Rodrigo Werle*; University of Wisconsin-Madison, Madison, WI (164)

The rapid evolution and widespread distribution of waterhemp (*Amaranthus tuberculatus* (Moq.) Sauer) populations resistant to several herbicide sites of action limit farmers' options for effective chemical control. Hence, the need for integrating innovative practices to reduce the reliance on herbicides for waterhemp control. The use of phytohormones to manipulate waterhemp seed germination is an area of research to be further explored. Thus, the objective of this study was to investigate the effects of the phytohormones gibberellic acid (GA3) and abscisic acid (ABA) on waterhemp and soybean emergence as a novel approach for integrated weed management by attempting to manipulate waterhemp seed germination/dormancy. Studies were conducted under greenhouse and field conditions. The greenhouse dose-response study evaluated different doses of GA3 (0.0, 14.0, 28.0, 42.0, 56.0, 84.0, and 112.0 g ha^{-1}) and ABA (0.0, 18.2, 36.5, 73.0, and 146.1 g ha^{-1}) sprayed over two soil types (clay loam [pH 6.4, sand 18%, silt 53%, and clay 30%] and sandy loam [pH 6.0, sand 82%, silt 7%, and clay 12%]) on the emergence of five waterhemp populations (W20, W25, W39, W89, and W21) collected from corn and soybean fields in WI. Approximately 130 seeds of each waterhemp population were seeded to the two soil types filled in plastic plots and the phytohormones sprayed over the soil immediately after waterhemp seeding. The % of waterhemp emergence was recorded at 21 DAE (days after establishment) and described by a linear regression model. The field study evaluated three treatments (check with no phytohormone application, GA3 [56.0 g ha^{-1}], and ABA [73.0 g ha^{-1}]) in two locations (Brooklyn and Lancaster, WI) on waterhemp and soybean emergence during the 2021 and 2022 growing seasons. Phytohormones were sprayed right after soybean planting and waterhemp density (plants m^{-2}) was recorded at 14, 21, and 28 DAE, and soybean stand (plants m^{-1}) at 28 DAE. It was hypothesized that GA3 and ABA would have a stimulative and a suppressive effect on waterhemp emergence, respectively. However, the greenhouse study did not show strong tendencies in waterhemp emergence response to GA3 and ABA doses across populations and soil types. Moreover, most coefficients of determination (R2) were close to zero indicating poor response to the increase in the phytohormones' doses. The field study was designed to test this concept under natural waterhemp emergence conditions and the phytohormones' effect on crop response. Corroborating the greenhouse observations, no significant differences in soybean and waterhemp emergence were observed in the field studies. These studies tested a novel concept that could have positive effects on integrated weed management by manipulating waterhemp seed germination/dormancy. However, non-significant responses were obtained with the adopted methodology, which simulated a way that practitioners could potentially adopt it (via a broadcast application). Nevertheless, our results can serve as a starting point for novel methodological approaches for future research on this concept. rwerle@wisc.edu

Integrative Effect of Row Spacing and Herbicide Programs on Photosynthetically Active Radiation (PAR) in Corn Infested with Multiple Herbicide-resistant Palmer Amaranth. Ramandeep Kaur*[1], Stevan Knezevic[2], Yeyin Shi[2], Nevin Lawrence[2], Parminder Chahal[2], Rachana Jhala[2], Amit J. Jhala[1]; [1]University of Nebraska-Lincoln, Lincoln, NE, [2]University of Nebraska, Lincoln, NE (165)

Multiple herbicide-resistant Palmer amaranth is among the most troublesome summer annual broadleaf weeds across the mid-south, southeastern, and north central United States. It uses C_4 photosynthetic pathway, thus having greater growth rate. It has evolved resistance to number of herbicide sites of action. Agronomic and weed management strategies are required that can alleviate the evolution and spread of multiple herbicide resistant weeds. Integration of narrow row spacing with herbicide application might augment herbicide efficacy for its management. The objectives of this study were to determine the effects of row spacing and herbicide programs on photosynthetically active radiation (PAR) in glyphosate/glufosinate-resistant corn infested with ALS-inhibitors, atrazine, and glyphosate-resistant Palmer amaranth. Field experiments were conducted during the summer 2020 and 2021 in a grower's field infested with ALS-inhibitors/atrazine/glyphosate-resistant Palmer amaranth near Carleton, NE. With 38cm row spacing, Acetochlor/flumetsulam/ clopyralid fb Glufosinate received highest interception of 90 to 97% as compared to 76cm row spacing, thus this treatment provided having higher yield and 95% to 99% season long- control of ALS inhibitors/ atrazine/ glyphosate -resistant Palmer amaranth. The results of this research illustrate that PAR interception with narrow spacing will act as integrative approach in combination with herbicide programs to manage the multiple herbicide resistant Palmer amaranth in glyphosate/glufosinate-resistant corn.

Cover Crops as a Weed Seed Bank Management Tool. Cynthia Sias[1], Bethany R. Wolters[2], Mark S. Reiter[3], Michael L. Flessner*[1]; [1]Virginia Tech, Blacksburg, VA, [2]University of Tennessee, Martin, Martin, TN, [3]Virginia Tech, Painter, VA (166)

The herbicide resistant weed epidemic and increasing pesticide regulations have led us in search of other forms of weed suppression. The ability to manipulate the weed seed bank environment through cover crops could be an effective integrated weed management tool that reduces weed seed survival, promotes dormancy, and/or discourages germination. Our objective is to connect literature on weed seed fate to literature on how cover crops alter the environment surrounding weed seeds. We stop short of weed establishment as this is covered in other works. Key findings of this review were as follows: Cover crops alter the weed seed bank environment by influencing survival, dormancy, and germination. Weed seedling density may be reduced as a result of decreased soil temperature and fluctuations thereof, light, and soil nitrogen. Weed seedling density may be increased by greater soil moisture, soil nitrogen, and oxygen. The complexity of differing responses due to location and environment deters from providing uniform recommendations for producers. For greatest potential, management should maximize cover crop biomass, decrease soil nitrogen, and delay termination. Future research on using cover crops for weed management should include measurements of weed seed banks, including dormancy status, predation, and germination. This review was recently published in the *Italian Journal of Agronomy:* doi: 10.4081/ija.2021.1852.

Evaluating Different Deep Learning Models for the Detection of Poison Ivy (*Toxicodendron radicans) in Images*. Renato Herrig Furlanetto*[1], Nathan Boyd[2], Ana Claudia Buzanini[1], Alex G. Rodriguez[3]; [1]University of Florida, Wimauma, FL, [2]University of Florida, Balm, FL, [3]University of Florida, Belle Glade, FL (168)

Poison ivy (*Toxicodendron radicans*) is the most common toxic plant encountered in Florida parks, woodlands, home landscapes and rural field and the contact with all parts of this plant can cause severe skin burning, followed by redness, swelling and watery blistering. The recognition and differentiation of Poison ivy (*Toxicodendron radicans*) in relation to other non-toxic species can be difficult by the similarity in leaf development and often misidentified, representing a potential health risk, especially for children and animals. With the advancement of real-time processing technology, machine learning and computer vision, these tools have become a great ally for the identification of different plant species quickly and accurately manner. Thus, in this work we evaluate and compare two convolutional neural networks for detection of Poison ivy (*Toxicodendron radicans*) leaves using images. Ten different versions of Yolov5 and six versions of Yolov7 were trained to identify leaves in images. The dataset used was composed of 1427 Poison ivy (*Toxicodendron radicans*) images taken from different conditions, in which the leaves were manually annotated, totaling approximately 6.000 annotations. The transfer learning and augmentation processing were applied to increase and simulate different environment conditions, being applied random noise, brightness, rotation and blur. The training was performed using the University of Florida supercomputer (HiPerGator) with 4 cpus, 31 gb RAM and 1 gpu Nvidia Ampere A100 with 80 gb. After training, the performance of the trained models were tested using random images to evaluate the detection performance. The metrics evaluated were accuracy, precision, recall and time to train the model. The results shown that Yolov7-E6 achieved the best accuracy between the models, showing 70.3% and 2.24 hours to train, followed by Yolov5x6 + TTA with accuracy of 69.7% and 1.18 hours to train. The worst models observed were for Yolov5n (63.6% and 0.76 hours) and Yolov5s (64.7% and 0.72 hours), respectively. The remaining models achieved accuracy between 67.1% to 69.2% and 0.82 hours to 3.52 hours to train. During the evaluation step, using different images to perform the Poison ivy (*Toxicodendron radicans*) leaves identification using the models trained, the results shown that Yolov5 models identified between 15 to 33 leaves and for Yolov7 were identified between 24 to 52 leaves, mis classification and under classification were observed for both models evaluated, indicating that are necessary more images to train the model to recognize leaves in different scenarios and development stages. Therefore, these results suggest that is possible to use these trained models to identify Poison ivy (*Toxicodendron radicans*) in real condition to avoid poisoning and potential skin injuries. The next step is to develop a mobile app to identify in real time the plant leaves.

Multi-year Herbicide Comparison for Italian Ryegrass (Lolium Perenne Ssp. Multiflorum) Control in Wheat. Wyatt J. Stutzman*, Kevin Bamber, Michael L. Flessner; Virginia Tech, Blacksburg, VA (169)

With the ever-present possibility of herbicide resistance, the constant need to diversify weed management techniques and creating variety in herbicide modes of action, an experiment was completed spanning the years 2018 to 2021 to evaluate herbicide control methods for Italian ryegrass (*Lolium multiflorum* Lam) in wheat production systems. A total of 7 site years were completed in Blacksburg and Blackstone, evaluating eleven combinations of herbicide treatment and application timings. Applications conducted include seven days pre-planting (PRE), seven days after planting (DPRE), fall post and spring post. Trials were conducted in a RCBD design with four replications and nontreated checks in each replication. Visual ratings evaluated plots for Italian ryegrass control at subsequent weekly intervals after application. Interactions were observed between treatments and weeks after application (WAA). In all years, all treatments were significantly different from the untreated check ($a \leq 0.05$). In all years except for 2020, the greatest control was observed in pyroxasulfone DPRE, and pyroxasulfone and carfentrazone applied seven days pre-planting, with up to 95 and 85 % control respectively. This research shows that there is observed control in PRE and DPRE applications and differing combinations of POST applications. Future research should continue to evaluate treatment combinations of PRE and POST herbicide treatments together over time.

Weed Suppression from Frost-seeded *Brassicaceae* Cover Crops. Huong Nguyen*, Olivia Fisher, Amy Fox, Kristen Loria, Kathryn Marini, Chris J. Pelzer, Adam Sharifi, Domenic Varma, Sandra Wayman, Matthew R. Ryan; Cornell University, Ithaca, NY (170)

Cover crops can suppress weeds through a variety of mechanisms including competition for resources. Most research on cover crops has focused on grass and legume cover crops that are seeded in the fall. A field experiment was conducted in central New York in 2022 that compared cover crops in the Brassicaceae family that were seeded in the spring. A randomized complete block design with 4 replications was used to compare 10 treatments (10 brassica cover crop species) and two control treatments (red clover and no cover crop). The ten *Brassicaceae* species and red clover were frost-seeded on March 23 into plots within a field where organic soybean (*Glycine max* (L.) Merr.) was no-till planted into rolled-crimped cereal rye (*Secale cereale* L.) in 2021. Ground cover, cover crop biomass, and weed biomass were sampled in early June. Results show a negative relationship between weed biomass and cover crop biomass and that collard (*Brassica oleracea* L., var. viridis) provided among the greatest ground cover and weed suppression. However, the biomass of all cover crops was low, indicating that frost seeding may not be an appropriate seeding method when soil is covered with soybean residue and mulch from cereal rye.

Evaluation of Two-pass Herbicide Programs for Broad-spectrum Weed Control in Conventional Corn in Wisconsin Atrazine Prohibition Areas. Ahmadreza Mobli*[1], Ryan P. DeWerff[2], Nicholas J. Arneson[1], Rodrigo Werle[1]; [1]University of Wisconsin-Madison, Madison, WI, [2]University of Wisconsin, Madison, WI (171)

Effective herbicide selection is of great importance to Wisconsin conventional corn growers farming in atrazine prohibition areas. Per our Extension clientele's request, a field study was conducted at Arlington and Brooklyn, Wisconsin, in 2021 and 2022 (4 site–years) to evaluate the weed control efficacy of eleven comprehensive two-pass (PRE followed by POST) herbicide program treatments, without atrazine, in conventional corn production. The most abundant weed species in Arlington were common ragweed (Ambrosia artemisiifolia L.), velvetleaf (Abutilon theophrasti Medik.), and giant foxtail (Setaria faberi Herrm.), and in Brooklyn, waterhemp (Amaranthus tuberculatus [Moq.] J.D.Sauer) and velvetleaf. Nine and eight out of the 11 two-pass treatments investigated provided >90% giant foxtail and broadleaf weed control in 2021, respectively; however, only three treatments provided >90% broadleaf control and 79% was the highest giant foxtail control in 2022. Higher end-of-season broadleaf and grass weed control was observed in 2021 compared to 2022 due to high weed pressure and lack of timely early-season activating rainfall for soil residual herbicides in 2022. The two-pass treatments Harness® MAX fb Diflexx® Duo (94%-100%,) and Surestart® II fb Accent® Q + Status® (85%-100%) provided the most consistent and effective broadleaf control across all site-years. These results demonstrate that a well-designed and timely deployed two-pass herbicide program can provide effective control of common broadleaf and grass weeds in conventional corn production assuming moderate weed pressure and timely rainfall for soil residual herbicide activation, otherwise effective weed control, particularly of grass weeds, in the absence of atrazine and glyphosate can be challenging. Although this weed control study was designed for conventional corn in Wisconsin atrazine prohibition areas, our results are also translatable to support weed management decisions for glyphosate-resistant corn hybrids in non-atrazine prohibition areas.

A Multi-state Evaluation of Weed Seed Fate During Soybean Harvest. Jacob E. Vaughn*[1], Joseph T. Ikley[2], Vipan Kumar[3], Sarah Lancaster[4], Christy L. Sprague[5], Rodrigo Werle[6], Nicholas J. Arneson[6], Kevin W. Bradley[1]; [1]University of Missouri, Columbia, MO, [2]North Dakota State University, Fargo, ND, [3]Kansas State University, Hays, KS, [4]Kansas State University, Manhattan, KS, [5]Michigan State University, East Lansing, MI, [6]University of Wisconsin-Madison, Madison, WI (172)

The increasing prevalence of herbicide-resistant weeds in U.S. agriculture has led to increased interest in non-conventional methods of weed control, including harvest weed seed destruction. Harvest weed seed destruction refers to any method of destroying or managing weed seed that has been produced by escaped weeds, such as windrowing, burning, or grinding chaff as it exits the combine harvester. The use of combine-mounted impact mills as a means to achieve seed destruction began in Australia to control resistant ryegrass *(Lolium spp.*) in wheat (*Triticum aestivum*). Research has been conducted in the U.S. to determine the efficacy of commercially available impact mills such as the Seed Terminator™ or Redekop™ implements, but the ultimate efficacy of such mills is dependent on the percentage of total seeds that reach them. In 2020 and 2021, an experiment was conducted in Illinois, Indiana, Iowa, Kansas, Nebraska, and Missouri to determine the fate of the seed of common weed species during soybean harvest. Header loss, threshing loss, and grain tank contamination of waterhemp (*Amaranthus tuberculatas*), palmer amaranth (*Amaranthus palmeri*), common ragweed (*Ambrosia artemisiifolia*), common lambsquarters (*Chenopodium album)* and giant foxtail (*Setaria faberi*) were evaluated across the experimental locations. Depending on location and weed species, between 1 and 42% of the total weed seed collected were lost at the header and did not enter the combine due to shatter as the reel contacted the weed seedheads. A greater fraction of the total weed seed collected (58 to 99%) passed through the machine and was accounted for as threshing loss as it exited the rear of the combine with the chaff fraction. Grain tank contamination was also highly variable by year and weed species. Pigweed species were found to be present from 51,600 to 630,000 seeds per bushel of soybean harvested, while common ragweed occurred in the range of 2,700 to 81,000 seeds per bushel of soybean harvested. The results from these experiments indicate that while the majority of weed seeds could enter the combine and be available for destruction with on-combine impact mills, there are significant fractions that shatter before entering the machine, or enter the grain tank.

Cover Crop Residue Management Strategies Impact Deposition and Washoff of Soil-applied Residual Herbicides. Cody Smith[1], John M. Wallace*[2], Mark VanGessel[3], Michael L. Flessner[4]; [1]Penn State University, University Park Pa, PA, [2]Penn State University, University Park, PA, [3]University of Delaware, Georgetown, DE, [4]Virginia Tech, Blacksburg, VA (173)

Use of cereal rye surface mulch remains a promising integrated weed management (IWM) tactic for early-season suppression of small-seeded summer annual weeds. However, previous research has demonstrated that use of surface mulch can result in reduced soil bioavailability of residual herbicides. Further research is necessary to optimize the benefits of cultural and chemical weed control tactics while minimizing drawbacks. Cover crop management decisions that likely influence soil bioavailability of residual herbicides include (1) the growth stage of cover crops at the time of application, which influences total residue mass; (2) taxonomic traits of cover crops and use of mixtures, which may influence cover crop quality; (3) use of residue management tools such as the roller-crimper prior to application, which influences residue architecture; (4) and the timing of residual herbicide application relative to cover crop termination, which influences the absorption and adsorption rates of herbicides. A field experiment was replicated at Rocks Springs, PA, Georgetown, DE, and Blacksburg, VA in 2022 to investigate cover cropping management effects on the soil bioavailability of residual herbicides using a two-factor randomized complete block design with a split plot treatment structure. The main plot included alternative cover crop management tactics, including cereal terminated and left standing at the (1) flagleaf stage; (2) heading stage; (3) and anthesis stage, and (4) cereal rye; (5) crimson clover + cereal rye; and (6) hairy vetch + cereal rye that was roll-crimped at cereal rye anthesis. Split plot treatments including application of a test residual herbicide (pyroxasulfone; 0.09 lb ai/ac) applied (1) at the time of cover crop termination (glyphosate 1.12 lb ae/ac; 0 DAT); or (2) as an EPOST treatment (21 DAT). Field assays were employed to measure deposition and washoff processes relative to a no cover crop control by placing 50 g of sieved field soil in a petri dish positioned at the soil surface and beneath cover crop residues. Soil samples were collected 2 to 4 hr after residual herbicide application (deposition assay) and replaced with another set of petri dishes that were collected immediately after an incorporating rainfall (washoff assay). Pyroxasulfone (ppb) was extracted from soil samples using acetone as the solvent and analyzed using HPLC methodology. Pennsylvania results are reported: A significant interaction ($F = 3.3$; $p = 0.01$) was observed between cover crop strategy and residual application timing. When applied 0 DAT (green), pyroxasulfone recovery (deposition + washoff) was lower for treatments applied at the anthesis stage (< 33%) compared to early stages (55-78%), but no differences among cover crop treatments were observed at the 21 DAT treatment (43 to 63%). The washoff process contributed a greater proportion of total herbicide recovered in roll-crimped treatments (45 to 80%) compared to standing treatments 0 DAT, with hairy vetch + cereal rye resulting in the lowest deposition (4% of control) and greatest washoff (80% of total recovery). Preliminary results suggest that delaying cover crop termination to the anthesis stage and use of a roll-crimper reduces the bioavailability of pyroxasulfone at the time of an incorporating rainfall. Delaying application (21 DAT) of residuals may increase soil bioavailability relative to 0 DAT. These experiments will be replicated in 2023, resulting in six site-years to further identify trends in cover crop by residual herbicide interactions on bioavailability of residual herbicides.

Goosegrass (*Eleusine indica)* Control Options in Bentgrass Putting Greens. Bridgette C. Johnson*, Joseph S. McElroy; Auburn University, Auburn, AL (174)

US-Herbicide Resistance Action Committee (HRAC): Who We Are and What We Do? Zahoor A. Ganie*[1], Lowell Sandell[2], Greg Elmore[3], Chandrashekar Aradhya[4]; [1]FMC, Newark, DE, [2]Valent, Ashland, NE, [3]Bayer Crop Science, Saint Louis, MO, [4]Bayer Crop Science, Chesterfield, MO (175)

United States-Herbicide Resistance Action Committee (US-HRAC) is an industry group operated by technical members of the companies represented in CropLife America. US-HRAC has a mission to protect crop yields and quality in the US by supporting fight against herbicide resistant weeds, and to promote proper product stewardship required to delay the evolution of new herbicide resistance issues. Diverse weed management approaches can help growers to mitigate the evolution and spread of herbicide-resistant weeds. We support development and distribution of research based technical information on diverse weed management strategies that are effective, reliable, practical, and economical. Our primary audience are Grower Influencers. These are industry experts (personnel, ag advisors, extension experts, retail agronomists etc.) who work directly with farmers. To achieve its goals USHRAC has three strategic focus areas that includes regulatory engagement, stakeholder engagement and communication.

Light Quality Alteration is a Mechanism of Weed Suppression by Cereal Rye Cover Crop. Gustavo Camargo Silva*, Muthukumar V. Bagavathiannan; Texas A&M University, College Station, TX (176)

Cover crops have become one of the main non-chemical weed control methods in recent years. The most popular cover crop species is cereal rye, which apart from offering soil and water conservation benefits, has great potential for weed resistance management. Cereal rye controls weeds through several mechanisms, including mulching and competition. However, there is evidence that cereal rye can also influence the red to far-red ratio (R/FR) and induce shade-avoidance response in weeds. This project seeks to determine the effect of living cereal rye on the R/FR around the weeds and its influence on the growth and development of weeds. Cereal rye was planted at four seeding rates (0, 20, 40, and 80 kg ha^{-1}) and terminated at three timings (6, 4, and 2 weeks before planting the cash crop), totaling 12 treatments. Cotton was planted into the cereal rye residues with a no-till drill. Cereal rye biomass production was determined at termination, which was carried out using glyphosate (870 g ae ha^{-1}). Light quality was measured prior to termination with a R/FR sensor. The growth and development of individual *Amaranthus palmeri* and *Acalypha ostryifolia* plants were tracked during the growing season to see the response of the weeds to cereal rye residue. Preliminary results show that high seeding rates of cereal rye reduced the R/FR at the soil surface by up to 50%. Cereal rye also decreased the height of the weeds by up to 20 cm at the early season, while increasing the internode-stem diameter ratio in five times, a sign of competition avoidance response. This experiment shows that cereal rye can hinder weed growth and development by changing the light quality environment, benefitting the cash crop during the growing season.

Building an Agriculture Weed ID Network for New York: Development of Web Based Resources to Interface with Growers and Beyond. Sharon Bachman*[1], Caroline Marschner[2], Antonio DiTommaso[2], Isabella Colucci[2]; [1]Cornell Cooperative Extension, East Aurora, NY, [2]Cornell University, Ithaca, NY (177)

Weeds are a primary cause of yield loss in North American crops, and successful management requires accurate weed identification, a critical skill for New York State farmers and their technical service providers. Our Weed ID Network for New York involves Extension educators, crop advisors, other nonprofit technical service providers, regional invasive species partner organizations, and the Weed Ecology & Management Lab at Cornell. In addition to the comprehensive web resources developed targeting the diverse crops grown in New York and the Northeast Region, the project team offered a range of virtual and in-person trainings, developed printable weed ID resources for different commodities, accepted submissions of unknown weeds for identification, and created an outreach toolkit for 4-H youth audiences. Our Weed ID Network resources are available at blogs.cornell.edu/weedid.

Weed Control and Rice Response to Florpyrauxifen-benzyl. Rachelle Berger*; University of Florida, Belle Glade, FL (178)

Economical production of rice in Florida requires efficacious herbicide weed control programs. Florpyrauxifen-benzyl, a new arylpicolinate synthetic auxin herbicide with activity on grasses, sedges, and broadleaf weed species can provide solutions for weed control in Florida rice. Florpyrauxifen-benzyl has a 3-month rotational restriction and would fit in the Florida rice cropping system that involves its rotation with sugarcane, rice, or leafy vegetables. Auxin herbicides are of concern when used in proximity to broadleaf crops because of injury from off-target movement. The objectives of the study were 1) to evaluate florpyrauxifen-benzyl for postemergence weed control in Florida rice compared to currently used herbicides, and 2) to evaluate sensitivity of broadleaf crops (endive, escarole, and radish) to low concentrations of florpyrauxifen-benzyl. Field and greenhouse experiments for objectives 1 and 2, respectively were conducted at the University of Florida Everglades Research and Education Center (EREC) in 2022. The field experiment was a randomized complete block design with 11 treatments and an untreated control. Florpyrauxifen-benzyl (30 g ha^{-1}) applied alone and in combination with cyhalofop (313 g ha^{-1}), bentazon (840 g ha^{-1}), halosulfuron (40 g ha^{-1}), and carfentrazone-ethyl (22 g ha^{-1}) were compared to cyhalofop, bentazon, halosulfuron, and carfentrazone-ethyl applied alone or to tank-mixes of cyhalofop + bentazon + carfentrazone or cyhalofop + halosulfuron + carfentrazone. The herbicides were applied at the 4-leaf stage of rice followed by permanent flooding after treatment. Weed species evaluated included fall panicum, yellow nutsedge, and spiny amaranth. At 21 days after treatment (DAT), rice stunting (<9%) was observed in florpyrauxifen-benzyl treatments. The tank-mix of florpyrauxifen-benzyl with cyhalofop, bentazon, halosulfuron, and carfentrazone-ethyl controlled fall panicum 93% to 100% at 21 DAT. Fall panicum control with florpyrauxifen-benzyl was not significantly different from cyhalofop treatments. Addition of bentazon, carfentrazone-ethyl, cyhalofop or halosulfuron to florpyrauxifen-benzyl neither antagonized nor synergized fall panicum control. Yellow nutsedge and spiny amaranth control with florpyrauxifen- was 91% and 98%, respectively at 21 DAT. Yellow nutsedge control with florpyrauxifen-benzyl was not significantly different from treatments containing halosulfuron. However, florpyrauxifen-benzyl provided significantly better yellow nutsedge control than treatments containing bentazon. Spiny amaranth control with florpyrauxifen-benzyl was not significantly different from control with carfentrazone-ethyl. Tank-mixing the herbicides with florpyrauxifen-benzyl neither antagonized nor synergized spiny amaranth control. Rough rice yield was not significantly different between florpyrauxifen-benzyl treatments and cyhalofop tank-mixes. Florpyrauxifen-benzyl was applied at 1/512 to 1/2 of its 1X rate of 30 g ha^{-1} to endive, escarole, and radish in a greenhouse experiment arranged in a completely randomized design for crop. The higher rates of florpyrauxifen-benzyl (1/4 to

1/2X rates) resulted in complete death of all crops at 28 DAT. At 1/512X rate, up to 25% injury on the crops from florpyrauxifen was observed. Injury was expressed as severe leaf cupping, leaf and stem epinasty, and stunting followed by chlorosis, necrosis and death at the highest rates. At the lowest rates (<1/128X), symptomology was expressed as minor stunting, leaf cupping, and leaf and stem epinasty. Crop biomass response to florpyrauxifen-benzyl was like that observed for injury. Florpyrauxifen rates that resulted in 10% visual injury and biomass reduction of endive and escarole was estimated to be >1/128X compared to >1/32X rate for radish showing that endive and escarole were more sensitive to the herbicide. These results indicate that florpyrauxifen-benzyl can provide broad spectrum weed control in the Florida rice and would be a suitable tank-mix partner with commonly used herbicides due to lack of antagonism. However, because of injury potential from off-target movement to sensitive crops, care must be taken to mitigate injury on these crops by following florpyrauxifen-benzyl label instructions. Additional research will be conducted in 2023 to corroborate these results. Berger555@ufl.edu, dcodero@ufl.edu

To Bury or Uproot? Modelling Selectivity of Physical Weed Control in Sensitive Crops. Daniel C. Brainard*, Noelle Connors; Michigan State University, East Lansing, MI (179)

Selectivity, the ability to kill weeds without killing the crop, is a challenge for in-row mechanical cultivation, especially in slow-growing crops like carrots. To gain insight into the optimal tool type and timing for cultivation given different weed species, we adapted a model from Kurstjens et al. (2004) to predict potential efficacy (PE)—the greatest weed mortality attainable at a given level of crop mortality assuming an idealized tool — based on early growth characteristics of carrots and weeds. We parametrized the baseline model using anchorage force and height data of carrots and five weed species grown in a greenhouse and used it to predict PE for idealized tools that precisely uproot or bury weeds. We also used the model to predict the impact on PE of cultural practices (e.g. stale seed bedding and cultivar choice) which increase carrot height and anchorage force relative to weeds at the time of cultivation. Under baseline model assumptions, the predicted PE of tools depended on timing, weed species and tool mode of action. Grass weeds were predicted to be most susceptible to burial at 1-2 weeks after planting (WAP). In contrast, broadleaf weeds were predicted to be more susceptible to uprooting at 1-2 WAP, and more susceptible to burial thereafter. Stale seed bedding and carrot cultivar choice were predicted to increase PE, with the magnitude of those effects varying considerably based on weed species and tool. The model provides a useful tool for generating hypotheses to facilitate efficient identification of effective approaches for improving selectivity of mechanical cultivation.

Effect of Tillage, Cover Crop, and Herbicide on Giant Ragweed (*Ambrosia trifida* L.) Management in Soybean. Venkatanaga Shiva Datta Kumar Sharma Chiruvelli*[1], Thomas J. Peters[2], Gregg A. Johnson[1], Debalin Sarangi[1]; [1]University of Minnesota, St Paul, MN, [2]North Dakota State University, Fargo, ND (181)

Overuse of herbicides from a single site of action caused the evolution of herbicide-resistant weeds. Integrated weed management approaches that include non-chemical weed control tactics such as cover cropping and tillage along with herbicides are considered among the best management practices to mitigate herbicide resistance. Giant ragweed is an early-emerging annual weed in Minnesota and the herbicide-resistant biotypes of this weed are difficult to control in soybean. The objective of this study was to compare different integrated weed management approaches for giant ragweed control and soybean yield. A field study was conducted in 2022 at the Rosemount Research and Outreach Center near Rosemount, MN, using a split-plot design. The main plot treatments included pre-plant weed management tactics as:1) cereal rye (*Secale cereale* L.) cover crop terminated at soybean planting, 2)cereal rye terminated at 7 d after planting (DAP), 3) reduced tillage, 4) conventional tillage,5) no-tillage plus burndown herbicide, and 6) no-tillage control. The subplot treatments included herbicide programs as POST-only, PRE followed by (*fb*) POST, and a nontreated control. At 21 d after PRE, no-tillage plus burndown, and cover crop terminated at 7 DAP resulted in = 84% control of giant ragweed. Application of PRE improved giant ragweed control by 13% compared to treatments without PRE. At 21 d after POST application, the POST-only and PRE *fb* POST herbicide programs in all pre-plant weed management tactics (except for the no-tillage control) showed > 95% giant ragweed control and reduced the weed density to < 2 plants m^{-2}. Cover crop terminated at soybean planting or 7 DAP combined with PRE *fb* POST herbicide program resulted in highest soybean yield (= 4,030 kg ha^{-1}), which was comparable to the PRE *fb* POST program in no-tillage plus burndown, and POST-only treatments in both cover crop treatments. An integrated management approach that includes cover crop and PRE *fb* POST herbicide program should be used for better control of giant ragweed and soybean yield. chiru005@umn.edu

Weed Detection Using Convolutional Neural Networks Under Scenarios of Occlusion. Daniel J. Ginn*[1], Ramon G. Leon[2], Chris Reberg-Horton[2], Steven Brian Mirsky[3], Muthukumar V. Bagavathiannan[1]; [1]Texas A&M University, College Station, TX, [2]North Carolina State University, Raleigh, NC, [3]USDA ARS, Beltsville, MD (182)

A Multi-Tactic Waterhemp Management Approach in Soybean-Sugar Beet Rotation. Navjot Singh*[1], Thomas J. Peters[2], Seth L. Naeve[1], Ryan P. Miller[3], Debalin Sarangi[4]; [1]University of Minnesota, Saint Paul, MN, [2]North Dakota State University, Fargo, ND, [3]University of Minnesota, Rochester, MN, [4]University of Minnesota, St. Paul, MN (183)

Because of its competitiveness and prolific seed production ability, herbicide-resistant waterhemp [*Amaranthus tuberculatus* (Moq.) J.D. Sauer] is the most problematic weed in agronomic crops in Minnesota. Waterhemp control is further challenging in soybean-sugar beet rotation due to the crop rotation interval restrictions of many soybean PRE herbicides and limited POST herbicide choices in sugar beet. Therefore, integrating herbicide programs with cultural and mechanical weed management tactics can facilitate better waterhemp control in soybean and reduce weed seedbank for the subsequent sugar beet crop. Field experiments were conducted in Franklin and Rosemount, MN, in 2021 and 2022 to evaluate the effect of soybean row spacing and herbicide programs on waterhemp control, seed production, and soybean yield. The herbicide programs included low-input, high-input herbicide treatments, and a high-input herbicide with harvest-time weed seed control (HWSC) treatment. Narrow-row soybean closed the canopy earlier and the differences in cumulative growing degree days (GDD) required for 90% soybean canopy closure between wide- and narrow-row were 418 and 166 in Franklin and Rosemount, respectively. Soybean row spacing did not impact waterhemp control at 28 d after late-POST (DALP) and waterhemp seed production at harvest. High-input herbicide programs provided = 94% waterhemp control across the research sites and reduced waterhemp density by 98% at 28 DALP. Waterhemp produced over 1.5 million seeds m^{-2} in nontreated control, and high-input herbicide treatments reduced the seed production to < 4,500 seeds m^{-2}. Narrow-row spacing increased soybean yield by 339 and 562 kg ha^{-1} than the wide-row in Franklin and Rosemount, respectively. Email: sing0807@umn.edu

Increased Soil Temperature Effects on Weed Seed Mortality Vary Across Species. Ashley N. Isaacson*[1], Richard G. Smith[2], Chris Stamplis[3], Alix Contosta[2], Carolyn J. Lowry[4]; [1]Penn State University, State College, PA, [2]University of New Hampshire, Durham, NH, [3]Pennsylvania State University, University Park, PA, [4]The Pennsylvania State University, University Park, PA (284)

The warmer temperatures associated with climate change will likely have profound effects on the persistence of seeds in the soil seedbank. However, it is not yet known to what extent warmer temperatures affect seed persistence, and whether this varies across species of common agricultural weeds. Therefore, our goal was to determine how warmer soils affected weed seed persistence in the soil seedbank within a perennial forage mixture. We utilized open-top chambers (OTCs) to manipulate soil temperature *in situ* by passively trapping solar radiance within the chamber, warming the air within the chamber and subsequently the soil. The OTCs increased soil temperatures by an average of 0.38ºC. After 19 months, weed seed persistence in the warmed soil decreased in two (*Amaranthus retroflexus* and *Ambrosia artemisiifolia*) out of four weed species. However, *Digitaria sanguinalis* had greater seed viability, and therefore greater weed seed persistence, in the warmer soils. Our study suggests that the effect of warmer soils on weed seed persistence varies by species, which may cause weed communities to shift towards weed species that are less affected by warmer soil temperatures after a perennial phase in a crop rotation.

Steam Disinfestation of Soil with and without Accelerants. Donald E. Fox*[1], Emma Volk[1], Mark Hoffmann[1], Steve Fennimore[2], Joe C. Neal[1]; [1]North Carolina State University, Raleigh, NC, [2]University of California - Davis, Salinas, CA (285)

Steam Disinfestation of Soil with and without Accelerants. D. Fox*, M. Hoffmann, E. Volk, S. Fennimore, and J. Neal, North Carolina State University, Raleigh, NC (1/2023) WSSA/NEWSS 2023 Abstract Methyl bromide is a fumigant that was widely used for disease and weed control in specialty crop production but most uses were phased out by 2015. Although other fumigants are available, none are as effective as methyl bromide on weed propagules, and many growers would prefer control measures consistent with organic production standards. Steam has been shown to provide effective control of soil-borne pathogens, arthropod pests and weed propagules but strategies to reduce the thermal inputs and time required for steam sterilization are needed. Exothermic reactions have been shown to reach higher temperatures for a longer duration in soil. The objective of this study was to evaluate exothermic-reactive additives (accelerants) combined with steam to increase the efficacy of treatments for weed control. The experiment was conducted at the Central Crops Research Station, Clayton, NC, in 2021 and repeated in 2022. The field was cultivated in May of each year and treatments were applied on June 1, 2021 and June 1, 2022 to plots measuring 1 m by 0.5 m. Treatments were arranged in a randomized complete block design with four replicates and included non-treated control, steam, sodium peroxide (172 g/m^2), steam + sodium peroxide, quicklime (1715 g/m^2), and steam + quicklime. Quicklime and sodium peroxide were manually incorporated into the soil immediately before steam injection. To ensure uniform soil conditions in all treatments, the soil was turned in the same manner in all plots. Four bioassay species, Italian ryegrass (Lolium perenne ssp. multiflorum), common vetch (Vicia sativa*)*, prickly sida (Sida spinosa), and yellow nutsedge (Cyperus esculentus) were placed into 250 µm nylon mesh sachets and buried in soil 2.5, 12.5, 25, and 37.5 cm from steam injection points, each at a depth of 10 cm (corresponding to the depth of steam injection). In 2022 thermocouples were placed adjacent to seeds to record soil temperatures. Steam was injected into soil for 30 minutes. Plots were individually covered with plastic during steam injection; plastic was removed 15 minutes after steam-run ceased. Sachets were removed the next morning, seeds were spread onto the surface of peat/perlite substrate in 4 pots, covered with minimal amount of substrate/vermiculite, and then pots were placed in a greenhouse. Seed germination was recorded weekly until no new germination was observed. Germination data was subjected to analysis of variance using the GLIMMIX procedure, orthogonal contrasts, and Dunnett's Multiple Comparison test in SAS. Accelerants alone did not affect germination. Also, at distances of 25 and 37.5 cm from the steam injectors there was no reduction in weed propagule germination. At 2.5 cm from the injection point all steam treatments provided nearly complete control of all species. Differences among steam treatments were observed at 12.5 cm distance from the steam injectors. In 2021, the addition of accelerants improved control of weed propagules compared to steam alone for all species but not in 2022. In 2022 results among replicates of the same treatments was much more variable than in 2021. Poorer control of weed propagules appears to be associated with lower peak soil temperatures. Weed germination was greater where maximum soil temperatures were less than 70° C. Results from this study indicate that exothermic agents added to steam soil sterilization may improve control of weed propagules, but solutions to variable / inconsistent soil heating are still needed.

Acreage Allocation for Variable Density Planting to Optimize Crop Yield and Weed Suppression. Sandra R. Ethridge*[1], Wesley Everman[1], Anna M. Locke[2], David L. Jordan[1], Ramon G. Leon[1]; [1]North Carolina State University, Raleigh, NC, [2]USDA-ARS, Raleigh, NC (286)

The use of high crop densities is effective to suppress weeds in row crops. However, growers may be reluctant to use this method due to increased seed cost. Because weeds tend to cluster in aggregated patches throughout fields, the weedy and less- or non-weedy areas may be managed differently. Therefore, in the weed-free areas of a field, low-density planting can be used to offset seed costs. When using high- and low-density plantings in combination, it is critical to optimize each planting density to consider not only seed costs, but also the productivity per plant. In this study, the growth and yield of maize, cotton, and soybean were characterized under several densities, and a bioeconomic model was created to optimize area allocation to high- and low-density plantings and increase weed suppression without increasing seed costs for the producer. Maize and cotton showed the most plasticity in yield per planted seed and the area with low-density required to compensate for the area with high-density planting. It would be possible to plant less acreage in a low-density planting of these crops because plants were able to increase yield (in g seed^{-1}), while also increasing return and profit. Soybeans, although morphologically plastic, did not show the same area optimization. Soybeans maintained the 1:1 ratio for area optimization and maintained return and profit. The use of this optimization model could be used for large scale agriculture, in order to maintain or reduce seed cost, while maintaining yields and increasing weed suppression.

Degradation of Spring Applied Methiozolin and its Effects on Smooth Crabgrass and Goosegrass Control. John M. Peppers*, Shawn Askew; Virginia Tech, Blacksburg, VA (287)

Persistence of preemergence herbicides often determine their viability for use. Methiozolin is a newly registered herbicide typically utilized for PRE and POST control of annual bluegrass (*Poa annua*). Methiozolin is also labelled for control of goosegrass (*Eleusine indica*) and smooth crabgrass (*Digitaria ischaemum*), but usage directions are unclear on the label and no peer-reviewed literature exists on the subject. Four field and two greenhouse studies were conducted in Blacksburg, Virginia in order to determine if methiozolin can effectively control goosegrass and smooth crabgrass in turfgrass systems. Two greenhouse studies were conducted in fall of 2021 to evaluate goosegrass and smooth crabgrass response to methiozolin rate based on weed stage of growth. Treatments were arranged as a six by five by two factorial with six levels of methiozolin rate (0, 125, 250, 500, 1000, and 2000 g ai ha^{-1}), five levels of weed growth stage (three to four tillers, one to two tillers, three to four leaves, one to two leaves, and PRE) and two weed species (goosegrass and smooth crabgrass) for a total of 60 treatments. Weed seed planting was staggered in order for all treatments to be applied at once. Visual weed control and above-ground biomass was assessed 28 days after treatment (DAT). Four field studies were conducted between 2020 and 2022 at the Turfgrass Research Center in Blacksburg, VA to evaluate spring-applied methiozolin control of goosegrass and smooth crabgrass. Treatments were arranged as a randomized complete block design and included a nontreated check, methiozolin applied once (500 g ai ha^{-1}), methiozolin applied twice at a 6 to 8 wk interval, bensulide (8967 g ai ha^{-1}), oxadiazon plus bensulide (1681 plus 6725 g ai ha^{-1}) applied twice at a 6 to 8 wk interval, and four biweekly applications of siduron (3363 g ai ha^{-1}). Weed control was assessed via weed counts at the end of the summer (~September 10) for all trials. In the greenhouse study, methiozolin controlled smooth crabgrass and goosegrass 90% at 125 and 102 g ai ha^{-1}, respectively, when applied PRE. Averaged across growth stage and rate, methiozolin controlled smooth crabgrass more effectively than goosegrass when applied POST. In the field, methiozolin applied once controlled smooth crabgrass 33%, which was less than all other treatments. Two applications of methiozolin controlled smooth crabgrass 64%, which was similar to oxadiazon plus bensulide, but less than all other herbicide treatments. Single applications of methiozolin controlled goosegrass 32%, which was similar to bensulide and siduron. A sequential application of methiozolin increased goosegrass control to 55%, and was less effective than oxadiazon plus bensulide. Results from this study indicate that although methiozolin is highly efficacious against smooth crabgrass and goosegrass at below-labelled rates, methiozolin degrades in the field too rapidly to be utilized as a preemergence herbicide applied at traditional timings. These results led to ongoing research evaluating frequent, biweekly applications of low methiozolin rates to control smooth crabgrass and goosegrass on golf greens and collars.

Weed Risk Assessment of *Brassica carinata* in the Southeastern U.S. April M. Dobbs*, Chris Reberg-Horton, Lori Unruh Snyder, Ramon G. Leon; North Carolina State University, Raleigh, NC (288)

Carinata (Brassica carinata (A.) Braun) is a promising winter oilseed crop in the southeastern US. Because it is a new crop, assessing its agricultural and ecological safety is critical for growers. This study evaluated weediness potential for carinata in the southeastern US. A field study was conducted in Goldsboro and Clayton, North Carolina comparing seedling emergence and survival of volunteer plants with and without predator exclusion. Cumulative emergence at both locations was highest for buried seeds with predator exclusion (42% and 15%) and lowest in unburied seeds without predator exclusion (16% and 1%). Survival 90 days after planting (DAP) at both locations was highest for buried seeds with predator exclusion (10% and 5%) and lowest in unburied seeds without predator exclusion (3% and <1%). Frost damage contributed to predator damage, increasing mortality of established plants to 100% 120 DAP. In addition to the field study, we evaluated carinata using the Australian Weed Risk Assessment and Plant Risk Evaluation Tool. The cumulative Australian WRA score was -1 (low risk), which was below the minimum score of 6 for rejection of introduction. The agricultural and environmental scores were -5 and -2, corresponding to a low risk in agricultural and non-agricultural settings. The cumulative Plant Risk Evaluation score was 6 (low risk), which was below the minimum score of 13 for rejection. Based on the field study and risk assessments, carinata has a low risk for becoming weedy/invasive in the southeastern US. New varieties will need to be assessed as they are developed.

Response of Italian Ryegrass Accessions from South-Central North Carolina to Spring Burndown Herbicides. Jose H. de Sanctis*, Charlie W. Cahoon, Wesley Everman, Travis Gannon; North Carolina State University, Raleigh, NC (289)

Italian ryegrass (*Lolium perenne ssp. multiflorum*) is a troublesome weed commonly found across the US. Furthermore, this weed has developed resistance to 7 herbicide modes of action in the US. In North Carolina, biotypes resistant to herbicides from groups 1, 2, and 9 have already been confirmed. Recently, paraquat (group 22) resistant populations were confirmed in the south-central region of the state. Because of multiple reports of Italian ryegrass escapes from paraquat and the historical issues with herbicide-resistant Italian ryegrass populations in the area, an assessment of Italian ryegrass response to commonly used burndown herbicides from different modes of action was established. The objective of the study was to characterize the distribution of multiple herbicide-resistant Italian ryegrass biotypes from the south-central region of North Carolina to commonly used burndown herbicides. Greenhouse studies were conducted with 38 accessions from Union and Stanly counties. Seed samples were collected from fields naturally infested with Italian ryegrass and each sample was composed by seed heads from multiple plants from each location. Herbicide treatments consisted of clethodim (271 g ai ha^{-1}), glyphosate (1260 g ae ha^{-1}), glufosinate (880 g ai ha^{-1}), nicosulfuron (63 g ai ha^{-1}), and paraquat (840 g ai ha^{-1}), plus a nontreated control. The trial was conducted as a RCBD, with five replications and two runs, where each individual plant was considered a replicate. Italian ryegrass populations were considered resistant if visual ratings averaged =75% control. When averaged over populations, nicosulfuron and glyphosate resulted in similar Italian ryegrass control, with 74% and 75%, respectively. Clethodim was the most effective and resulted in 98% control. Furthermore, 52% and 47% of populations tested exhibited =75% control for nicosulfuron and glyphosate, respectively. Paraquat, glufosinate, and clethodim exhibited 16%, 5%, and 2% of populations with =75% control, respectively.

Optimizing Application Time of Residual Herbicides in a Planted Green Soybean System. Megan R. Czekaj*[1], John M. Wallace[2], Mark VanGessel[3], Michael L. Flessner[4]; [1]Penn State University- Agricultural and Environmental Plant Science, State College, PA, [2]Penn State University, University Park, PA, [3]University of Delaware, Georgetown, DE, [4]Virginia Tech, Blacksburg, VA (290)

Use of cereal rye surface mulch for early-season suppression of small-seeded summer annual weeds is a promising integrated weed management (IWM) to reduce weed population impacts on crop yields and increase environmental sustainability by facilitating reduced herbicide use and mitigate risks of herbicide resistance evolution. Currently, there is a significant knowledge gap in understanding how to position the use of soil-applied residual herbicides in high-residue cover crop systems. High-residue cover crops have the potential to interfere with decomposition and wash off processes that influence soil bioavailability of residual herbicides. Alternative tactics to optimize use of residual herbicides include application at early cereal rye growth stages where residue interception is minimized or delaying application to after terminating the cover crop when losses to plant uptake are minimized. We conducted a field experiment in three states (Rock Springs, PA, Georgetown, DE, and Blacksburg, VA) to examine how residual herbicide application timing relative to cereal rye termination and soybean planting affects populations dynamics of water hemp (*Amaranthus tuberculatus*; PA) and Palmer amaranth (*Amaranthus palmeri*; DE and VA) in no-till soybean (Enlist) production. The experiment was a two-factor randomized complete block with a split-plot treatment structure. Main plots were soybean planting date, based on cereal rye growth stage and termination timing, including boot (14 DPP), boot (-1 DAP) and anthesis (-1 DAP). Split plot treatments were residual herbicide application timing relative to cereal rye and soybean growth stages, including boot, anthesis, MPOST (4-6 WAP), anthesis + MPOST (4-6 WAP) and an untreated control. Pyroxasulfone (0.15 kg ai/ha) was used as the test residual herbicide. All treatments received a 2,4-D + glufosinate application at the MPOST timing. Results showed that, across all three locations, there was no residual herbicide application timing (boot, anthesis, control) and cover crop management interaction on *Amaranthus* weed density at the MPOST application timing. However, there was a significant herbicide treatment main effect, where pyroxasulfone application at cereal rye anthesis reduced Amaranthus populations (94%) compared to application at the boot (88%). These results suggest that weed emergence germination timing is a stronger driver than cover cropping impacts on herbicide bioavailability. Application of pyroxasulfone at the MPOST stage resulted in higher probability of weed free conditions compared to earlier application. The 2-pass residual treatment (anthesis + MPOST) resulted in 99% control. At 2 of 3 locations there was a soybean yield advantage when planting green early relative to standard planting.

Do Plant Growth Regulator Admixtures Influence Annual Bluegrass Control by Methiozolin? Navdeep Godara*, Shawn Askew, John M. Peppers; Virginia Tech, Blacksburg, VA (291)

Annual bluegrass (*Poa annua* L.) is the most troublesome weed of managed turf, causing reduced aesthetics and affecting playability on putting greens. Plant growth regulators (PGRs) are utilized routinely on creeping bentgrass (*Agrostis stolonifera* L.) putting greens to suppress annual bluegrass. Methiozolin (PoaCure®) selectively controls annual bluegrass preemergence and postemergence. However, the methiozolin label prohibits all PGRs except trinexapac-ethyl from use concurrent with methiozolin treatments due to suspected antagonistic interactions. A field experiment was conducted on in-play greens in Radford, VA, in 2021 to evaluate creeping bentgrass and annual bluegrass response to methiozolin applied alone and in conjunction with four PGR admixtures. The experiment was designed as a randomized complete block with treatments arranged in a 2x4 factorial with methiozolin (with and without) and PGRs (no PGR, trinexapac-ethyl at 53 g ai/ha, paclobutrazol at 280 g ai/ha, flurprimidol at 280 g ai/ha, and prohexadione calcium at 154 g ai/ha) replicated four times. PGRs were applied biweekly with a total of four consecutive applications. Methiozolin was applied in conjunction with only the last two applications of PGRs. Only prohexadione calcium and paclobutrazol significantly injured creeping bentgrass 20 and 14%, respectively, at 2 weeks after initial treatment (WAIT). Creeping bentgrass was stunted and discolored by these two PGRs after the first treatments but had fully recovered prior to the third treatment. At 24 WAIT, two and three times as much annual bluegrass coverage was present in plots treated with methiozolin + paclobutrazol compared to methiozolin + trinexapac ethyl and methiozolin alone, respectively. Furthermore, all PGR + methiozolin treatment programs controlled annual bluegrass >70% except methiozolin + paclobutrazol, which controlled annual bluegrass 55%. These data support current label restrictions on PGR admixtures with methiozolin. With additional repetitions of this trial that were initiated in fall 2022, we hope to make more definitive conclusions.

Evaluation of Various Preemergent Herbicides at Different Rates to Control False-Green Kyllinga (*Kyllinga gracillima*) from Seed. Matthew T. Elmore[1], Brandon C. McNally[1], Aaron J. Patton[2], Daniel P. Tuck[1], Trevor S. Watson*[1]; [1]Rutgers University, New Brunswick, NJ, [2]Purdue University, West Lafayette, IN (292)

Greenhouse experiments were conducted to determine efficacy of herbicides used in cool season turfgrass against false-green kyllinga (*Kyllinga gracillima*) emergence from seed in soil. A replicated preliminary experiment evaluated various herbicides at 100 and 10% typical use rates for cool season turfgrass. Herbicides and 100% use rates included bensulide (11,200 g ha^{-1}), corn gluten meal (880,000 g ha^{-1}), dimethenamid (1,680 g ha^{-1}), dithiopyr (560 g ha^{-1}), isoxaben (1,120 g ha^{-1}), mesotrione (280 g ha^{-1}), methiozolin (516 g ha^{-1}), oxadiazon (4,500 g ha^{-1}), pendimethalin (3,360 g ha^{-1}), prodiamine (840 g ha^{-1}), siduron (13,500 g ha^{-1}), and sulfentrazone (280 g ha^{-1}). The 100 and 10% use rate applications of corn gluten meal, pendimethalin, methiozolin, and siduron were less effective than sulfentrazone, dithiopyr, isoxaben, oxadiazon, prodiamine, dimethenamid, and bensulide against false-green kyllinga from seed.Herbicides deemed effective in the first experiment were evaluated in a greenhouse dose-response experiment. Treatments were a complete factorial of weed species (false-green kyllinga and smooth crabgrass), herbicide, and herbicide rate. Herbicides were applied at 100, 50, 25, 12.5, 6.3, and 3.1% of typical use rates for cool season turfgrass. Herbicides evaluated were bensulide, dimethenamid, dithiopyr, isoxaben, oxadiazon, prodiamine, and sulfentrazone. Local collections of false-green kyllinga seed were sowed to 10 by 10 cm pots filled with local field soil. Smooth crabgrass (*Digitaria ischaemum*) was seeded to separate pots as a standard for comparison. Treatments were replicated four times. Seedlings were counted at 42 days after application and expressed as a percent of the non-treated control. Data were analyzed as a three-way complete factorial using the GLIMMIX procedure in SAS ($P= 0.05$). Fisher's protected LSD test was used to separate means. Main effect interactions were detected ($P < 0.05$) for seedling count expressed as a percent of the non-treated control 6 weeks after treatment. Bensulide, dimethenamid, and oxadiazon provided >94% false-green kyllinga control at all rates. Dithiopyr, prodiamine, sulfentrazone and isoxaben provided less false-green kyllinga control than bensulide and oxadiazon at 3.1 and 6.3% rates. Prodiamine, isoxaben, and oxadiazon provided more smooth crabgrass control than false-green kyllinga control at 6.3, and 3.1% rates. Sulfentrazone controlled more false-green kyllinga than smooth crabgrass at the 12.5% rate. Prodiamine, pendimethalin, and dithiopyr are less effective against false-green kyllinga seed than smooth crabgrass. Sulfentrazone applied preemergence is more effective against false-green kyllinga than smooth crabgrass. Bensulide and oxadiazon were the most effective herbicides evaluated for preemergence control of false-green kyllinga seed in a controlled environment.

Discrimination Between PPO-inhibiting Herbicide-resistant Redroot Pigweed (Amaranthus Retroflexus) Utilizing Spectral Reflectance. Eric A. Jones*, Robert Austin, Diego J. Contreras, Jeffrey Dunne, Charlie W. Cahoon, Ramon G. Leon, Wesley Everman; North Carolina State University, Raleigh, NC (293)

Chemical weed control options in soybean production are limited due to overreliance on a small number of herbicides, which has selected for herbicide-resistant weeds. The current assays to confirm herbicide resistance can be time and labor intensive (dose-response) or require a technical equipment/skillset (molecular sequencing). Farmers are in need for a rapid assay to confirm herbicide-resistant weeds to ensure economically and ecologically sustainable crop production. Since PPO-inhibiting herbicides cease chlorophyll production causing necrosis, we propose a new, rapid assay utilizing spectral reflectance to confirm resistance. Leaf discs were excised from two PPO-inhibiting herbicide-resistant (target site and non-target site) and –susceptible *Amaranthus retroflexus* populations and placed into a 24-well plate containing different concentrations (0-10 mM) of fomesafen for 48 hours. A multispectral sensor captured images from the red, blue, green, and red-edge wavebands after a 48-hour incubation period. The green leaf index (GLI) was utilized to determine spectral reflectance ratios of the treated leaf discs. Differences of spectral reflectance were observed between the PPO-inhibiting herbicide-resistant and –susceptible populations with the 10 mM concentration for the green waveband in the dose-response assay. Clear differences of spectral reflectance was observed in the red-edge waveband for all populations with the 10 mM concentration in the dose-response assays. Leaf discs from the aforementioned *A. retroflexus* populations and two additional susceptible population were subjected to a similar assay with the discriminating concentration (10 mM). Spectral reflectance was different between the PPO-inhibiting herbicide-resistant and –susceptible populations in the red, blue, and green wavebands. Spectral reflectance was not distinctive between the populations in the red-edge waveband and the GLI. The results provide a basis for rapidly (~48 hrs) detecting PPO-inhibiting herbicide-resistant *A. retroflexus* via spectral reflectance. Discrimination between target- and non-target site-resistant populations was possible only in the dose-response assay, but the assay still has utility in distinguishing herbicide-resistant plants from herbicide-susceptible plants.

Influence of Tall Fescue Canopy Architecture on Weed Emergence. Daewon Koo*[1], Clebson G. Goncalves[2], Shawn Askew[1]; [1]Virginia Tech, Blacksburg, VA, [2]University of California, Agriculture and Natural Resources, Ukiah, CA (294)

With increasing resistance and limited availability of herbicides, annual bluegrass control has become increasingly reliant on integrated approaches that include cultural practices that maximize turfgrass competitiveness. Greenhouse trials were conducted at Virginia Tech's Glade Road Research Facility in Blacksburg, VA to examine the impact of three tall fescue mowing heights on canopy architecture, soil-surface temperature, light penetration, and seedling establishment of six weeds. 'Falcon III' tall fescue mown at 3.8 cm for one-year reduced penetration of photosynthetically active radiation (PAR) at 1 cm above soil level by 43%. For each cm increase in mowing height, PAR decreased an additional 4.4% such that PAR was reduced by 90% at the highest mowing treatment of 11.4 cm. Canopy coverage, assessed via digital analysis of images collected from a 0.6-cm tall micro camera, decrease with mowing height in trends that mirrored that of PAR penetration. A concomitant linear decrease in temperature was also related to increased mowing height. Although many weed seedlings were observed at 14 d after seeding, most died and were degraded by 28 d after seeding. Seedling counts at 28 days after seeding were reduced by tall fescue competition for all species but only annual bluegrass, dandelion (*Taraxacum officinale* F.H. Wigg.), and white clover (*Trifolium repens* L.) seedling counts were dependent on mowing height. Percentage reduction in 28-d seedling counts increased linearly with mowing height and were 77 to 88% for annual bluegrass and 43 to 77% for white clover. Dandelion seedling counts were at least 25% reduced at tall fescue mowing heights of 3.8 or 7.6 cm and 100% reduced by tall fescue mown at 11.4 cm.

Control of Common Row Crop Weeds with Dicamba and Glufosinate Applied Alone, Mixed, and Sequentially. Jackson W. Alsdorf*, Eric A. Jones, Diego J. Contreras, Charlie W. Cahoon, Ramon G. Leon, Wesley Everman; North Carolina State University, Raleigh, NC (295)

Dicamba and glufosinate are commonly used herbicides and remain effective in various row crop systems in the Southeastern United States. Field trials were conducted to determine the control of dicamba and glufosinate individually, mixed, and sequentially on common ragweed (Ambrosia artemisiifolia), Palmer amaranth (Amaranthus palmeri), sicklepod (Senna obtusifolia), and goosegrass (Eleusine indica) at two different sizes. Sequential herbicide treatments were applied seven days after the initial treatment. Weed size did not affect control with the exception of Palmer amaranth where control decreased as size increased. Goosegrass was best controlled following two applications of glufosinate or glufosinate + dicamba. There was more control when glufosinate was included in the initial treatment for goosegrass, while control was less when dicamba was the initial application. Palmer was best controlled after at least two applications while no application followed by a treatment was less effective. As expected, control was higher for each treatment when the plants were treated at the smaller size. Application order did not affect common ragweed or sicklepod control. This research provides that dicamba and glufosinate effectively control common weed species, but the species present may dictate how the herbicides are applied.

Effect of Increased Soil Temperatures on Weed Seed Mortality Vary Across Species. Ashley N. Isaacson*[1], Carolyn J. Lowry[2]; [1]Penn State University, State College, PA, [2]The Pennsylvania State University, University Park, PA (296)

Multi-state Survey of Waterhemp (Amaranthus Tuberculatus) and Palmer Amaranth (Amaranthus Palmeri) in the USA on Sensitivity to Multiple Herbicides. Daljit Singh*[1], Alejandro Perez-Jones[2], Drew Tyre[3], Chandrashekar Aradhya[2]; [1]Bayer, Chesterfield, MO, [2]Bayer Crop Science, Chesterfield, MO, [3]Bayer Crop Science, Lincoln, NE (418)

Herbicide resistance in Palmer amaranth (*Amaranthus palmeri*) and waterhemp (*A. tuberculatus*) are on the rise and pose great concern to growers in the United States (US). A multi-state survey was conducted for these two weed species in the US to assess their sensitivity to glufosinate, dicamba and 2,4-D herbicides. The survey was designed to understand the weed sensitivity landscape and potential emerging trends in resistance evolution by testing them at the field recommended label rate. The testing was conducted in a controlled environment at the Bayer Crop Science research facility in Chesterfield, Missouri. Treatments consisted of dicamba, glufosinate and 2,4-D applied at the recommended label rate. Overall, the efficacy of glufosinate, dicamba and 2,4-D was excellent in all these populations with greater than 90% average injury level. The survey provided a snapshot of the distribution of herbicide sensitivities in Palmer amaranth and waterhemp populations in the US and highlights the value of glufosinate, dicamba and 2,4-D herbicides in managing troublesome weeds as part of a diversified weed control program integrated with other chemical, mechanical, and cultural practices.

SYMPOSIUM

Symposium Welcome. Steve Young*; USDA-ARS, Beltsville, MD (184)

Current State of Cover Crops and Weed Management – What Do We Know and How Do We Know It? Erin Haramoto*; University of Kentucky, Lexington, KY (185)

Foundational mechanistic research has advanced our understanding of how cover crops suppress weeds. Likewise, many (many) field experiments across multiple crops, geographies, climates, and with varying management strategies have clarified the role that cover crops can play in integrated weed management systems. Across all of these systems and regions, the use of cover crops for weed management can arguably be distilled into a few goals: reduce and delay weed emergence, slow weed growth rates, and lower the number of viable propagules produced and persisting in the soil. Achieving these goals leads to reduced competition with crops and improved efficacy of reactive strategies including herbicides and cultivation. They may also lessen the selection pressure of these reactive strategies, improving the longer-term sustainability of weed management.While cover crops are growing, they can suppress the cohort of weeds in the field concurrently through competition and allelopathy. Cover crops are often temporally separated from cash crops, so this interference may not directly impact weed management in the cash crop phase. However, consider the importance of the facultative winter annuals horseweed (Conyza canadensis) and annual ryegrass (Lolium multiflorum) to summer annual cropping systems. Our research has shown that fall- and early spring-emerging horseweed cohorts are managed effectively with a cereal rye (Secale cereale) cover crop, demonstrating the utility of cover crops for managing these facultative species.Cover crops also suppress weeds after termination through residue-mediated effects. Cover crop residues, particularly when left as surface mulches in no-till production, alter the biotic and abiotic environment, with resulting impacts on weed seed survival and germination, seedling emergence and establishment, and plant growth and reproduction. Residues change availability of abiotic factors such as light, water, and nutrients, and provide a physical barrier to seedling emergence. Residues may also provide habitat for seed predator insects and pathogenic organisms, thus reducing seedbank density and seedling survival. While residues rarely result in season-long suppression, nor reduce weed density to agronomically acceptable levels, they can effectively delay emergence, slow growth, and lengthen the window for POST herbicide applications.In addition to these mechanisms, cover crops also contribute to general weed suppression and disruption through increasing an agroecosystem's diversity – rotational, spatial, and temporal. This allows implementation of different practices at different times that exert variable pressures on weeds – mowing, harvest time, different herbicides, etc. Overall, this reduces weeds' ability to adapt. In the soil, cover crops can also alter seedbank dynamics and may induce seed mortality, changes in dormancy status, seedling disease, etc., though these mechanisms are harder to show experimentally. Using cover crops should still be considered only part of an integrated weed management system so weeds do not adapt to the practice on an individual or community level.

Cover Crop Options and Opportunities in Australian Cropping Systems. Michael Walsh*[1], Asad Shabbir[2], Saliya Gurusinghe[3], Leslie A. Weston[4]; [1]University of Sydney, Narrabri, Australia, [2]The University of Sydney, Camden, Australia, [3]Charles Sturt University, Bathurst, Australia, [4]Charles Sturt University, Wagga Wagga, Australia (186)

Australian grain growers are interested in using cover crops primarily to increase soil surface residues that lead to improved soil moisture and nutrient conservation. In Australian broadacre cropping the improved production capacity realised by the adoption of no-till cropping systems is largely due to the soil moisture and nutrient benefits from the retention of crop residues. Soil moisture is particularly critical to crop production and for this reason there is some hesitancy by growers in adopting cover crops as the current perception is that their use will reduce soil water storage. The weed control opportunity associated with cover crops is a secondary but increasingly important consideration due to the increasing frequency of herbicide resistance in Australian cropping systems. Despite rainfall patterns varying between winter and summer dominance across the diverse southern Australian grains region, grower focus on the production of winter crops continues, especially wheat. Consequently, potential for cover crop integration is during the winter growing season as break crops and as short-duration crops during the summer fallow phase between winter crops. Recent research efforts have, therefore, centred on identifying both suitable winter and summer growing cover crop species exhibiting rapid growth and weed suppressive ability. Studies have identified grazing oats and tillage radish as the most productive (6.5 to 8 t ha-1 of biomass) winter cover crop species following short duration (80 d) early season growth. The most productive (~6 t ha-1) short-term (40 d) summer cover crop species included Japanese millet, sorghum and sudangrass. These covers were also highly competitive, generally reducing weed biomass by 95 to 98%. Ongoing research is underway to evaluate the weed suppressive ability of these and other crop species as well as their impact on soil moisture storage. The cost efficacy and implications of these results on the role of winter and summer cover crops in Australian cropping systems will be further described.

What is the Potential for Actually Managing Weeds with Cover Crops? A Reality Check. Richard G. Smith*; University of New Hampshire, Durham, NH (187)

The weed science literature is replete with studies demonstrating that cover crops suppress weeds and subsequent conclusions that cover crops are key to successful IWM in row crops. However, while there is no doubt that a relatively productive cover crop will suppress weeds, sometimes to levels approaching 100%, if cover crops do not result in meaningful reductions in weed abundance during the cash crop phase their utility for IWM will be limited. Unfortunately, available evidence demonstrating that even the most weed-suppressive cover cropping scenarios are unlikely to lead to agronomically meaningful reductions in weed populations suggests that simply integrating cover cropping into a business as usual row cropping system and expecting this will eventually enable crop production in bare, weed-free rows with minimal herbicide use is unrealistic. If cover crops are to play anything more than a marginal role in IWM, they must be able to prevent or delay weed emergence during the cash crop growth period. Growing evidence suggests that alternative strategies for implementing cover crops, either as living or killed mulches, represent the most promising opportunities for capitalizing on the weed-suppressive potential of cover crops. However, to do so successfully will require investments in research and likely changes to our cropping practices.

Opportunities and Challenges of Growing Cover Crops in the Semi-arid Great Plains. John Holman*[1], Augustine Obour[2]; [1]Kansas State University, Garden City, KS, [2]Kansas State University, Hays, KS (188)

Weighing Cover Crop Mediated Weed Management Benefits with Other Risks. Erin Haramoto*; University of Kentucky, Lexington, KY (189)

Cover crops suppress weeds and are an important tool in the integrated weed management toolbox. But ill-planned and implemented cover crop plantings can create challenges for growers, necessitating additional management and often expense. These challenges can include, but are not limited to, contaminated seed bringing in undesirable plant species; improper termination leading to the cover crop becoming a weed in the subsequent cash crop; planting a species or mixture not well-suited to the rotation and available equipment; unintended pest consequences; changes in early-season moisture and nutrient availability; and cover crop residues interfering with cash crop planting. Some of these challenges are exacerbated in high-residue systems that optimize weed suppression. Moreover, a disconnect between cover crop reality and expectations might make producers decide cover crops are not worth the time and expense. Planting cover crops too late can lead to shortened establishment and growth windows and reduce resulting ecosystem services, including weed suppression. Improper planting can lead to poor establishment with similar consequences. Across much of North America, and despite our best efforts to develop flexible planting options, a short cover crop growing season also reduces their ability to accomplish goals. Lastly, optimal establishment must be planned around residual herbicides applied to the cash crop – a critical piece to deal with herbicide resistant biotypes.We can advise growers on these challenges, helping them think through their rotations, their equipment, their timelines, and, importantly, their goals for using a cover crop. We can help growers evaluate their cover crop expectations and plans to achieve them. Importantly, we can also help growers navigate increased management and trade-offs that might result from optimizing cover crops for weed suppression vs. other agroecosystem services.Given the added complexity that cover crops introduce, we should also critically evaluate the weed management benefits that they bring and be realistic with expectations around reducing herbicide use and managing herbicide resistance. While it is clear that cover crop residues can reduce and delay weed emergence in no-till systems, it is not as clear in what cases, if any, this results in reduced herbicide use or weed management costs. Given the complexity of agroecosystems, and the importance of initial conditions (i.e., the seedbank), it is unlikely that we will be able to provide these prescriptions. Rather, we can consider and emphasize other weed management benefits that cover crops and their residues bring to integrated weed management systems, and work with growers to develop adaptive management strategies to optimize services provided. Lastly, modeling impacts of realistic cover crop benefits (with or without reduced herbicide use) on the evolution and spread of herbicide resistance is important and may streamline messaging emphasizing but not over-selling the realistic benefits of cover crops for weed management, balanced with their challenges.

The Highs and Lows of Working with a Perennial Cover Crop for Weed Suppression. Cerruti Rr Hooks*, Alan W. Leslie, Veronica Yurchak; University of Maryland, College Park, MD (190)

During the past few decades, growers have increasingly adopted cover crops into their crop rotation systems. On US cropland, cover crop usage increased by 50% during a five-year period between 2012 and 2017. Many producers sow annual cover crops in the fall after harvesting their cash crops for future land quality improvements. Thus, cover crops are primarily grown during the fallow period and terminated prior to cash crop planting. Notwithstanding, annual and perennial cover crops can be grown as an interplanted living mulch within the cash crop during the growing season. However, limited research has been directed to studying services provided by perennial cover crops in a living mulch-cash crop system. Such a system can be deployed to manipulate multiple pest complexes. Some of the findings related to our research in this area as well as the highs and lows of deploying a perennial living mulch-cash crop system to manage weeds will be presented.

Cover Crop-Based Integrated Weed Management: Communicating an Adaptive Management Framework. John M. Wallace*; Penn State University, University Park, PA (191)

Foundational studies focusing on the contribution of cover crops to weed management in sustainable agricultural systems have concluded that weed control by cover crop residues increases with increasing residue biomass, is species specific, and is incomplete (Teasdale 1996). More recent field-based research and meta-analyses have done little to alter these conclusions, yet cover cropping is still emerging as an essential feature of integrated weed management (IWM) in no-till production systems. Three trends support the call for advancing cover crop-based IWM using an adaptive management framework. First, emerging interest in planting green by no-till producers effectively increases cover crop biomass potential and consequently, weed suppression potential. However, early adopters of planting green are principally motivated by soil health considerations and adaptively manage cover crop termination to preserve soil moisture conditions at the time of cash crop planting. As a result, weed scientists should refrain from prescriptive weed management recommendations based on cover crop biomass thresholds, as such targets are not easily realized due to varying growing season conditions and windows, as well as competing management decision points. Second, use of high biomass cover crops has a disproportionate population-level effect on small-seeded annual species, including multiple herbicide resistant species (Conyza canadensis, Amaranthus tuberculatus, A. palmerii), which have a disproportionate effect on weed control costs and net returns in many no-till systems. Future studies should identify weed community shifts that are likely to occur from sustained use of cover crop residues and the role of other control tactics in managing weeds at a community-level. Finally, because weed control by cover crops is incomplete, it is imperative to optimize the benefits and minimize the drawbacks of integrating cover crops with herbicide-based weed control programs. Future studies should identify cover crop management tactics that minimize interference with herbicide-based weed control tactics. Decision support tools are needed to guide herbicide use in cover crop- based IWM systems and two research directions are needed for advancement. First, weed scientists should engage in multi-disciplinary research that adequately addresses the range of conservation and production tradeoffs that growers must negotiate when managing cover crops. For example, cover crop termination timing will increasingly become a water management tool under increasing spring precipitation scenarios, which influences the number of days suitable for field operations, including sprayable days, as well as early-season weed recruitment dynamics. And second, precision agriculture technologies, including discrimination of vegetation at high spatial resolutions, has the potential to better identify direct and indirect factors that mediate the relationship between cover crop biomass and weed suppression potential, including fine-scale spatial variability in surface layers that influences weed recruitment and herbicide fate, as well as functional trait responses to varying residue levels and co-varying soil-environment conditions. In summary, leaning into systems-level research and precision agriculture technologies are likely to hasten the adoption of cover crop- based IWM.

Emerging Opportunities to Understand Cover Crop x Weed Management Interactions from Data-Model Mashups. Armen R. Kemanian*, Yuning Shi, Curt A. McConnell; Penn State University, State College, PA (192)

In this presentation cover crops are conceived as benign weeds, i.e., a community of plants that can compete for resources with cash crops but that is expected to provide multiple ecosystem services. Live cover crops and the resulting changes in biophysical and biochemical properties of the soil-plant system can interact significantly with weed dynamics. The goals of this presentation are to: 1) illustrate the potential of mechanistic models to analyze certain biophysical properties that affect weed emergence and growth, particularly the light, temperature, and moisture environment at the soil surface but also through the canopy-soil profile; 2) explore data mining approaches that can help us extract general principles from multiple realizations of on-farm and plot trials; and 3) discuss results from simulations in counties in the Chesapeake Bay that can help us refine expensive experiments. Through the presentation, I will provide examples of the soil surface diel pattern of temperature, radiation load, and vapor pressure deficit in systems with partial canopy or residue cover, which in the context of this symposium require better description. For example, the vapor pressure deficit at the height of an emerging cotyledon above a dry soil surface at 45 °C can be 7 kPa, and it is likely lethal for an emerging weed or cover crop. However, a wet soil surface, partial canopy or residue cover may produce multiple environments that can promote or suppress weed emergence and establishment (and other biotic processes including insect dynamics). Frost damage is difficult to simulate with accuracy when temperatures oscillate around the threshold for damage. Hourly simulation models can provide substantial biophysical information of this nature, but do we know how to use it? A supplemental approach is to use data mining to detect latent patterns in cover crop x weed management interactions. These approaches should combine the power of simulations models and that of remote sensing techniques to create rich data panels to explore on-farm data. The end point would be improved mechanistic models and more targeted research at in field plots and growth chambers. A partial example of such approach is exemplified by the creation of a large synthetic data panel in a study in the Chesapeake Bay addressing the use of cover crops in corn for grain. We simulated a combination of conventionally seeded or interseeded cover crops consisting of a winter annual grass, a legume blend, or a mixture of a grass and legumes. After analyzing the results, we concluded that research should focus in three bins: establishing legume cover crops in high-yielding fields with low residual mineral nitrogen, grass cover crops in manured or low-yielding fields with residual mineral nitrogen, and a mixture of both when nitrogen supply is uncertain. The need to understand and manage cover crops x weed management interactions provide a clear focal point to deploy these tools.

Assessing Regional Variations of Weed Suppression with Cover Crops. Steven Brian Mirsky*[1], Victoria Ackroyd[2], Muthukumar V. Bagavathiannan[3], Kevin W. Bradley[4], William S. Curran[5], Adam S. Davis[6], Jeffrey Evans[7], Wesley Everman[8], Michael L. Flessner[9], Nicholas R. Jordan[10], Eugene P. Law[11], Lauren M. Lazaro[12], Ramon G. Leon[8], John Lindquist[13], Jason K. Norsworthy[14], Lovreet S. Shergill[15], Mark VanGessel[16]; [1]USDA ARS, Beltsville, MD, [2]University of Maryland, College Park, MD, [3]Texas A&M University, College Station, TX, [4]University of Missouri, Columbia, MO, [5]Penn State University, Bozeman, MT, [6]University of Illinois, Urbana, IL, [7]Five Rivers Conservation Trust, Concord, NH, [8]North Carolina State University, Raleigh, NC, [9]Virginia Tech, Blacksburg, VA, [10]University of Minnesota, St. Paul, MN, [11]University of Delaware, Beltsville, MD, [12]Blue River Technology, Sunnyvale, CA, [13]University of Nebraska, Lincoln, NE, [14]University of Arkansas, Fayetteville, AR, [15]Montana State University, Huntley, MT, [16]University of Delaware, Georgetown, DE (193)

Panel Discussion. Steve Young*; USDA-ARS, Beltsville, MD (194)

Introduction to Workshop, Objectives and Format. William J. Chism*; US EPA, Point Of Rocks, MD (271)

Keynote. Jake Li*; US EPA - Deputy Assistant Adminsitrator for Pesticide Programs, Washington, DC (272)

Keynote. Kyle Kunkler*; American Soybean Association, St Louis, MO (273)

Discuss WSSA Comments to EPA Endangered Species Workplan. William J. Chism*; US EPA, Point Of Rocks, MD (274)

Description of Workshop Topics. William J. Chism*; US EPA, Point Of Rocks, MD (275)

Workgroup Discussion Sessions. William J. Chism*; US EPA, Point Of Rocks, MD (276)

Post Workshop Discussion. William J. Chism*; US EPA, Point Of Rocks, MD (277)

Our Changing Climate: Science, Impacts, and Vulnerabilities. Donald J. Wuebbles*; University of Illinois, Urbana, IL (297)

The science is clear: Climate change is happening now and it is happening throughout the world. These conclusions are based on observations in combination with scientific understanding of the processes driving our climate system. This has been the warmest decade on record, but surface temperature is just one of many indicators of our changing climate. Certain types of extreme weather events – such as heat waves and heavy precipitation – are increasing in severity and frequency, leading to droughts in some locations and floods in others. Sea levels are rising as the oceans warm and land ice melts. We can see these changes play out in the news – whether its wildfires in the American west or Australia, flooding in Germany, or heatwaves in the United Kingdom and Italy. Observations show that the climate is changing about ten times more rapidly than natural changes in climate based on paleoclimatic observations of the changes that occurred since the end of the last ice age. The evidence is clear that the climate changes over the last half-century are primarily due to the burning of fossil fuels in tandem with the clearing of forests and other land-use changes. Natural mechanisms cannot explain the observed changes in climate. There are extensive impacts on humanity and nature resulting from the changing climate. Climate change along with land use change are especially important factors in the increasing concerns about biodiversity. Changes in the characteristics of extreme events are particularly important for human safety, infrastructure, agriculture, water quality and quantity, and natural ecosystems. Some extremes have already become more frequent, intense, or of longer duration, and many extremes are expected to continue to increase or worsen, presenting substantial challenges. The Earth's climate is projected to continue to change over this century and beyond. I feel confident that we can slow climate change and reduce its magnitude, but it will take a concerted worldwide effort to greatly reduce the human-related emissions that are driving these changes. All nations need to significantly transition away from emissions associated with fossil fuels and they also need to be more proactive in protecting their forests and other natural resources because of their importance in storing carbon. A variety of analyses show this can be done without harming the economy. The clock is ticking. We need to face the challenges of addressing our changing climate. Meaningful solutions rest on technological, educational, social, and cultural actions. We need to come together to address what is clearly the biggest challenge of our time.

Modeling Invasive Weed Distributions Under Future Climate Scenarios. Antonio DiTommaso*[1], Sophie Westbrook[1], Emma K. Nikkel[2], David R. Clements[3], Caroline Marschner[1]; [1]Cornell University, Ithaca, NY, [2]University of British Columbia, Vancouver, BC, Canada, [3]Trinity Western University, Langley, BC, Canada (298)

Invasive weeds threaten agricultural productivity, biodiversity, and ecosystem function. Climate change may increase these threats by altering the distributions and impacts of invasive weeds. To minimize harm, policymakers and land managers require accurate predictions about how climate change will affect local weed communities. Such predictions are difficult to develop. Factors associated with climate change have different effects on different weed species, and sometimes have mixed effects on individual weed species. In many species, physiological responses to climate factors are variable and not fully understood. The potential for adaptive evolution adds another layer of complexity. Other factors, such as land management, are important determinants of invasion risk at local scales, but are seldom incorporated into large-scale weed modeling projects. We illustrate some of these challenges and potential solutions using examples from agricultural and natural systems, including examples focused on swallow-wort (*Vincetoxicum*) species and emerging weeds poised to invade the northeastern United States. Overall, we emphasize the power of both correlative and mechanistic modeling approaches to support anticipatory management of invasive weeds.

Effect of CO_2 on Weedy Rice Growth, Seed Shattering, and Seed Dormancy. Luis A. Avila*; Universidade Federal de Pelotas, Pelotas, Brazil (299)

Weedy rice (*Oryza sativa* f. *spontanea*) is one of the most troublesome global weeds in cultivated rice. The seed shattering and dormancy traits allows its long-term reinfestation and persistence in rice fields. Seed shattering traits are genetically fine-tuned controlled, being affected by different environmental conditions including changes in carbon dioxide concentration [CO_2]. High [CO_2] could alter shattering and dormancy, with subsequent effects on weed seed input and competition. The aim of this study was to evaluate the effect of atmospheric [CO_2] on weedy rice development, seedbank longevity and shattering profile. We also evaluated the effect of [CO_2] and the transcriptional regulation of seed shattering-related genes in four weedy rice biotypes and three cultivated rice. All the analyzed traits were evaluated in two atmospheric [CO_2]: ambient {a[CO_2]} and enhanced {e[CO_2]} concentration, 400 and 700 µmol mol^{-1} respectively. Our results indicated that e[CO_2] increased weedy rice growth and biomass, seed shattering and lengthened seed viability within the seedbank. The transcriptional regulation of the seed shattering-related genes (*OsCPL1*, *qSH1*, *Sh4*, *SHAT1*, *OsXTH8*, *OSH15,* and *SH5*) was also affected by the

$e[CO_2]$, with high variability observed among genotypes. The weedy rice genotypes showed the highest upregulation level of these genes. In addition, $e[CO_2]$ positively affects panicle number in cultivated and weedy rice and grain yield mainly in cultivated rice. Our results suggest that weedy rice seed bank persistence and potential competition may be exacerbated with rising $[CO_2]$. Weedy rice seed shattering profile is greatly affected by $[CO_2]$ by modulation of seed shattering-related genes.

Impact of High Temperature Stress on Crops, Weeds, and Herbicide Efficacy. Pv V. Prasad*; Kansas State University, Manhattan, KS (300)

Crop production is extremely sensitive to changing environmental conditions, particularly high temperature. Projections of climate change show increases in season-long mean temperatures and greater frequency of short-episodes of high temperature during crop growing seasons. These changes will have significant negative impacts not only on yield of major grain crops, but also on weeds. Current knowledge on the impacts of season-long high temperature stress; and impacts of short-episodes of high temperature stress on various physiological, growth and yield processes of selected crops and weeds will be presented and discussed. Results indicate that above optimum temperatures will have negative impacts on physiological processes (e.g., thylakoid membrane damage, loss of leaf chlorophyll, leaf photosynthesis and green leaf area duration) and reproductive processes (such as the viability of gametes, fertilization, grain numbers and individual grain weight) resulting in lower harvest index and grain yield of major food grain crops (e.g., wheat, sorghum, soybean). Grain crops are most sensitive to high temperature stress during gametogenesis, flowering, and grain-filling stages. High temperature stress during gametogenesis and flowering leads to loss of gamete fertility, poor pollination, decreased fertilization and embryo abortion resulting in fewer grain numbers. High temperature stress during grain-filling periods decreases grain-size due to shorter grain-filling period. Similarly, high temperature stress impacts physiology and growth of weeds, and can also alter the level of resistance to herbicides in weeds, and hence, their management. Impacts of temperature stress on efficacy and resistance level of mesotrione, glyphosate, dicamba, and 2,4-D in important weeds of Midwestern agriculture (e.g., common waterhemp, Palmer amaranth, kochia, giant or common ragweed were evaluated. Warmer temperatures reduce the control of both 2,4-D-resistant and susceptible common waterhemp and Palmer amaranth, primarily because of the rapid metabolism of 2,4-D at high-temperature stress. Similarly, Palmer amaranth was more sensitive to mesotrione at low than high temperatures because of the rapid metabolism of herbicide at high temperature. At high temperatures, poor control of kochia when treated with dicamba was observed due to reduced herbicide translocation. Whereas reduced absorption of glyphosate was found to contribute to poor control of kochia at high temperatures. Improved herbicide efficacy and weed (resistant or susceptible) control were found when are applied at cooler than warmer temperatures. Climate change can influence herbicide performance and potentially increase the incidence of herbicide-resistant biotypes. Development of high temperature tolerant genotypes will be of prime importance for adaptation to climate change and climate variability. Genetic variability exists for high temperature tolerance in grain crops for key physiological and biochemical traits. In addition, it is critical to study crop-weed interaction and timing of herbicide application to ensure greater efficacy. Continued collaboration between physiologists, breeders, molecular biologists, agronomists and weed physiology and ecologist is essential for developing high temperature stress-tolerant genotypes and best managing practices to combat negative effects of weeds and climate change on productivity of crops under current and future climates.

Climate Change and Pest Treatment Costs: What is Happening? Bruce A. McCarl*; Texas A & M University, College Station, TX (301)

Climate change is altering the pest-crop environment and in turn, causing alterations in the ways that pest management is carried out. Here we examine some of the underlying forces and do a statistical analysis on climate impacts of treatment costs by crop and three classes of pest treatment methods, in particular herbicides, insecticides, and fungicides. The presentation will first review evidence on how crop incidence is changing as that will alter spatial needs for pest treatment Then we present results from a statistical examination of how cost pest management is being altered by climate change. In our analysis, we examined how current climate affects the distribution of planting incidence and pesticide costs for five crops: corn, cotton, soybean, potato, and winter wheat. In our statistical analysis, we construct interrelationships between cost and climate, individually and jointly for herbicides, insecticides, and fungicides herbicides, insecticides, and fungicides. Empirically we find that solar radiation and temperature have heterogeneous impacts on the pesticide cost of these five crops. Total precipitation has an inverted U-shaped impact on cotton, potato, and soybean herbicide cost, but its impact varies for insecticides and fungicides. Moreover, extreme climate events such as drought show an inverted U-shaped effect on pesticide costs for corn, cotton, and soybean, indicating that pesticide expenditure rises as the frequency and magnitude of drought increases past a critical level. Meanwhile, pesticide costs decrease as more genetically engineered (GMO) varieties are adopted for corn, cotton, and soybean. Our projected result shows climate change by the end of this century stimulates an increase in pesticide costs ranging from 2% to 15% on the national level, with solar radiation and precipitation contributing the most to the change.

Weeds and Weed Management in a Warmer, Wetter World. Carolyn J. Lowry*[1], Ashley N. Isaacson[2], Alix Contosta[3], Erin Haramoto[4], Richard G. Smith[3], John M. Wallace[5]; [1]The Pennsylvania State University,

University Park, PA, [2]Penn State University, State College, PA, [3]University of New Hampshire, Durham, NH, [4]University of Kentucky, Lexington, KY, [5]Penn State University, University Park, PA (302)

Throughout the Northeast, anthropogenic climate change is causing warmer temperatures and an increase in the number of extreme precipitation events, both of which may have profound effects on weed communities, as well as weed management efficacy. We will present results from two separate studies that aim to improve our understanding of how weeds and weed management efficacy may be affected by climate change. First, we examined how extreme precipitation events affect residual herbicide efficacy, and whether cover crop surface residues can increase or decrease weed control when extreme rain events occurs. Simulated precipitation treatments were applied using rainfall simulators, and included: 1. *Ambient precipitation* consisted 0 cm added precipitation; 2. *Very Extreme precipitation* consisted of 12.5 cm added precipitation in a single day event; or 3. *Frequent Extreme precipitation* consisted of 2 events of 6.25 cm added precipitation). We examined precipitation effects on weed control efficacy of residual group 15 herbicides (None, s-metolachlor, dimethenamid-P, and pyroxasulfone); and cover crop (with or without a cereal rye cover crop) surface residues in a no-till system. In the absence of herbicides, we found that *Frequent Extreme* events increased the emergence density of redroot pigweed (*Amaranthus retroflexus* L.) and giant foxtail (*Setaria faberi* Herrm.). Regardless of precipitation treatment, all group 15 herbicides reduced pigweed and giant foxtail density compared to the no herbicide control. However, efficacy of the group 15 herbicides tended to be reduced when extreme precipitation events occurred, and these effects were often exacerbated by the presence of cereal rye surface residues. Second, within a fall-planted alfalfa (*Medicago sativa* L.)-orchardgrass (*Dactylis glomerata* L.) forage mixture, we examined how increasing temperatures affected forage productivity and weed suppression of common chickweed (*Stellaria media* L.). To simulate warmer temperatures in the field, we used passively heated hexagonal open-top chambers (OTC) and compared three treatments: 1) ambient temperature; 2) Constant Warming (OTC consistently in place); and 3) Fluctuating Warming (OTC periodically removed in winter to exacerbate temperature fluctuations). We found that increasing temperature fluctuations in the winter led to greater alfalfa winter injury but was not enough to affect alfalfa yields. Warmer temperatures increased alfalfa yields early in the season in 2021 when precipitation was greater but warming had no effect in late summer of 2021 or throughout 2022 when conditions were drier. Warming also had no effect on chickweed biomass. These results suggest that warming may promote alfalfa productivity only when water is not limiting.

Does Environmental Change Increase the Impact of Weeds? Montserrat Vilà*; Estación Biológica de Doñana (EBD-CSIC), Sevilla, Spain (303)

Crops worldwide are simultaneously affected by weeds, which reduce yield, and by climate change, which can negatively or positively affect both crop and weed species. While the individual effects of environmental change and of weeds on crop yield have been assessed, the combined effects have not been broadly characterized. To explore the simultaneous impacts of weeds with changes in climate-related environmental conditions on future food production, we conducted a meta-analysis of 171 observations measuring the individual and combined effects of weeds and elevated CO_2, drought or warming on 23 crop species. The combined effect of weeds and environmental change tended to be additive. On average, weeds reduced crop yield by 28%, a value that was not significantly different from the simultaneous effect of weeds and environmental change (27%), due to increased variability when acting together. The negative effect of weeds on crop yield was mitigated by elevated CO_2 and warming, but added to the negative effect of drought. The impact of weeds with environmental change was also dependent on the photosynthetic pathway of the weed/crop pair and on crop identity. Native and non-native weeds had similarly negative effects on yield, with or without environmental change. Weed impact with environmental change was also independent of whether the crop was infested with a single or multiple weed species. Since weed impacts remain negative under environmental change, our results highlight the need to evaluate the efficacy of different weed management practices under climate change. Understanding that the effects of environmental change and weeds are, on average, additive brings us closer to developing useful forecasts of future crop performance.

Combined Effects of Climate Change and Intensification on Potential Yield. Jonathan Storkey*; Rothamsted Research, Harpenden, United Kingdom (304)

The efficacy and widespread use of herbicides since the 1960's has meant that changes in potential yield loss from weeds ((weed free yield – weedy yield) / weed free yield) are difficult to study. However, in the context of evolved herbicide resistance, understanding the impact of environmental and management change on weed competition will be important for designing more resilient cropping systems. Potential yield loss may change over time because of shifts in the weed flora or differential responses of crops and weeds to the environment or management. The Broadbalk winter wheat experiment at Rothamsted has a unique time series of crop yield data from plots with and without herbicides along a gradient of fertiliser inputs. These data were analysed to study the impact of environmental and management change on potential yield loss between 1969 to 2014. Over this time period, average air temperatures across the growing season of the weeds increased by 1.8°C and the wheat cultivars grown have become progressively shorter. Potential yield loss was found to increase along the nitrogen fertiliser and also with time. Differences between the fertiliser treatments was explained by shifts in the weed flora towards more competitive species. Temporal trends were related to the cultivation of shorter crop cultivars and warming temperatures associated with climate change. I interpret these results using knowledge of the ecology of the weeds and discuss their relevance for informing more sustainable weed management approaches.

Crops and Weeds Running to Stand Still in the Red Queen's Climate Change. David R. Clements*; Trinity Western University, Langley, BC, Canada (305)

In Lewis Carroll's fairytale Through the Looking Glass, Alice complains to the Red Queen that she is tired out from running but still at the tree where she and the Queen started. The Red Queen retorts: Now, here, you see, it takes all the running you can do to keep in the same place. If you want to get somewhere else, you must run at least twice as fast as that. This is the basis for Leigh Van Valen's Red Queen Hypothesis that co-evolving species must keep evolving to keep up with each other. Since the agricultural revolution, crops and weeds have been running an evolutionary race, and now both are also running against the winds of climate change. Climate change acts as a selection pressure on weeds to alter life history characteristics or physiological traits (e.g., stress tolerance, metabolism, or herbicide resistance), sometimes quite rapidly. Meanwhile, crop breeders must also work quickly to maintain high yielding crops despite climate change. Although crop breeders possess increasingly sophisticated tools, weeds are inherently more responsive to environmental change so it is hard to guess which will run faster. Clearly crop improvement faces the urgent need for speed described by the Red Queen. Meanwhile one of the strongest evolutionary mechanisms weeds possess is the ability to steal genes from closely related crops. Many crops have feral counterparts including rice, sorghum, maize, canola, radish, oats, and sunflower which stand to benefit from genetic improvements to their relatives. Or more optimistically, can humanity harness rapid weed evolution to harvest genetic material to improve the crops the weeds are racing? The Red Queen Hypothesis has stimulated many advancements in the study of evolution. With the effects of climate change on agroecosystems ever more apparent, we need to up our agricultural chess game with more evolutionary thinking.

Climate Change, Weed Biology and Public Health. Lewis H. Ziska*; Columbia University, Baltimore, MD (306)

Although the response of weedy species and corresponding changes in their management has been, justifiably, the focus of research in the context of climate change, there are additional concerns regarding subsequent consequences for public health. Such consequences reflect a number of intersections from contact dermatitis to aerobiology and from toxicology, to pesticide exposure. I will provide an overview of these concerns with an especial focus on aerobiology, providing a history of research that transitions from laboratory to continental to hemispheric effects that emphasizes changes in timing, pollen loads and allergenicity. It is clear that climate change, specifically the ongoing increase in carbon dioxide and temperature will, through a weedy lens, alter a number of public health concerns. However, at present, little research is addressing these issues.

Introductions and Objectives. James J. Kells*; Michigan State University, East Lansing, MI (400)

Summary and Interpretation of WSSA Research Priorities Survey. Daniel C. Brainard*[1], Erin Haramoto[2], Ramon G. Leon[3], James J. Kells[1], Lee Van Wychen[4], Pratap Devkota[5], Mithila Jugulam[6], Jacob Barney[7]; [1]Michigan State University, East Lansing, MI, [2]University of Kentucky, Lexington, KY, [3]North Carolina State University, Raleigh, NC, [4]Weed Science Society of America, Alexandria, VA, [5]University of Florida, Jay, FL, [6]Kansas State University, Manhattan, KS, [7]Virginia Tech, Blacksburg, VA (401)

Question and Answer Session on Survey Results. Daniel C. Brainard*; Michigan State University, East Lansing, MI (402)

Research Panel Discussion. Stephen O. Duke*; University of Mississippi, Oxford, MS (403)

Management Panel Discussion. Lee Van Wychen*; Weed Science Society of America, Alexandria, VA (404)

Closing Comments and Next Steps. James J. Kells*; Michigan State University, East Lansing, MI (405)

Novel Technology for Weed Management - Symposium Introduction. Lynn M. Sosnoskie*[1], Marcelo L. Moretti[2]; [1]Cornell University, Geneva, NY, [2]Oregon State University, Corvallis, OR (423)

Weeds and invasive plants are significant threats to food and fiber production, the ecological functioning of ecosystems, and human and animal health. Weeds can reduce crop yields through competitive interactions, facilitate pest damage, alter microclimates to support disease development, disrupt natural habitats, displace native vegetation, impede the flow of water in canals and streams, trigger allergic responses, create or enhance fire and safety hazards, and reduce the aesthetics and enjoyment of both managed and natural areas, among other impacts. Herbicides have been the predominant tool for weed management for many producers, although there is growing interest in alternate control strategies due to the evolution of herbicide resistance, crop safety concerns, changes in public perceptions about pesticide use, and regulatory impacts on pesticide registration and use. The topic of this symposium directly aligns with the WSSA's commitment to enhancing the diversity in and sustainability of weed control practices as highlighted by several of the society's most recent research priorities (https://wssa.net/wp-content/uploads/Weed-Science-Research-Priorities_NIFA-2018.pdf. This symposium addresses novel strategies for managing weeds while reducing or eliminating herbicide use and covers a variety of technologies including thermal weeding (e.g., steam, electric), interrow mowing, vision-guided spray technology, and harvest weed seed control. Speakers in the symposium work in a diverse array of crops ranging from vegetables to agronomic row crops to berries, grapes, and tree nuts. The symposium speakers also are based in different regions of the country, including the Southeastern US, Mid-Atlantic states, Northeastern US, Mid-Western states, and the Western US, which will allow for discussion about how environmental factors and weed community composition influence observed responses. Additionally, all participants work closely with stakeholders and can speak to both the benefits of and barriers to adoption of novel technology.

Identifying and Overcoming Barriers of Adoption for Robotic Weed Control Implementation in Specialty Crops. Matthew A. Cutulle*[1], Lynn M. Sosnoskie[2], Thierry E. Besancon[3], Nathan Boyd[4], Sushila Chaudhari[5], Matthew B. Bertucci[6], Joe Maja[7], Tara Wade[8], Felipe Silva[9], Laura Pineda-Bermudez[2]; [1]Clemson University, Charleston, SC, [2]Cornell University, Geneva, NY, [3]Rutgers University, Chatsworth, NJ, [4]University of Florida, Balm, FL, [5]Michigan State University, East Lansing, MI, [6]University of Arkansas, Fayetteville, AR, [7]Clemson University, Blackville, SC, [8]University of Florida, Gainesville, FL, [9]Clemson University, Clemson, SC (424)

Both conventional and organic specialty crop producers are experiencing labor shortages and higher manual weeding costs. Robotic weeders are promising new weed control tools for specialty crops because they are cheaper to develop and, with fewer environmental and human health risks, are less regulated than herbicides. However, accessibility to robotic weed control platforms is not equal among all farm demographics and regions. Outside of Arizona and California there is limited use of robotic weed control platforms in specialty crops in the US. Furthermore, access to robotic weed control platforms by small, diversified crop growers in certain regions of the country is non-existent. Preliminary surveys were conducted to identify grower issues and interests with respect to robotic our autonomous weed control. More than 100 respondents from across the country provided answers to our survey. Seventy six percent of the respondents were classified as small growers by USDA standards and were typically diversified specialty crop growers. The most common method of controlling weeds was noted to be hand weeding while growers were split on the importance of autonomy. With respect to the decision-making process for acquiring new equipment, respondents indicated that purchase price, availability of parts and service, compatibility with existing equipment, battery life, versatility, and the increasing cost of labor were the most important factors driving adoption. Future research will focus on growers' willingness to pay (WTP) for novel robotic weed control technologies.

Smart Weed Control in Vegetables with Band Steaming. Steve Fennimore*; University of California - Davis, Salinas, CA (425)

Steam injected into the soil, raising soil temperatures to >70 °C for 20 minutes, i.e., dwell time, will control weed seed and soilborne pathogens. The effect of this reduction in the weed seedbank viability results in weed control in the treated zone that can persist for several weeks or months. The effect of steam pasteurization of soil on weed seeds, produces results similar to a preemergence herbicide. In our study, steam was applied to the soil to control weed seed and propagules of *Sclerotinia minor* and *Pythium* spp. Field trials in carrot, and lettuce were conducted using band steam applicators during 2020 to 2022 at Salinas , CA. Data collected were soil temperatures after steam application, weed control, hand weeding times, and crop yields. Post-steam soil temperature dwell times >70 °C in the top 10 cm of the soil were >20 minutes. Steam reduced weed densities by 64 to 100% and lowered hand weeding times by 23 to 94%. Lettuce head diameters in steamed soils were 10 to 24% larger compared with the no steam control. Carrots grown in the steamed treated soil had a 10% greater root diameter than the no steam control. Steam increased lettuce yields in two trials 22% to 28% compared to the no steam control. Gross revenues for the steam treated lettuce were \$3,231 ha^{-1} higher than in the no steam control. The data suggest that band steam is a viable soil pest control treatment for vegetable crops.

Electrical Weed Control in Perennial Cropping Systems. Marcelo L. Moretti*; Oregon State University, Corvallis, OR (426)

Electrical Weed Control in Perennial Cropping Systems Marcelo L Moretti Oregon State University, Horticulture Department Electrical weed control (EWC) is an emerging thermal weed control technology. Electric currents are applied to plants, increasing their temperature as the electric current passes through the tissue leading to plant death. A tractor-powered mobile unit produces the current, which is delivered to the targeted plant's foliage by electrodes. Crop tolerance to this management practice largely depends on the placement of the electrodes and the plant's resistance to electricity. The benefit of EWC is that the current moves throughout the plant, affecting below-ground parts without soil disturbance. One negative aspect is that the high-voltage current can risk fire. The source of electricity is a high-voltage system delivering alternate or direct current. The latter has been employed in more recent versions of commercially available units with auto-adjustment of the energy applied based on site conditions to improve energy efficiency and reduce risk of fire. The need for non-chemical weed control alternatives in perennial crops that do not disturb the soil stimulated our interest in EWC. Non-chemical alternatives are needed to manage herbicide-resistant weed biotypes and for organic production systems. Herbicide resistance is a problem now dominating the landscape in multiple cropping systems. An alternative to repeated tillage for weed control will support soil health and is essential to systems for which tillage is incompatible. We conducted field evaluations of electric weeders using alternating current and 30 kW (EH-30 Thor, Zasso™) treating swaths 1.2 m wide in hazelnut orchards, or using direct current and 6 kW (Raiden, Zasso TM) treating swaths 0.4m wide in blueberry fields. In both evaluations, electricity was applied at speeds ranging from 0.4 to 6 km h^{-1}. Treated plants wilted within minutes after treatment, and wilting was noticed sooner at lower speeds and after treatment with higher-powered equipment (EH-30). Weed control peaked 7 to 14 days after treatment, reaching 70-95%. Greater control was exerted by applications at speeds below 3.2 km h^{-1}. In an orchard infested with Italian ryegrass, electric current reduced shoot dry weight by up to 93% and inflorescence weight by 61 to 93% by 56 days after a single application. Weed regrowth and new germination after treatment suggest that repeated applications at greater speeds are the more effective strategy. No injury was observed to hazelnuts or blueberries in these experiments. These initial data suggest that electricity is an effective and safe weed control option in perennial crops.

Electrical Weeding and Inter-row Mowing in Annual Crops. Annika Rowland*[1], Lynn M. Sosnoskie[2], Chris J. Pelzer[1], Matthew R. Ryan[1]; [1]Cornell University, Ithaca, NY, [2]Cornell University, Geneva, NY (427)

Weed management in soybean [*Glycine max* (L.) Merr.] production typically involves herbicides or soil tillage, both of which can lead to environmental problems. No-till planting soybean into mulch from rolled-crimped cover crops, such as cereal rye (*Secale cereale* L.) is an alternative production strategy that has been shown to effectively suppress weeds, especially small-seeded annual weed species. However, if inadequate cereal rye mulch biomass is produced, weeds will not be suppressed. Integrating cultural and physical weed management practices into the no-till soybean system may enhance weed suppression, but little research has been conducted on interactions between practices. Two physical weed management practices, inter-row mowing and electric weed control, were tested in a field experiment in central New York in 2021 and 2022. The practices, along with the cultural weed management practice of increasing the soybean seeding rate (i.e., planting 42% more seed) were implemented alone and in all possible combinations in soybean managed without herbicides or inter-row cultivation. A custom-made inter-row mower was used to target mid-season weeds between 76-cm soybean rows. Electric weed control was used to target weeds growing above the soybean canopy. Inter-row mowing reduced weed biomass ($P < 0.05$) and trait-based analysis revealed that mowing targeted annual broadleaf weeds. Electric weed control did not reduce weed biomass, as few weeds grew above the soybean canopy before canopy closure prohibited use. The combination of the high soybean seeding rate and inter-row mowing, as well as the combination of all three tactics increased soybean yield compared to the control ($P < 0.05$). However, there were no synergistic interactions between tactics for weed biomass or soybean yield. Weed seed production was reduced when combining all three tactics for four of six dominant weed species. Additional work could be completed to investigate weed seed production and seed viability post-weed management. Reduction of weed biomass through inter-row mowing suggests ease of soybean harvestability and anecdotal evidence suggests similar results when the Weed Zapper is used to manage early emerging and/or fast-growing weed species. Moreover, inter-row mowing and weed electrocution have the potential to reduce weed biomass and weed seed production in systems without herbicides or cultivation. Additional opportunities exist to understand tactic profitability, effect on weed seed viability, and contribution to harvest efficiency.

The Future of Precision Weed Management in Specialty Crops. Nathan Boyd*[1], Arnold W. Schumann[2], Ana Claudia Buzanini[3]; [1]University of Florida, Balm, FL, [2]University of Florida, Lake Alfred, FL, [3]University of Florida, Wimauma, FL (428)

Recent advances in deep learning, computational hardware, and artificial intelligence have advanced the potential for precision weed management. Weed detection and identification accuracy are rapidly improving, and the use of synthetic data and image augmentation have made the acquisition of labelled data sets more efficient. There is an on-going need for end-effectors that can be used in conjunction with pest detection that accurately target weeds without negatively affecting surrounding vegetation. This could include chemical or mechanical mechanisms of weed removal with non-chemical options preferred in many crops where few herbicides are registered. Equipment design will need to consider the relatively small acreage fields typical of many specialty crops, the limitations and advantages of autonomous vehicles, and the extensive number of crop species and production systems. Adoption of precision technologies in specialty crops offers unique challenges and opportunities but is likely to transform the future of weed management.

Weed Control Efficacy and Perennial Crop Tolerance Using an Herbicide Sprayer Based on Chlorophyll Fluorescence Sensing in Perennial Crops. Thierry E. Besancon*[1], Lynn M. Sosnoskie[2], Wesley M. Bouchelle[1], Elizabeth C. Maloney[2], Bethany M. Schulteis[3]; [1]Rutgers University, Chatsworth, NJ, [2]Cornell University, Geneva, NY, [3]Cornell University, Ithaca, NY (429)

Research findings in agronomic systems suggest that precision sprayers have the potential to reduce herbicide use while providing acceptable control of unwanted vegetation and maintaining yields in annual crops. However, there is limited research describing the performance and safety of smart-sprayers in perennial cropping systems. While herbicides are an effective tool for managing weeds, the number of herbicides labeled for use in perennial crops such as blueberry remains limited because of crop tolerance concerns. Concurrently, the development and spread of herbicide resistant weed biotypes require the integration of new modes of action to perennial crops weed management programs. In 2021 and 2022, field trials were conducted at the Rutgers P.E. Marucci Research Center (Chatsworth, NJ) to evaluate the performance of a vision-guided precision spray system for controlling emerged weeds while maintaining excellent crop tolerance in mature highbush blueberry fields. Vision-guided precision sprayer applications were made using a Weed-It Quadro system equipped with 4 chlorophyll fluorescence sensing units and were compared to a banded boom-application. Postemergence herbicides included glufosinate (Rely 280) at 105 g ha^{-1}, 2,4-D choline (Embed Extra) at 100 g ai ha^{-1}, fluroxypyr (Starane Ultra) at 17 and 34 g ai ha^{-1}, florpyrauxifen-benzyl at 2 and 4 g ai ha^{-1}, and diquat (Reglone) at 35 and 70 g ha^{-1}. Two applications were made to control emerged weeds less than 4 in late spring and 30 days later. Averaged across herbicides, the vision guided sprayer allowed better weed control of blueberry ridge top with 41% weed coverage 28 days after treatment (DAT) as compared to 58% for banded application whereas weed coverage of the ridge side was similar (10% on average) for both systems. In 2022, control of glyphosate-resistant horseweed 14 DAT was significantly improved with systemic herbicides (florpyrauxifen, fluroxypyr, and 2,4-D choline) applied with the vision guided sprayer (89%, 73%, and 75%, respectively) as compared to the banded application (68%, 60%, and 62%, respectively). However, broadcast application of diquat and glufosinate performed better horseweed control (90% and 80%, respectively) than vision-guided applications (66% on average). In 2021, blueberry injury 14 DAT following the second application of readily translocated herbicides (2,4-D and florpyrauxifen) was lower with the vision guided (12%) compared to broadcast application (20%). Similar observation was made for glufosinate with crop injury following the second application reduced by 50% with the vision guided sprayer compared to banded application. In 2022, the amount of injury on blueberry new canes was similar for both types of application. Berry yield was not affected by treatments or method of application both years. The volume of herbicide sprayed per nozzle was significantly lower with the vision-guided sprayer (292 ml per application per treatment) compared to the broadcast application (384 ml per application per treatment). However, since there are 4 nozzles on the vision-guided sprayer for covering the same spray width as for the backpack sprayer (2 nozzles), the volume of spray solution required by the vision-guided sprayer was 1.5 times greater than for the broadcast application.

See &Amp; Spray™ Ultimate - A New Tool for Weed Management in Agronomic Crops. Lauren M. Lazaro*[1], William L. Patzoldt[1], Scott Brown[1], Jesaelen Gizotti de Moraes[1], Michael M. Houston[2], Gaylia Ostermier[3], Chad Yagow[3]; [1]Blue River Technology, Sunnyvale, CA, [2]Blue River Technology, Stoneville, MS, [3]John Deere, Urbandale, IA (430)

See & Spray Ultimate™ is the first factory-integrated targeted application sprayer commercialized for in-crop weed management. See & Spray Ultimate is a collection of new application technology improvements that include: 1) cameras and computers that identify and spray weeds in real-time, 2) a dual tank system allowing for simultaneous broadcast and targeted applications, 3) a new lighter and more stable boom for optimal spray accuracy, and 4) fully integrated in-cab controls and data uploading to John Deere Operations Center generating weed and as-applied maps. Small-plot equipment, the Agronomy Test Machine, has been designed to mimic See & Spray Ultimate performance to expand research capabilities and allow for rapid iteration of concepts. These concepts include model sensitivity by application timing, single vs dual tank programs, nozzle selection, and reduction of antagonism. In general, regardless of research trial, no significant difference was observed between See & Spray treatments and a broadcast standard for end of season weed control or crop yield. While there are similar outcomes today, less sprayed areas allow for better stewardship in the future. Furthermore, it is imperative that users recognize that the fundamentals of weed science still apply to new technologies.

Advances in Unmanned Aerial Systems-based Spray Applications, 2023. Vijay Singh*[1], Dhiraj Srivastava[2], Vipin Kumar[2], Daniel E. Martin[3], William Reynolds[4]; [1]Virginia Tech, Painter, VA, [2]Virginia Polytechnic Institute and State University, Painter, VA, [3]United States Department of Agriculture, College Station, TX, [4]LeadingEdge Aerial Technologies, New Smyrna Beach, FL (431)

Unmanned Aerial Systems are widely used for mapping and classification of weed species these days. Along with sensing and collecting image data, a novel idea of pest management through UAS-based site-specific and broadcast pesticide-spray applications has emerged as a potential crop protection tool. Such interventions are necessary for reducing expenses, labor, and chemical load on the agroecosystems and managing pest-resistance issues. However, not much research has been conducted on these aerial platforms and data on their efficiency for controlling major pests (weeds) in diverse cropping systems are not available. Several studies were conducted in 2020-2022 at Painter, VA to evaluate the efficacy of UAS-based blanket and spot spray applications in corn, cotton and soybean. UAS-based herbicide applications were conducted at two different volumes at 18.7 (UAS-2) and 37.4 L ha^{-1} (UAS-4) and compared with a CO_2 pressurized backpack sprayer at 140 L ha^{-1} (BP-15). Water sensitive strips, and six-cell trays filled with seedlings of Palmer amaranth and large crabgrass were placed randomly in the test plots. All crop studies were replicated four times and repeated after one-month. UAS-based POST spray applications provided 90-100% control of weeds in all crop studies which was either similar to weed control efficacy of backpack sprayer or better with UAS when applied late-POST. Additional studies were conducted to evaluate the site-specificity of UAS-based herbicide applications and results have indicated that this technology can perform automatic spray operations with overall weed control efficacy of 94%. However, one of the major challenges of spot spray is to manually locate or extract geo-coordinates of weed spots from orthomosaic image and converting it to shapefile. Studies were conducted to evaluate if machine learning based object detection method can be used to automate the process of creating shapefile. Digital library of weeds was used to detect weeds using machine learning algorithms and center of the boundary box provided the coordinates of weed in the frame. This information was imported as CSV file and further converted into ESRI shapefile using GeoPandas python library. The process used reduced time as compared with traditional approach and improved the efficiency of spot sprays. Results have indicated that current UAS technology can spot spray with high precision, but more studies and technological improvements are required for standardization.

Current Status of Harvest Weed Seed Control Research in the Eastern United States. Michael L. Flessner*[1], Prashant Jha[2], Lovreet S. Shergill[3], Lynn M. Sosnoskie[4], Muthukumar V. Bagavathiannan[5], Wesley Everman[6], Eugene P. Law[7], Lauren M. Lazaro[8], Steven Brian Mirsky[9], Mark VanGessel[10]; [1]Virginia Tech, Blacksburg, VA, [2]Iowa State University, Ames, IA, [3]Montana State University, Huntley, MT, [4]Cornell University, Geneva, NY, [5]Texas A&M University, College Station, TX, [6]North Carolina State University, Raleigh, NC, [7]University of Delaware, Beltsville, MD, [8]Blue River Technology, Sunnyvale, CA, [9]USDA ARS, Beltsville, MD, [10]University of Delaware, Georgetown, DE (432)

Recent publications have established tremendous potential for harvest weed seed control (HWSC) in US soybean and wheat production to address herbicide resistance and improve sustainability of weed control practices by reducing soil seed banks. For HWSC to be successful, seeds must be retained on the weedy plant at harvest. Initial research found that seed retention at soybean harvest from the most troublesome herbicide resistant species is very high (>90% for most broadleaf weed species). Subsequent field research evaluated HWSC in small plots. Recently completed work in the eastern US indicates that soybean chaff is about 13% of yield by weight, which enables further research in chaff lining since this amount was not previously known. This work also reveals a so-called hidden cost of HWSC, which is the nutrient value in harvest residue fractions: 7.46 and 1.58 USD ha-1 for HWSC methods that concentrate or remove soybean straw and chaff or just chaff, respectively. Other small plot chaff lining work indicates that chaff inhibits subsequent weed and crop emergence. Greater chaff amounts lead to greater reductions in crop emergence. However, crop yield reductions were not observed at the field scale, other than at one site-year. Future work should examine weed seed fate in chaff lines. Ongoing work is examining weed seed fate in commercial combines. Weed seeds that exit in the straw fraction are not subjected to HWSC methods like seed impact mills or chaff lining. Preliminary data shows that >94% of weeds seeds that enter the combine are removed from the field when using a seed impact mill. On-going seed kill testing using the Redekop Seed Control Unit indicates >90% kill for Italian ryegrass (Lolium perenne L. ssp. multiflorum (Lam.) Husnot) and >98% kill of other species tested. Future work will also include seed kill rate testing by chaff amount and moisture level and with the iHSD seed impact mill. Other on-going work includes collecting combine operating metrics when using a seed impact mill such as fuel usage, engine load, and travel speed in an effort to create an HWSC Cost Calculator. Multi-year studies evaluating weed responses to HWSC have been initiated in corn/soybean and crop rotations including wheat. After a single harvest using HWSC, weed density was reduced on average about 30% for both common ragweed (Ambrosia artemisiifolia L.) in soybean and Italian ryegrass in wheat; however, reductions were highly variable. Research and on-farm collaborations to date have revealed issues with HWSC. Seed impact mills require a lot of horsepower and may reduce harvest speed. Additionally, seed impact mills can clog during high moisture harvest conditions, halting harvest. Since all HWSC work to date uses a platform/grain header during harvest, work examining weed seed capture with a corn header and cotton harvesters has also been initiated. Newer forms of HWSC are being explored that utilize blue light and infrared radiation to kill weed seeds. As data accumulates, future work will likely include extension activities, optimizing combine settings for HWSC, and dealing with high moisture harvest conditions, where possible.

Implementing Harvest Weed Seed Control Strategies in the Midwestern and Southern U.S. Prashant Jha*[1], Jason K. Norsworthy[2], Vipan Kumar[3], Karla L. Gage[4], Steven Brian Mirsky[5], Muthukumar V. Bagavathiannan[6], Michael L. Flessner[7], Michael Walsh[8], Avery J. Bennett[1], Austin H. Schleich[1]; [1]Iowa State University, Ames, IA, [2]University of Arkansas, Fayetteville, AR, [3]Kansas State University, Hays, KS, [4]Southern Illinois University Carbondale, Carbondale, IL, [5]USDA ARS, Beltsville, MD, [6]Texas A&M University, College Station, TX, [7]Virginia Tech, Blacksburg, VA, [8]University of Sydney, Narrabri, Australia (433)

The evolution of multiple herbicide-resistant waterhemp and Palmer amaranth (9 sites of action) has created an urgent need to develop novel, IWM strategies in soybean-based production systems of the U.S. Harvest weed seed control strategies involve collection and/or destruction of weed seeds during crop harvest; thus, drastically reducing weed seed spread and inputs to the soil seedbank. The efficacy of harvest weed seed control methods in managing herbicide resistance depends on the propensity of some weed species to retain a majority of seeds until crop harvest. For instance, waterhemp and Palmer amaranth plants can retain 70 to 90% of seeds at the typical harvest dates of soybean in the Midwestern and southern U.S. Chaff lining involves making a simple chute to divert the weed seed-bearing chaff fraction (from the sieves) into a narrow windrow (45-55 cm wide) while the straw is chopped and spread at soybean harvest. In recent field studies conducted in Iowa, the chaff liner concentrated more than 95% of waterhemp seeds that entered the combine into a narrow band at the center of the 4.2 m header; hence, drastically reducing the herbicide-resistant weed seed spread across the field at the time of soybean harvest. Waterhemp emergence in the subsequent corn crop (soybean-corn rotation) was significantly higher inside vs. outside the chaff line area of the plot. Waterhemp had a delayed emergence and accumulated a lower biomass inside vs. outside the chaff line area in corn plots. An effective PRE followed by POST herbicide program (multiple sites of action) was much needed to control late-emerging waterhemp plants inside the chaff line in the corn crop; however, there was a reduced burden on the POST herbicide to control waterhemp outside the chaff line area of the plots. This was because of the very low density of waterhemp that emerged outside the chaff lines. Field experiments were conducted across IA, KS, and AR in 2021 and 2022 to evaluate the efficacy of the Redekop™ Seed Destructor in killing Palmer amaranth and waterhemp seeds that entered the combine at the time of soybean harvest. The combine head can shatter 20 to 30% of the waterhemp/Palmer amaranth seeds at the time of soybean harvest (header loss). Additional losses in weed seeds can occur through the grain tank and thresher. The Seed Destructor mechanically destroyed up to 95% of pigweed seeds (seed-kill efficacy) that entered the combine at the time of soybean harvest, thereby drastically reducing the viable weed seed inputs. Results indicate a great potential for adopting these two harvest weed seed control tactics to mitigate herbicide resistance in soybean-based production systems. On-going long-term studies funded by USDA-NIFA at IA, IL, AR, and KS would determine the reductions in weed seed banks and herbicide use and also cost-benefit of implementing these two harvest weed seed control tactics.

Current Status of Harvest Weed Seed Control Research in the Northern Great Plains and Pacific Northwest Region of the United States. Lovreet S. Shergill*[1], Todd A. Gaines[2], Ian Burke[3], Drew J. Lyon[3], Judit Barroso[4], Eric P. Westra[2]; [1]Montana State University, Huntley, MT, [2]Colorado State University, Fort Collins, CO, [3]Washington State University, Pullman, WA, [4]Oregon State University, Adams, OR (434)

Research in Northern Great Plains (NGP) and Pacific Northwest (PNW) regions has shown potential of using harvest weed seed control (HWSC) to manage problematic and herbicide resistant weeds in these wheat-based cropping systems. The initial research has focused on assessing seed shatter of the target weed species at crop maturity. Recently completed research recorded low seed retention (50% on average) by downy brome *(Bromus tectorum* L.*)*, Italian ryegrass [*Lolium perenne* L. ssp. *Multiflorum* (Lam.) Husnot], and rattail fescue [*Vulpia myuros* (L.) C.C. Gmel.] at wheat harvest in PNW. Preliminary data from NGP shows that *B. tectorum* and wild oats (*Avena fatua* L.) retained upto 90% seeds at winter and spring wheat physiological maturity, respectively, but they tend to shatter seed rapidly after the maturity. Therefore, early harvesting wheat is recommended to capture maximum weed seeds. Ongoing research is examining seed retention in various weed species in competition with winter wheat, spring wheat, corn, and soybean. Crop residues are critical for reducing soil erosion and moisture retention in this region; thus, current research efforts have been focused on assessing chaff lining and impact mill systems for weed management. Simulated chaff lining field trials in dryland cropping systems in Montana and Colorado indicates that inhibition of weed emergence is dependent on the quantity of the chaff. Greater weed emergence was observed in no chaff compared to simulated wheat chaff lines. Preliminary observations from the field and growth chamber experiment indicate that presence or absence of chaff may not influence germinability of weeds. Multi-year studies have been initiated across the region to evaluate weed responses to chaff lining and impact mill HWSC systems.

Northeastern Weed Science Society (NEWSS)
P.O. Box 25
Woodstown, NJ 08098
USA

ISBN 979-8-3313-0917-6

9 798331 309176

CURRAN ASSOCIATES INC.
proceedings.com